Emerging Needs and Opportunities for Human Factors Research

Raymond S. Nickerson, Editor

Committee on Human Factors
Commission on Behavioral and Social Sciences and Education
National Research Council

NATIONAL ACADEMY PRESS
Washington, D.C. 1995

National Academy Press • 2101 Constitution Avenue, N.W. • Washington, D.C.

NOTICE: The project that is the subject of this report was approved by the Governing Board of the National Research Council, whose members are drawn from the councils of the National Academy of Sciences, the National Academy of Engineering, and the Institute of Medicine. The members of the committee responsible for the report were chosen for their special competences and with regard for appropriate balance.

This report has been reviewed by a group other than the authors according to procedures approved by a Report Review Committee consisting of members of the National Academy of Sciences, the National Academy of Engineering, and the Institute of Medicine.

The National Academy of Sciences is a private, nonprofit, self-perpetuating society of distinguished scholars engaged in scientific and engineering research, dedicated to the furtherance of science and technology and to their use for the general welfare. Upon the authority of the charter granted to it by the Congress in 1863, the Academy has a mandate that requires it to advise the federal government on scientific and technical matters. Dr. Bruce Alberts is president of the National Academy of Sciences. The National Academy of Engineering was established in 1964, under the charter of the National Academy of Sciences, as a parallel organization of outstanding engineers. It is autonomous in its administration and in the selection of its members, sharing with the National Academy of Sciences the responsibility for advising the federal government. The National Academy of Engineering also sponsors engineering programs aimed at meeting national needs, encourages education and research, and recognizes the superior achievements of engineers. Dr. Harold Liebowitz is president of the National Academy of Engineering.

The Institute of Medicine was established in 1970 by the National Academy of Sciences to secure the service of eminent members of appropriate professions in the examination of policy matters pertaining to the health of the public. The Institute acts under the responsibility given to the National Academy of Sciences by its congressional charter to be an adviser to the federal government and upon its own initiative, to identify issues of medical care, research, and education. Dr. Kenneth I. Shine is president of the Institute of Medicine.

The National Research Council was established by the National Academy of Sciences in 1916 to associate the broad community of science and technology with the Academy's purposes of furthering knowledge and of advising the federal government. Functioning in accordance with the general policies determined by the Academy, the Council has become the principal operating agency of both the National Academy of Sciences and the National Academy of Engineering in providing services to the government, the public, and the scientific and engineering communities. The Council is administered jointly by both Academies and the Institute of Medicine. Dr. Bruce Alberts and Dr. Harold Liebowitz are chairman and vice chairman, respectively, of the National Research Council.

The work of the Committee on Human Factors is supported by Department of Army Contract No. DAAD05-92-C-0087 issued by the U.S. Aberdeen Proving Ground Support Activity. The views and opinions, and findings contained in this report are those of the author(s) and should not be construed as an official Department of Army position, policy, or decision, unless so designated by other official documentation.

Library of Congress Catalog Card No. 95-70762
International Standard Book Number 0-309-05276-9

Additional copies are available from National Academy Press, 2101 Constitution Ave., N.W., Box 285, Washington, D.C. 20055. Call 800-624-6242 or 202-334-3313 (in the Washington Metropolitan Area).

Copyright 1995 by the National Academy of Sciences.

Printed in the United States of America

COMMITTEE ON HUMAN FACTORS
1990 - 1994

RAYMOND S. NICKERSON (*Chair, 1991-1994*), Bolt, Beranek, and Newman Laboratories (retired), Cambridge, Massachusetts
DOUGLAS H. HARRIS (*Chair, 1988-1991*), Anacapa Sciences, Inc., Charlottesville, Virginia
PAUL A. ATTEWELL, Department of Sociology, City University of New York
M.M. AYOUB, Department of Industrial Engineering, Texas Tech University
JEROME I. ELKIND, Lexia Institute, Palo Alto, California
PAUL S. GOODMAN, Center for Management of Technology, Carnegie Mellon University
JOHN D. GOULD, IBM Corporation (retired), Yorktown Heights, New York
MIRIAN M. GRADDICK, AT&T Corporation, Basking Ridge, New Jersey
OSCAR GRUSKY, Department of Sociology, University of California, Los Angeles
ROBERT L. HELMREICH, NASA/UT Aerospace Crew Research Project, Austin, Texas
JULIAN HOCHBERG, Department of Psychology, Columbia University
WILLIAM C. HOWELL, American Psychological Association Science Directorate, Washington, D.C.
ROBERTA L. KLATZKY, Department of Psychology, Carnegie Mellon University
THOMAS K. LANDAUER, Department of Psychology, University of Colorado, Boulder
TOM B. LEAMON, Liberty Mutual Research Center, Hopkinton, Massachusetts
HERSCHEL W. LEIBOWITZ, Department of Psychology, Pennsylvania State University
ANN MAJCHRZAK, Human Factors Department, University of Southern California
NEVILLE P. MORAY, PERCOTEC-LAMIH (Laboratory for Automation, Mechanical, Industrial and Human Engineering), Université de Valenciennes, France
WILLIAM B. ROUSE (*Chair-Elect, 1994-1997*), Search Technology, Inc., Norcross, Georgia
JOYCE L. SHIELDS, HAY Management Consultants, Arlington, Virginia
LAWRENCE W. STARK, School of Optometry, University of California, Berkeley

CHRISTOPHER D. WICKENS, Aviation Research Laboratory, University of Illinois, Savoy
ROBERT C. WILLIGES, Department of Industrial Engineering and Operations Research, Virginia Polytechnic Institute and State University
J. FRANK YATES, Department of Psychology, University of Michigan
LAURENCE R. YOUNG, Man Vehicle Laboratory, Massachusetts Institute of Technology

ANNE S. MAVOR, *Study Director*
HAROLD P. VAN COTT, *Principal Staff Officer (until 1992)*
BEVERLY M. HUEY, *Senior Staff Officer*
JERRY KIDD, *Senior Adviser*
EVELYN E. SIMEON, *Senior Project Assistant*

CONTRIBUTORS TO PART II

PAUL A. ATTEWELL, Department of Sociology, City University of New York

M.M. AYOUB, Department of Industrial Engineering, Texas Tech University

JOSEPH B. CAVALLARO, HAY Management Consultants, Arlington, Virginia

JEROME I. ELKIND, Lexia Institute, Palo Alto, California

PAUL S. GOODMAN, Center for Management of Technology, Carnegie Mellon University

JOHN D. GOULD, IBM Corporation (retired), Yorktown Heights, New York

DOUGLAS H. HARRIS, Anacapa Sciences, Inc., Charlottesville, Virginia

ROBERT L. HELMREICH, NASA/UT Aerospace Crew Research Project, Austin, Texas

BEVERLY M. HUEY, Committee on Human Factors, National Research Council

ROBERTA L. KLATZKY, Department of Psychology, Carnegie Mellon University

HERSCHEL W. LEIBOWITZ, Department of Psychology, Pennsylvania State University

NEVILLE P. MORAY, Department of Mechanical and Industrial Engineering, University of Illinois, Urbana-Champaign

RAYMOND S. NICKERSON, Bolt, Beranek, and Newman Laboratories (retired), Cambridge, Massachusetts

D. ALFRED OWENS, Whitely Psychology Laboratory, Franklin and Marshall College

PENELOPE M. SANDERSON, Department of Mechanical and Industrial Engineering, University of Illinois, Urbana-Champaign

KAREN SEIDLER, U.S. West Advanced Technologies, Denver, Colorado

JOYCE L. SHIELDS, HAY Management Consultants, Arlington, Virginia

HAROLD P. VAN COTT, Committee on Human Factors, National Research Council (retired)

CHRISTOPHER D. WICKENS, Aviation Research Laboratory, University of Illinois, Savoy

ROBERT C. WILLIGES, Department of Industrial and Systems Engineering, Virginia Polytechnic Institute and State University

J. FRANK YATES, Department of Psychology, University of Michigan

CAROLYNN A. YOUNG, Department of Psychology, University of Michigan (deceased)

Note: Affiliations current as of the report preparation period.

Contents

PREFACE		ix
EXECUTIVE SUMMARY		1
PART I:	SUMMARY REPORT	11
PART II:	BACKGROUND PAPERS	69
1	Productivity in Organizations	71
2	Training and Education	86
3	Employment and Disabilities	106
4	Health Care	131
5	Environmental Change	158
6	Communication Technology and Telenetworking	177
7	Information Access and Usability	200
8	Emerging Technologies in Work Design	220
9	Transportation	241
10	Cognitive Performance Under Stress	262
11	Aiding Intellectual Work	291

Preface

The Committee on Human Factors was established in October 1980 by the Commission on Behavioral and Social Sciences and Education of the National Research Council. The principal objectives of the committee are to provide new perspectives on theoretical and methodological issues, to identify basic research needed to expand and strengthen the scientific basis of human factors, and to attract scientists inside and outside the field for interactive communication and performance of needed research.

Human factors issues arise in every domain in which humans interact with the products of a technological society. To perform its role effectively, the committee draws on experts from a wide range of scientific and engineering disciplines. Members of the committee include specialists in such fields as psychology, engineering, biomechanics, physiology, medicine, cognitive sciences, machine intelligence, computer sciences, sociology, education, and human factors engineering. Other disciplines are represented in the working groups, workshops, and symposia organized by the committee. Each of these disciplines contributes to the basic data, theory, and methods required to improve the scientific basis of human factors.

Since its inception in 1980, the Committee on Human Factors has issued more than a dozen reports regarding the state of knowledge and research needs on topics deemed important by the committee and its sponsors. This report is the product of a committee-initiated project. It identifies major problem areas in which human factors research can make an important contribution during the next few decades. Part I provides the committee's recommendations and conclusions. These conclusions are drawn from the background papers prepared by committee members, staff, and their colleagues and included in Part II of the report. The committee list is a complete membership roster covering the report preparation time. During this time, some members completed their terms and new members were added.

Throughout the project the committee received encouragement and support from a variety of sources. In particular, we would like to express our gratitude to our sponsors for their continuing interest and important insights. The sponsors include: the Air Force Office of Scientific Research, the Army Research Institute for the Behavioral and Social Sciences, the National Aeronautics and Space Administration, the Air Force Armstrong Aerospace Medical Research Laboratory, the Army Advanced Systems Research Office, the Army Human Engineering Laboratory, the Army Natick RD&E Center, the Federal Aviation Administration, the Nuclear Regulatory Commission, the Naval Training Systems Center, and the U.S. Coast Guard.

The committee would also like to thank staff of the National Research Council for their important contributions to our work. Harold P. Van Cott, study director until his retirement in 1992, was involved in shaping the early planning stages; Beverly M. Huey, acting study director from 1992 to 1994, supported the committee through its deliberations and the many stages of report drafting; and Anne Mavor, the current study director, helped bring the project to a successful conclusion. Finally, we gratefully acknowledge Barbara White for her fine editorial contribution and Evelyn Simeon for her excellent administrative assistance and hard work on the manuscript.

> Raymond S. Nickerson, Chair 1991-1994
> Committee on Human Factors

Emerging Needs and Opportunities for Human Factors Research

Executive Summary

This report identifies 11 areas considered by the National Research Council Committee on Human Factors as fertile ground for human factors research during the next few decades. The papers in Part II, written by committee members, staff, and colleagues and addressing these areas, provide the basis for the conclusions and recommendations articulated in Part I of the report.

The committee's first report, *Research Needs for Human Factors*, published in 1983, three years after the committee was established, focused on six topics that reflected interests of the sponsors at the time and were within the expertise of the committee members to address. Since the committee's establishment, both its sponsorship and the composition of its membership have broadened considerably. Technology has also advanced rapidly, and both the national and world situations have changed in significant and remarkable ways. Some of these changes create new challenges and opportunities for human factors research.

The committee therefore thought it appropriate to again attempt to identify research needs and opportunities for human factors. The process of selecting the topics to emphasize in the report was lengthy and deliberative. It was agreed at the outset that committee members should write the report, that the report should be forward-looking, that it should focus on problems that need attention rather than on the research that is currently being done, and that it should be selective rather than comprehensive. The process produced a consensus on 11 topics, which can be fairly easily grouped under three major headings—as follows:

National or global problems
- Productivity in organizations
- Training and education
- Employment and disabilities
- Health care
- Environmental change

Technology issues
- Communications technology and telenetworking
- Information access and usability
- Emerging technologies in work design
- Transportation

Human performance
- Cognitive performance under stress
- Aiding intellectual work

In keeping with the committee's decision to be forward-looking, relatively little past research is reviewed in the report, and that is done only to help provide an appropriate frame of reference for thinking about future possibilities. Some of the recommendations are relatively general; others, quite specific. Some follow traditional lines of research; others relate to problems that have received little attention from the human factors community so far. In the latter cases, the recommendations are offered more as points of departure for further discussion and planning than as items for a research agenda.

PRODUCTIVITY IN ORGANIZATIONS

Productivity is a major national and international concern. Economists see productivity as a primary determinant of competitiveness both for companies within an industry and for national economies. Productivity is also believed to be causally linked to standard of living, so increasing productivity globally is seen as the best hope for improving living conditions worldwide. Human factors researchers have given considerable attention to how to improve human performance in the workplace and thereby increase individual productivity; they have put relatively little effort into determining how individual productivity relates to the productivity of the groups, organizations, or industries in which the individuals' work is done.

Our focus is on the latter issue, in particular on identifying the conditions under which increases in individual productivity are likely to be reflected in increases at organizational levels. Research needs and opportunities include developing a better understanding of the linkages among within-job activities that have implications for total-job productivity, determining how individual jobs within an organization are causally linked to the productivity of the organization as a whole, and identifying structural differences among organizations that show how the productivity of higher organizational units depends on the productivity of their components.

TRAINING AND EDUCATION

It is widely recognized that education and training are of critical importance to the national economy and to the well-being of the nation and that the country's educational and training systems have not been meeting the nation's needs. The problem is compounded by the expectation that in the future the workforce will have to be more proficient—especially more versatile and adaptive—in order to cope effectively with the rapidity of technological change in the workplace. As a discipline that historically bridges psychology and engineering, human factors should be important in applying laboratory discoveries about learning to the design of educational and training systems.

Determining how best to do this is a major challenge for research. A closely related research challenge is the identification of training and educational requirements that arise from interactions among changing technologies, changing workforce demographics, changing organizational structures, and other developments that are likely to have implications for the design of effective educational and training systems. Other opportunities for human factors research in this area include evaluating technologically innovative approaches to education and training, applying user-centered design principles to educational and training systems, developing approaches to support lifelong learning within work settings, and anticipating technology-induced changes in job-skill requirements and their implications for educational and training needs.

EMPLOYMENT AND DISABILITIES

The population of people with one or another type of disability is large. The rate of unemployment for people with disabilities who would like to work is several times higher than the unemployment average nationwide, and the percentage of people who live below the poverty line is two to three times greater for people with a disability that interferes with their ability to work than for the total working population. These circumstances not only represent a severe problem for many people with disabilities, but they also create a significant economic burden for the nation. Most of the previous research on how to enhance employment opportunities for people with disabilities has tried to find ways to use technology—especially computer and communication technologies—to mitigate the limiting effects of disabilities in the workplace and elsewhere.

With more than half the workforce now engaged in information-oriented jobs, we need to continue our efforts to find ways to use computer and communication technologies to increase employment opportunities for people with disabilities. Another major challenge for research is to increase our understanding of how the capabilities and limitations of people with specific disabilities compare with the requirements of specific jobs; such an understanding is essential to the design of devices, systems, and procedures that will expand the range of jobs that people with disabilities can effectively perform. There is also a need to explore ways in which specific jobs might be designed or redesigned to accommodate people with disabilities.

HEALTH CARE

Many health risks, such as cancer and heart disease, stem at least in part from behavioral and occupational factors. Significant threats to health are even found within the increasingly complex technology used for the delivery of health care. Many of these threats arise from the possibility of human error in using health-care equipment and carrying out health-care procedures, for example in trauma centers and by people without medical training in their homes.

A major challenge to human factors research is to help improve the designs of medical devices and health care procedures so that they are less conducive to human error. Another challenge is that, as the percentage of our population that is elderly and very elderly continues to grow, it will become increasingly important to learn to design living environments that limit the risk of accidents without sacrificing the functionality that residents need or want; this effort must include work on the design of spaces that accommodate the types of health care devices that are more and more commonly utilized in the home. Other challenges include developing effective memory aids for helping people to take medicines in proper dosages on schedule; improving the interpretability of labels, warnings, and instructions on medicines and health care devices that are intended to be used by nonprofessionals; developing and evaluating practical ways to train people in using such devices, and devising ways to convey health care and health maintenance to consumers generally.

ENVIRONMENTAL CHANGE

The topic of detrimental environmental change is not usually associated with human factors research but is included in this report because it is a serious national and international problem that has elements that deserve attention from human factors researchers. Scientists, policy makers, and the general public have become increasingly concerned about the threat that certain changes in the environment pose. This threat has many dimensions, and human activities are among its causes. Because these activities are aimed at satisfying human needs and desires, it is reasonable to assume that they will increase as world population grows. Some research has been done on how to get people to modify behavior to make it more environmentally benign. A complementary effort is needed to find ways to modify technology so as to decrease the opportunities it affords for harming the environment without impairing its effectiveness in meeting human needs.

Because many of the most significant threats to the environment are direct consequences of the production and use of energy, a major objective should be improving the efficiency of energy use and substituting forms of energy that do not harm the environment for those that do. Computer and telecommunications technologies—including electronic mail, teleconferencing, computer-supported work facilities, and virtual reality systems—provide many opportunities to substitute energy-light and material-light resources for energy-heavy and material-heavy ones. The challenge to human factors researchers is to help ensure that such facilities are sufficiently well designed from the users' point of view that they will be used in ways that decrease overall energy demands. Other challenges include helping to make mass public transportation a more attractive alternative to private automobiles for transport in urban areas; improving, from a human factors perspective, approaches to recycling and waste management, including procedures for handling radioactive and toxic wastes; and increasing accessibility of information in environmental databases to users and potential users of that information.

COMMUNICATIONS TECHNOLOGY AND TELENETWORKING

Since the establishment of the first computer networks, network technology has advanced rapidly. The ARPANET began as a four-node system in 1969; its successor, the Internet, now connects computers around the world, and the number of users, already in the millions, has been doubling annually for some time. With the carrying capacity of networks continuing to increase, it is now possible to transmit enormous amounts of data (including digital voice and high-quality video) from almost anywhere to almost anywhere at relatively low cost. People have unprecedented access to information (in libraries, museums, news databases, and other repositories), to information-based tools and resources, and to other people. Computer-based communication should remain a high priority for human factors research, because in the future nearly everyone is likely to be a user of this technology, directly or indirectly, for work and for numerous other reasons.

Human-computer interaction has been the focus of considerable human factors research in the past. Many of the problems that have received attention will continue to be major challenges for research. These include issues relating to interface design, information finding and utilization, personal information management, and users' comprehension of the systems with which they interact. Speech as a means of communication deserves continuing human factors research, and "walk-around," three-dimensional representations of objects and environments with which users can interact—i.e., virtual reality—are innovations that will require the resolution of a variety of human factors issues before their potential can be fully realized. Other needs and opportunities for human factors research involve the development of tools to facilitate interaction with huge and complex databases and studying interpersonal communication through computer-based systems to compare it with more traditional communication and to learn how to shape its technology to enhance its effectiveness for the various human purposes it is intended to serve.

INFORMATION ACCESS AND USABILITY

Among the more striking characteristics of modern society—the "information society"—is the dependence of its many institutions on constant accessibility to accurate, up-to-date information. Associated with this increasing dependence on information is a greatly increased quantity of information available to people in all walks of life. Availability does not necessarily mean ready accessibility, however, and users, or potential users, of information that is available often experience frustration in locating, accessing, and interpreting the information they want. Any effort to make information more accessible to people who need or want to use it—including many people who are not technically trained or oriented—must devote attention to a variety of human factors issues relating to the ways in which people might interface with information repositories and tools that are intended to facilitate finding, using, and conveying information.

Much relevant work has focused on the design of interfaces and query languages for databases representing restricted domains, on the effects of alternative organizations or representations of databases for search and retrieval, and on the relative effectiveness of various information search techniques. These issues will continue to require research. As databases continue to increase in both size and complexity, effective "navigational" tools will become more important to the user and more difficult to develop. Evaluating the effectiveness of information systems is, in large measure, a human factors problem. To support the design of effective automated search and retrieval systems, or retrieval aids, progress is needed on quantifying the value of information as distinct from the amount of information retrieved. Assessing the effectiveness of systems that are intended to facilitate browsing, as opposed to structured search, is another difficult, and unsolved, problem. Related problems include the need for a better understanding of how to exploit the potential of computer-based information systems for presenting data and information in innovative visual forms and of how to use the spatial metaphor to construct data worlds that can be explored effectively by moving around within them.

EMERGING TECHNOLOGIES IN WORK DESIGN

Many U.S. industries, especially in manufacturing, have found it increasingly difficult to remain competitive in the global marketplace. Competing effectively in the future will depend on the ability of the country to use its material and human resources more efficiently, and especially on its being able to adapt quickly to rapidly changing technology and market conditions. This means, among other things, that firms must become sufficiently flexible to shift production quickly from one item to another and that they must be able to produce items economically in relatively small quantities. This in turn requires a different job design from that found on conventional twentieth-century assembly lines. Emerging technologies—automation, robotics, information technologies—will figure prominently in the workplace changes that will occur.

Jobs have always changed over time as new technologies have been introduced in the workplace, but the rate of change has accelerated considerably in the recent past with the pervasive infusion of computer-based technologies. Human factors researchers must devote more attention to the introduction and use of new technologies in the workplace. This need has become evident in such disparate problems as the physical difficulties (e.g., eyestrain and carpal tunnel syndrome) that are sometimes associated with the use of information technology devices and the claims of some investigators that these technologies are not providing the expected increase in productivity. Other challenges include reducing human error in the workplace, especially in information work in which errors that are difficult to identify can significantly affect the quality of work products; determining the skill requirements of new high-technology jobs and of ways to combine human and machine capabilities effectively; and developing a better understanding of the increasing demand for people to play the role of supervisory controllers, and of backups in the case of failure, in highly automated systems. A special challenge is helping to ensure that the work climate, culture, and environment are humane and fulfilling to the human beings who work within it.

TRANSPORTATION

The goal of all forms of transportation systems is the expeditious movement of people and goods from place to place with a minimum risk to life and property. These two objectives—maximizing speed and minimizing risk—make up only one of several trade-offs involved in the design and operation of any transportation system. Helping to resolve these trade-offs is, in part, a human factors problem because these trade-offs involve questions of (1) how well vehicles can be controlled in various situations given human capabilities and limitations and (2) considerations of human attitudes and preferences with respect to comfort, convenience, and risk acceptance.

Considerable research has been done on driver and pilot performance, and many sources of driver/pilot error have been identified, some of which have affected vehicle and cockpit design. There are, however, many remaining needs and opportunities for human factors research, especially in view of the ever-increasing technical complexity of vehicles and transportation systems. An example of a problem relating to driving safety that has received some attention but that needs further research is the dependence of driving performance on dynamic visual acuity. To receive a license, drivers are currently tested for static visual acuity, but there is evidence suggesting that dynamic visual acuity, for which tests are not given, is a more effective predictor of accident probability. More work is needed on this important question. We also need work leading to a better understanding of the roles of alcohol and drug intoxication in driving and of the effects of driver attitudes toward risk. The introduction of increasingly sophisticated equipment in automobiles and airplane cockpits, as well as the automation or semiautomation of more and more functions, raises many issues that human factors research should address.

COGNITIVE PERFORMANCE UNDER STRESS

Stress manifests itself in modern life in a variety of ways, sometimes tragically. Some occupations are generally viewed as especially stressful, air traffic controllers being a case in point. Here the reason for the stress—the possibly tragic consequences of a significant error—is obvious. Jobs can also be stressful for much more subtle reasons; they may cause low levels of relatively continuous stress, a situation that can also have major consequences for workers. Although much research has been done on the effects of stress on cognition, the need for research on this topic is likely to increase over the foreseeable future for several reasons, including threats to job security from increasing competitive pressures in many industries, the need to accommodate to increasingly rapid change in the workplace, the greater cognitive demands of many jobs, and the application of new methods of electronic monitoring and surveillance.

Laboratory studies of stress have been limited by well-grounded restrictions on exposing participants in experiments to genuinely threatening situations and the infeasibility of the long-term experimental studies that would be required to study chronic stress. One priority for future research must be a search for feasible ethical solutions to the problem of these limitations and for ways to study the influences of stressors that more closely approximate those that occur in real-world situations. Possibilities that have been used to some extent but that should be more fully and creatively exploited include analyses of naturally occurring incidents and the use of simulations. Virtual reality technology should extend the possibilities for simulating stressful situations with a high degree of realism without exposing people to physical threats; how best to exploit this technology for this purpose should be one goal of research. Other major goals of research should be to develop a better understanding of how the more detrimental effects of stress can be counteracted, how to predict individual differences in reactions to stress, how to train people to cope with stressful situations, and how to make person-machine systems more resistant to the detrimental effects of stress.

AIDING INTELLECTUAL WORK

As computer technology has become widely available in the workplace, many software tools have been developed to help people perform intellectual tasks. Efforts to develop additional and more powerful aids to intellectual work will undoubtedly continue for the foreseeable future. Human factors research is needed to evaluate the effectiveness of such aids, to provide—through task analyses—a better understanding of what further aids would be useful, and to participate in the design, implementation, and iterative improvement of such aids. Past research has demonstrated that the introduction of technological aids to intellectual work has not had the anticipated effects and has not always increased either productivity or people's satisfaction with their jobs.

Evaluative studies are needed to provide the insights required to improve upon existing aids and to ensure that they accomplish what is intended. There is also a need to study the demands of intellectual work in specific settings to identify how it could be aided by the development of new tools; any such studies should be done with a sensitivity to the fact that new tools often change the nature of a job by making it possible to do things that could not be done before. Human factors researchers can contribute to the design, implementation, and evaluation of new aids for intellectual work, especially by bringing to the process an iterative design philosophy and a focus on users' needs; these factors have proved to be important in the development of systems whose ultimate characteristics are impossible to specify in detail in advance. Other challenges for human factors research include (1) identifying nonobvious but important aspects of current ways of performing tasks (e.g., spatial arrangement of papers on a desk serving as reminders of task priorities) for which some substitution should be made in an electronic aid and (2) developing a better understanding of why people often fail to use aids that could improve the performance of their jobs.

CONCLUSION

The needs and opportunities for research discussed in this report are considered by the Committee on Human Factors to be among the more important challenges to the human factors research community for the immediate future. The list is not exhaustive, nor was it intended to be, but the committee believes the items on it to be important without exception. The committee recognizes too that none of the problems discussed is the exclusive province of human factors research; most of them are broad in scope and deserve attention—especially collaborative attention—from many disciplines, human factors among them.

PART I

Summary Report

Emerging Needs and Opportunities for Human Factors Research

INTRODUCTION

This summary report identifies the areas that the Committee on Human Factors believes represent new needs and opportunities for human factors research during the next few decades. It is organized as follows: after a background discussion, we describe the process by which the committee determined which topics to cover. Each topic is then considered in turn, and the committee's consensus regarding needed research is presented.

A set of papers addressing these topics appears as Part II of this volume. These papers were written by committee members, sometimes with assistance from colleagues, as part of the process of informing the committee's discussions of the topics addressed. The conclusions and recommendations articulated in this report draw heavily on these papers.

Background

The National Research Council established the Committee on Human Factors in 1980. The committee's original sponsors were the Office of Naval Research, the Air Force Office of Scientific Research, and the Army Research Institute. The National Aeronautics and Space Administration became a sponsor in 1981. The committee's charter was to identify basic research needs of the military services as they relate to human factors issues and to make recommendations for basic research that would improve the foundations of the discipline.

The committee's first report, *Research Needs for Human Factors*, published in 1983, focused on six topics: human decision making, eliciting expert judgment, supervisory control systems, user-computer interaction, population group differences, and applied methods. These topics were selected after committee discussions of research needs, tours of military laboratories, and solicitation of suggestions from the human factors community through an article in the *Human Factors Society Bulletin*. Topics were selected because they were germane to the sponsors' interests, within the expertise of the committee members to address, incompletely addressed by previous or ongoing research, and important vis-à-vis the committee's charter.

Since the publication of *Research Needs for Human Factors*, the committee has been responsible for numerous panels and workshops, many of them about problem areas discussed in that report. Figure 1 gives a complete list of the reports produced by these panels and workshops.

Since the committee's establishment, its sponsorship has broadened con-

1994	*Organizational Linkages: Understanding the Productivity Paradox*
1993	*Workload Transition: Implications for Individual and Team Performance*
1992	*Human Factors Specialists' Education and Utilization: Results of a Survey*
1990	*Application Principles for Multicolored Displays: A Workshop Report*
	Quantitative Modeling of Human Performance in Complex, Dynamic Systems
	Distributed Decision Making: Report of a Workshop
	Human Factors Research Needs for an Aging Population
1989	*Human Performance Models for Computer-Aided Engineering*
	Fundamental Issues in Human-Computer Interaction
1988	*Human Factors Research and Nuclear Safety*
	Ergonomic Models of Anthropometry, Human Biomechanics, and Operator-Equipment Interfaces: Proceedings of a Workshop
1987	*Human Factors in Automated and Robotic Space Systems: Proceedings of a Symposium*
	Mental Models in Human-Computer Interaction: Research Issues About What the User of Software Knows
1985	*Human Factors Aspects of Simulation*
	Methods for Designing Software to Fit Human Needs and Capabilities: Proceedings of the Workshop on Software Human Factors
1984	*Research Needs on the Interaction Between Information Systems and Their Users: Report of a Workshop*
	Research Issues in Simulator Sickness: Proceedings of a Workshop
1983	*Research and Modeling of Supervisory Control Behavior: Report of a Workshop*

FIGURE 1 Reports of the Committee on Human Factors, 1983-1994.

siderably, and new sponsors have come from both military and civilian sectors. The committee's composition has also broadened to include not only researchers who represent traditional human factors interests in equipment design and use but also some with more cognitive and social orientations.

Concurrently, technology has been advancing rapidly, and both the nation and the world have changed in significant and remarkable ways. Some of these changes offer new challenges and opportunities for human factors research.

In view of these developments, the committee found it appropriate to again address the general topic of research needs and opportunities for human factors. Committee members agreed that the report should be:

- reflective of the committee's views and opinions, and also informed by inputs from several sources;
- forward looking: short on reviewing old and current work and long on identifying problems and opportunities for the future;
- problem/opportunity-oriented: the subject matter being determined more by what the committee perceives the needs and opportunities to be than by its understanding of what the existing research activities of the human factors community are; and
- selectively focused—with no attempt to be comprehensive—on a few major topics that the committee believes to be among the more important problem/opportunity areas for the near future.

Committee members took the lead in writing the papers that constitute Part II and that provide the material on which Part I is based.

Topic Selection

The process of selecting topics for emphasis was lengthy and deliberative. The decision was made early to cast a wide net for ideas and opinions and then, through committee discussion and debate, to attempt to reach a consensus regarding a subset on which to focus. Suggestions of major research needs were solicited from current and former members of the committee, from committee sponsors, and from numerous other members of the human factors research community, including several human factors leaders in Europe. The committee also reviewed the topics generated as candidates for inclusion in the original *Research Needs* report, the list of technical groups of the Human Factors and Ergonomics Society, the list of titles of Committee on Human Factors reports published or in process, and recent project suggestions from sponsors. This yielded a long list of suggested research topics.

As the list grew, the committee discussed it at several meetings. At one such meeting, an analysis of responses to a written request for suggestions from past and current committee members was presented and discussed. A complete list of all suggestions from all sources was then distributed to committee members, and at a later meeting, the members engaged in a structured exercise designed (1) to permit every member to identify all the suggestions he or she considered worthy of further consideration, (2) to facilitate grouping the suggestions to eliminate redundancies and replace similar suggestions with a more general category, and (3) to prioritize the resulting list through a quasi-formal, iterative voting procedure.

At all of these meetings, the discussions were long and spirited. All members saw the goal of identifying human factors research needs and opportunities as very important. However, the task of selecting a few problems or problem areas for special attention was difficult because selecting some areas for inclusion meant excluding others.

The committee's deliberations eventually led to a consensus on the topics that should be covered in its report. Continuing discussion resulted in some minor changes in terminology in the interest of clarity and the merging of some topics because of content overlap. Some of the selected topics are well within the mainstream of traditional human factors research; others are not. The committee intentionally took a relatively broad view of human factors and did not exclude a problem area simply because it has not traditionally been a major focus of the discipline. As it happens, the topics are fairly easily grouped under three major headings—national or global problems, technology issues, and human performance—as follows:

National or global problems
- Productivity in organizations
- Training and education
- Employment and disabilities
- Health care
- Environmental change

Technology issues
- Communication technology and telenetworking
- Information access and usability
- Emerging technologies in work design
- Transportation

Human performance
- Cognitive performance under stress
- Aiding intellectual work

Two topics the committee considers to be priority research areas—human performance modeling and human error—are not covered in this volume, because committee panels were established on both. A report on human performance modeling was recently published (Baron et al., 1990) and one on human error management in high-hazard systems is currently in preparation. With these exceptions, the topics on which the committee finally settled constitute the titles of sections of this summary report and of the papers that make up Part II.

The committee believes that these areas are, without exception, very important. However, failure of a topic to appear as a heading should not be taken as evidence that the committee considers the topic unimportant. The committee recognizes that many more human factors problems and problem areas are deserving of research than can be covered in a report of this sort.

This report focuses on needs and opportunities for human factors *research*; however, the committee also wishes to note that one of the most pressing needs at present, and probably during the near future, is to get the results of human factors research *applied* to equipment, procedures, systems, and situations. It continues to be easy to find equipment and operating procedures being designed in ways that are inconsistent with well-documented human factors research results. How to improve the dissemination of the results of human factors research, how to make this information readily available to users in a helpful form, and how to make potential users aware of its existence, are continuing major challenges to human factors researchers and to everyone who recognizes the potential usefulness of the results of human factors research.

The Report

Intended Audience

This report is intended for people who identify research funding priorities and sponsor research programs and projects and for those who perform human factors research. People responsible for identifying research funding priorities and sponsoring research efforts will find here the collective opinions of the committee on how needs and opportunities for human factors research relate to several problems of major national concern. Researchers will find specific questions that the committee believes constitute challenges to research that could contribute significantly to the solution or amelioration of these problems.

Contents

In keeping with the committee's decision to be forward-looking, little effort was made to review research extensively in problem areas that are or have been major foci. A very modest amount of reviewing is done in this summary report and somewhat more in Part II, but, in both cases, only to help provide an appropriate frame of reference for thinking about future needs.

Some of the committee's recommendations are relatively general; others, quite specific. Some follow traditional lines of research; others relate to problems that have not received much attention from the human factors community in the past. In the latter cases, the recommendations are offered primarily as points of departure for further discussion and planning rather than as items for a research agenda. The justification for including them is the perceived seriousness, from either a national or a global perspective, of the problem areas involved.

The opinions expressed in Part II are those of the authors of the papers, but their shaping has been strongly influenced by inputs from other members of the Committee on Human Factors and the other sources of ideas mentioned above. The recommendations that appear in this summary report are made by the committee as a whole.

Organization

The remainder of this summary report is organized by topic. A section is devoted to each of the topics selected by the committee for emphasis. In each case, the section describes the problem, briefly discusses representative previous work that relates to the problem, and gives a set of recommendations for research.

PRODUCTIVITY IN ORGANIZATIONS

The Problem

Productivity is a major national and international concern. Economists see productivity as a primary determinant of competitiveness both among companies within an industry and among national economies. Because it is also believed to be causally linked to standard of living, increasing productivity globally is seen to be the best hope of improving living conditions worldwide.

Although the United States is the most productive country in the world, its annual rate of increase in productivity has been considerably smaller than that of several other industrialized countries during the last few decades (Bureau of Labor Statistics, 1988a, 1988b). As a consequence, our national competitiveness has decreased significantly in the automobile, steel, shipbuilding, and textile industries and appears to be slipping in electronics, computers, robotics, and biotechnology as well. The U.S. share of the total world economy decreased from 35 percent in 1965 to 28 percent in 1985 (Johnston and Packer, 1987).

This trend can be attributed to many factors, such as the diffusion of technology, the globalization of work, and the development of new work processes and new work structures throughout the world. Many economists, however, are concerned that productivity in the United States is not what it should be and that the nation's competitiveness will continue to decline if ways are not found to accelerate the rate of productivity increase.

Human factors researchers have given considerable attention to the question of how to improve human performance in the workplace and thereby increase individual productivity. Yet they have made relatively little effort to determine how individual productivity relates to the productivity of the groups, organizations, or industries within which the individuals' work is done. Our focus is on organizational productivity and how the human factors community can contribute to improving it.

The prevailing assumption appears to be that increases in individual productivity automatically translate into increases in productivity at higher levels of organization, but this assumption gets little support from research. There is little evidence that increases in productivity at one level, say the level of the individual worker, automatically translate into increases in productivity at a higher level, say that of a work team, a corporation, or an industry. Similarly, there is very little theory or research that helps us understand how observed changes in organizational productivity relate to changes at the individual or group level.

The challenge is to better understand organizational productivity. We want human factors researchers to think about their work in terms of improvement in organizational productivity.

Previous Research

Most of the research on productivity has dealt with productivity at a particular level—individual, group, corporation. The research that has been done by applied psychologists and human factors specialists has dealt primarily with the individual. Training has been identified as an important determinant of individual productivity (Guzzo, 1988), along with goal setting (Locke and Latham, 1990) and the details of the design of specific tasks (Guzzo, 1988).

At the level of work teams or small groups, self-management has received some attention as a determinant of productivity (Goodman et al., 1988; Hackman, 1990). Several studies of larger systems have focused on the effects of the introduction of automation technologies in the workplace and, interestingly, have not found that greater automation invariably means higher productivity at the system level. In particular, technological change often has had little or no positive effect on system productivity unless appropriate organizational change has accompanied it (Goodman, 1979; MacDuffie and Krafcik, 1990).

Although the previous research has provided some knowledge of the determinants of productivity at specific levels of organization, very little of it has been directed at determining how improvements at one level do or do not effect changes at higher levels of operation. In particular, we know something about how to increase the productivity of individuals, but we are unable to specify the conditions under which improvements at this level will translate into improvements in the productivity of an organization as a whole.

Research Opportunities

Workplace interventions that appear to have the potential to increase productivity not only at the level of the individual worker but also at that of the organization as a whole are sometimes not implemented successfully (Goodman and Griffith, 1991). There is a need for better understanding of the implementation process and, in particular, of how to ensure that innovations that would increase individual and organization productivity will be adopted and used.

One opportunity for research is the exploration of how changes in organizations or in technology are related to improved productivity. Billions of dollars have been spent on new forms of technological and organizational change, many of which have clear or inherent benefits. There is, however, growing research evidence that many innovations are not successfully implemented or are implemented on a temporary basis. As a result, the objective benefits of the proposed technological and organizational changes are canceled out. If we cannot successfully implement improvements at the individual, group, or organizational level, we cannot expect to see productivity changes at the organizational level.

The challenge, then, is to find new research approaches to successful implementation and institutionalization. Much of the work over the past decade is on how different variables (e.g., top management support) facilitate the change process. This work suggests that new developments will come from identifying the critical processes that drive the implementation. Furthermore, a change in methodology is required—for example, there are very few multivariate and/or longitudinal studies capturing the implementation of new technologies or organizational intervention.

Another opportunity for research is the linkage between individual and organizational productivity. Focusing on single levels of analysis (e.g., individual, group) will not increase our understanding of organizational productivity. We assume that successful implementations create productivity increases at the individual level and that these changes will eventually appear at the organizational level. This assumption is incorrect—changes at one level may have no effect on other levels of analysis. There is neither good theory nor research about this linkage issue.

The challenge for the human factor researcher is first to acknowledge and understand the issues associated with these linkages. What are the factors that prevent productivity changes at the individual level from increasing productivity at the individual level? What are some of the enablers? Concepts such as organizational slack and different forms of organizational interdependence are starting points of our analysis. Increasing productivity on some jobs can simply create slack time that does not translate into increased organizational interdependence. The degree and form of interdependence between jobs and between organizational units may to a large degree determine whether changes in one unit will lead to positive, negative, or no changes in other units. Horizontal and vertical linkages are critical in determining whether changes in productivity at one level get translated to another level.

A third research opportunity is the congruency among technology, people, and organizational factors. It seems inevitable that organizational productivity must depend, in part, on the ways these factors interrelate. The empirical evidence supports the view that focusing solely on one factor (e.g., productivity) will not increase organizational productivity; rather it is the simultaneous restructuring of organizational, technological, and people factors that is key to increases in organizational productivity.

Although we know that congruency among these three factors is important, much of our research evidence is after the fact. We do not have any ex ante models about how combinations of organizational, technological, and people factors lead to increased organizational productivity. This is an important issue because the workplace is changing rapidly. All indications are that rapid change in technology, organizations, and characteristics of the workforce will continue. The opportunity for human factors researchers is to provide new insights into how combinations of changes in technology, organization, and people factors contribute to organizational productivity.

A fourth opportunity for research concerns internal and external integration. New innovations in design processes and linking production processes in manufacturing and retail/distribution organizations signal the importance of internal integrations. At the same time, there have been dramatic changes in relationships with suppliers and customers. "Arm's-length" transactions are a thing of the past. Boundaries between customers, suppliers, and the focal organization have become very permeable.

Although these forms of integration are initiated with the expectations of improving organizational productivity, little research is available. Consider the following: in our analysis of organizational productivity, we generally focus on linkages among levels (e.g., individual and group) within the organization. But let us assume that greater levels of integration between the organization and its customer increase productivity for the customer. The interesting question is: How do changes in productivity for the customer feed back and affect productivity of the focal organization?

TRAINING AND EDUCATION

The Problem

The critical importance of worker training to the national economy and to the general well-being of the nation is widely recognized. It is also recognized that the educational level of prospective trainees strongly influences the efficiency and effectiveness of training. Unfortunately, neither our education nor our training systems, as currently structured, appear to be meeting the country's needs. A key symptom of the problem is the growing number of programs in which employers attempt to provide remedial education for incoming employees as a way to build a base upon which specific, job-oriented training can proceed.

The seriousness of the education and training problem has been documented by numerous published studies, including the widely cited *A Nation at Risk* (National Commission on Excellence in Education, 1983) and subsequent reports by the Hudson Institute (Johnston and Packer, 1987), the William T. Grant Foundation (1988), the Commission on Workforce Quality and Labor Market Efficiency (1989), the Commission on the Skills of the American Workforce (1990), the Office of Technology Assessment (1990), and the National Council on Education Standards and Testing (1992).

The problem could become worse in the future because the workforce will need to be even more proficient—especially more versatile and adaptive—than it is now. This expectation is rooted in the rapidity of technological change and the increasingly tight interconnectedness of the economies of the world into a single global ensemble. Technological change means changes in job requirements. The ability to satisfy changing, and not entirely predictable, job requirements in a complex, culturally diverse, and constantly evolving environment will require a literate workforce that has good problem-solving and learning skills. The challenge to education and training will be amplified by expected changes in the demographics of the workforce. In the aggregate, these changes mean that an increasing percentage of the labor needs will be met by nontraditional sources; in some cases, special training will be required to tap these sources effectively. Furthermore, because advances in technology tend not only to create new jobs but also to increase the speed with which many existing jobs become obsolete, the need for worker retraining will become increasingly common.

Previous Research

A great deal of research on education and training has been done by psychologists and educational researchers. Inasmuch as our focus is primarily on postsecondary and work-based education and training, it suffices to mention here the following publications that, taken together, provide extensive coverage of research on the design, delivery, and evaluation of training methods and systems: Druckman and Bjork (1991), Druckman and Swets (1988), Goldstein (1988, 1992), Wexley (1984), Latham (1988), and Tannenbaum and Yukl (1991).

Unfortunately, many of the basic research findings on learning and skill development do not cross the boundary between the laboratory and the workplace. As a discipline that historically bridges psychology and engineering, human factors should be able to help transfer what has been discovered in the laboratory to the design of education and training systems. How best to do this is itself a question that will require intensive study.

Training and human factors are sometimes characterized as complementary ways of getting human-machine systems to function smoothly, the idea being that the objective of human factors research should be to determine how to design systems so that little, if any, training is required for the system to function effectively. Here we take the broader view that human factors encompasses whatever enhances the performance of human-machine systems, including training. However, even within the traditional, narrower characterization of human factors, many questions arise regarding the design and evaluation of training materials, methods, and systems. The research challenges that are identified in the following section arise from both modes of characterizing the field.

Research Needs and Opportunities

Training systems tend to be complex both functionally and structurally; they have to work under numerous constraints to achieve multiple objectives. Many variables are involved in the operation of such systems. Understanding the interactions among these variables requires using a systems-analytic approach that human factors researchers have effectively applied to the study of many other kinds of complex systems. The challenge is to apply such methods to the study of training when outcome requirements arise from interactions among changing technologies, changing workforce demographics, and changing organizational structures.

Technology has the potential to change the methods of education and training drastically. Computer and communication technologies, in particular, provide the possibility of new approaches involving interactive graphics, process simulation, individualized adaptive training, embedded training, and a host of other innovations. Each of these methods, however, is valuable only to the extent that it contributes to accomplishing the teaching/learning objectives of the training system into which it has been incorporated. There is a need for evaluations of technologically innovative approaches to education and training that reveal not only how effective the approaches are but also how they might be improved. Human factors researchers can perform evaluations of the type required.

An example of a type of evaluation that is much needed is a study to determine how much (and what type of) physical fidelity is required in a system simulation generated by virtual reality. Inasmuch as small increases in physical fidelity can mean large increases in costs, it is important to know how much fidelity is enough in specific cases.

The human factors community has promoted user-centered design—designing systems to match the capabilities and limitations of their intended users—and has widely applied it to the design of many kinds of systems, especially, but not exclusively, in military contexts. User-centered design principles that have proved to be effective should be applied to the design of education and training systems as well. The challenge is to devise, evaluate, and perfect ways to transfer the methods, data, and principles of human-centered design to the development of training systems and to ensure that feedback from applications closes the loop between research and practice.

Technology has had two particularly striking effects on jobs during the last few decades: (1) a decrease in the manual component of many jobs accompanied by an increase in the cognitive component and (2) acceleration of the pace of change in the workplace. The probability that these trends will continue establishes a need for a greater research focus on how individuals and organizations can best learn to adapt to rapid changes in work procedures and organizational structures while continuing to be engaged in production processes. There is need for the development of theory, tools, and techniques that will support lifelong learning in many occupations.

Because of the increasing rapidity with which jobs and the skill requirements of jobs for them are changing, it is also becoming increasingly important to have effective methods for anticipating future job requirements so that education and training programs can be designed to provide workers with the requisite skills in a timely way. Formal procedures used by human factors researchers to identify or analyze the skills, knowledge, and abilities that are requisite to the adequate performance of specific tasks should be useful in establishing education and training goals, but there is also a need for the development of more effective methods of identifying cognitive, as opposed to psychomotor, task demands.

Despite the widely acknowledged imperfection of current education and training systems and programs and the considerable amount of money and effort that continues to be spent on developing them, the technology for evaluating their effectiveness is still primitive and not very sensitive. A major practical need is for significant improvement in the means for performance assessment. Human factors researchers have experience in evaluating the effectiveness of complex systems of many types and thus should be in a position to help meet this need.

EMPLOYMENT AND DISABILITIES

The Problem

The population of people with one or another type of physical or cognitive disability is very large. According to the U.S. Census Bureau, there are about 75 million people with some sort of disability in the United States, about 30 percent of the entire population (Bureau of the Census, 1987, 1989). Not all of these disabilities are sufficiently severe to interfere with work or other activities, but, if we count only those that are, the number is still large, probably as many as 30 million in the United States (Elkind, 1990).

The rate of unemployment among people with disabilities who would like to work is several times higher than the unemployment average nationwide (Kraus and Stoddard, 1989). And the percentage of people who live below the poverty line is between two and three times greater for people with a disability that interferes with their ability to work than for the total working population (Vachon, 1990).

Failure to find more effective ways to make employment available to people with disabilities is costly to the national economy in two ways. The cost of maintenance support programs is approximately $100 billion per year and is increasing rapidly (Vanderheiden, 1990), and the cost of the country's lost opportunity to utilize the knowledge and skills of people with disabilities, although difficult to quantify, is surely also great.

Previous Research

Most of the previous research that directly relates to enhancing employment opportunities for people with disabilities has aimed at finding ways to use technology to mitigate the limiting effects of disabilities. The objective in many cases was the development of a device of some sort to help people compensate for the loss of sight, hearing, mobility, or other specific functions.

For at least two reasons, much of this work has focused on computer and communication technologies. First, the versatility of the computer provides opportunities for creative approaches to the augmentation of abilities that have been diminished by physical or cognitive disorders, disease, or accident; this has been amply demonstrated by the innovative ways that people with disabilities have adapted these technologies for their own use (Bowe, 1984). Second, because more than half the workforce is now engaged in information-oriented jobs (Strassman, 1985) and the percentage is expected to increase in the future (Kraut, 1987), it is also natural to look to computer and communication technologies as a source of job opportunities for people with disabilities.

Some emphasis on computer and communication technologies to enhance work opportunities for people with disabilities in the future is warranted and for the same reasons. We believe that these technologies offer great, and largely unrealized, potential to help remove barriers that have kept people with disabilities from jobs. We also believe that many, if not most, of the new job opportunities that could be readily made available to people with disabilities are likely to be in areas that make heavy use of these technologies.

Unfortunately, most of the research on how the requirements of specific jobs relate to the capabilities and limitations of prospective jobholders has excluded people with disabilities from consideration. Data have almost invariably been collected only on able-bodied people; people with disabilities are not even statistically represented in the results. As a consequence, little is known of how the capabilities and limitations of people with specific disabilities compare with the requirements of specific jobs. A requirement in this context is something that is essential to getting a job done; job requirements are to be distinguished from the architectural, social, and other barriers that hinder a person from getting to a job location or situation but are not essential aspects of the job itself.

Research Needs and Opportunities

Most disabilities are impediments only to certain activities, and typically many fewer activities are affected than are not. Specifically, any given type of disability usually impedes only *some* of the activities required by *some* jobs, and often even those impediments can be removed or overcome by restructuring the job or using assistive technology. A major need is for a much better understanding of precisely how the capabilities and limitations associated with specific disabilities relate to the functional requirements of specific jobs. This need suggests the following two objectives:

(1) development of a database of information on the performance capabilities of people with the more prevalent types of disabilities, with a view

to incorporating such information in design references, design texts, and in human performance models, thereby enhancing the ability to design for people with disabilities; and

(2) systematic analysis of the skill requirements of jobs, especially jobs in the information sector, to match requirements with the performance capabilities of people with disabilities. A clear distinction should be made between essential job functions and nonessential aspects of job situations that may have evolved as conveniences to able-bodied individuals who typically perform them.

Some jobs cannot be done by people with certain types of disabilities in the same way in which they are standardly performed by able-bodied people. It is important to know whether those jobs could be performed just as effectively in alternative ways that would be manageable by people with disabilities. A useful first step would be an extensive study of previous efforts to redesign jobs to accommodate people with specific disabilities; such a study should identify differences between efforts that have worked and those that have not.

As was noted already, the development of devices to assist people with disabilities in overcoming specific functional limitations has received considerable attention, especially from scientists and engineers who may be keenly aware of the kind of assistance that is needed because they are disabled in some way or because someone close to them is. Human factors researchers have been somewhat involved in these efforts, but this has not been a major focus of the discipline as a whole. The discipline has much to contribute to this design challenge and much to learn from involvement in this type of work.

Unfortunately, knowledge of much of what has been done to apply technology to benefit people with disabilities is not readily accessible, because the work was not all done by people who are members of the same professional community for which the channels of communication are well established and known. Much of the information exists in relatively inaccessible documents as opposed to mainstream technical journals. It can be very hard to obtain detailed information on a particular assistive device or type of device—or even to discover whether a device to serve a specified purpose exists. A very useful objective would be to develop a set of computer-based information resources on assistive devices, training programs, job opportunities, and job requirements for people with disabilities. Such resources could also serve the research community, making information on past and ongoing research much more readily accessible than it currently is.

HEALTH CARE

The Problem

Health risks stem from many sources. Prominent among them are behavioral and occupational factors. Cancer and heart disease, for example, are often attributable to smoking, diet, or other lifestyle variables (Newell and Vogel, 1988; Williams, 1991). Long-term occupational exposure to radiation or toxic substances is known to be a significant cause of certain types of illness, as is chronic job-related stress (Levi, 1990; Smith, 1987; Swanson, 1988).

Paradoxically, significant threats to health are to be found even within the technology for the delivery of health care. Many of these threats arise from the possibility of human error in using health care equipment and carrying out health care procedures. Such errors can have tragic consequences. For example, Bogner (1991) has cited the substantial number of potentially preventable incidents in which anesthesia has resulted in brain damage or death.

Much of the equipment in modern hospitals and trauma centers is complex, and even highly trained people can inadvertently use it incorrectly with injurious effects. In addition, more and more commonly, people without medical training are using complicated equipment, including ventilators, infusion pumps, and kidney dialysis machines, in their homes; these devices can extend and enhance the quality of life when used correctly, but they can cause irreparable harm when used erroneously.

Injurious errors can also occur in the administration of drugs, either by administration of the wrong substances or incorrect doses of the right ones. This problem too is compounded by the fact that patients are often responsible for their own drug regimens and are sometimes confused as a consequence of the effects of illness or age. An important challenge to research is developing ways to reduce the incidence of human error in the delivery of health care; this issue is likely to grow in importance as the technology that is available for health care delivery becomes ever more complex.

Previous Research

Although there have been attempts to call attention to the many challenges that medicine and health care delivery provide for human factors researchers (e.g., Pickett and Triggs, 1974), health care has not traditionally been a major focus of human factors research.

The human factors community has done considerable research both to determine the causes of human error in a variety of situations in which people interact with machines and to identify ways to decrease the probability of such errors or reduce harm when they do occur (Reason, 1990; Senders and Moray, 1991). Relatively little of this work has been done in medical or health care contexts.

Innovative work has been done on evaluating the relative effectiveness of medical imaging devices (Swets et al., 1979; Swets, 1988) and on human factors aspects of the design of imaging workstations (O'Malley and Ricca, 1990). Some human factors evaluations have been done of glucose measuring devices of the type used by people with diabetes (Kelly et al., 1990; McDonald, 1984; Moss and Delawter, 1986). Such studies have revealed a number of problems with the use of even these devices, which are relatively simple compared with many that are likely to become common for home use in coming years.

Research Needs and Opportunities

Accident prevention and wellness maintenance are important aspects of health care. As the population continues to age and as more and more elderly and very elderly people are able to live independently, an important objective of human factors research will be to determine how to design living environments that limit the risk of accidents without sacrificing the functionality that residents need or want. An aspect of this challenge that deserves increasing attention is designing living spaces for the elderly that can accommodate the types of health care devices that are more and more commonly utilized in the home.

Research in several other areas would also enhance the opportunities of people to live relatively independently, despite various types of sensory, psychomotor, or cognitive limitations. These include developing and evaluating memory aids to help people take medicines in proper dosages on schedule; improving the interpretability of labels, warnings, and instructions on medicines and on health care devices that are intended to be used by nonprofessionals in the home; and developing and evaluating practical approaches to training in the use of such devices.

Important questions for research are how to convey to nonspecialists accurate and understandable representations of (1) the health risks (or benefits) associated with specific behaviors and (2) the prospects for the various possible outcomes from specific types of treatments of their medical problems. More generally, the challenge is to find effective ways both to convey the information that people need in order to make truly informed decisions about their health care and to ensure that the information conveyed has been understood.

The need for group programs to increase positive health practices offers an opportunity for human factors involvement in health-related work. Programs to reduce smoking, decrease fat intake and increase fiber, reduce recreational sun exposure, and induce participation in screening programs for such diseases as breast and colon cancer should have beneficial results. Human factors, broadly defined, could help both in designing such programs and in evaluating their effectiveness.

The designs of medical devices need to be evaluated for usability and safety, just as do the designs of other devices. Such evaluations are especially important for medical devices because a patient's well-being often depends on the device's being used effectively.

Information systems—medical databases, diagnostic and decision aids, expert systems, and computer-based advisers—are becoming increasingly available for medical practitioners of all types. These tools differ greatly in how well they meet adequate design standards from a human factors point of view, and the amount of use they get varies greatly among potential users. Inclusion of the design features that have traditionally been considered essential to usability for a computer-based device or system appears not to be sufficient to ensure acceptance or use by health care providers. Research is needed to learn why potentially helpful computer-based systems are or are not used.

Among the many spectacular developments in medical technology during the recent past is the wealth of new imaging techniques (e.g., computed axial tomography, positron emission tomography, magnetic resonance imaging) that permit noninvasive visual exploration of the body's internal organs and structures. In part because these techniques are very expensive, it is important to understand exactly what advantages they offer and how useful they are for diagnosis, compared with one another and with more traditional diagnostic techniques. The appropriate cost-benefit comparisons require the use of psychophysical and statistical approaches of the type that have been widely used by human factors researchers and applied psychologists on similar comparison problems and on this problem as well (Swets et al., 1979; Swets, 1988).

Another challenge to the human factors community is helping find ways to reduce the frequency and severity of the injuries suffered by health care professionals in the performance of their jobs. Back injury, for example, appears to be an occupational hazard for nurses and nursing aides, who frequently put unsafe levels of stress on the spine in performing their work (Gagnon et al., 1986; Stubbs et al., 1983). Both training and the development of more effective devices to assist in lifting tasks could help remediate this problem.

Human factors, especially that part of the field that focuses on biomechanics, has much to offer not only in preventing the chronic pain and disabling injuries that can result from inappropriate physical stresses in work situations, but also in rehabilitating people after an injury and preventing its recurrence (Khalil et al., 1990; Rosomoff et al., 1981). Developing better biomechanical models could be an important part of this effort.

ENVIRONMENTAL CHANGE

The Problem

Although the problems of harmful environmental change are not usually associated with human factors research, the topic is included in this report because the committee considers it a serious national and international problem and believes that human factors research could contribute to a better understanding of, and perhaps solutions to, aspects of it.

Concern about the threat that certain changes in the environment pose for the future has been growing among scientists, policy makers, and the public. Aspects of this threat include the possible impact on the world's climate of an increasing concentration of carbon dioxide and other "greenhouse gases" in the atmosphere (Houghton and Woodwell, 1989; National Research Council, 1983); the acidification of precipitation and its effects on lakes and streams, forests, and materials (Baker et al., 1991; Mohnen, 1988; Schwartz, 1989); air pollution and urban smog (Gray and Alson, 1989; National Research Council, 1991; Office of Technology Assessment, 1988); thinning of ozone in the stratosphere (Stolarski, 1988; Stolarski et al., 1992); contamination and depletion of fresh water supplies (la Riviere, 1989; National Research Council, 1977; Postel, 1985); depletion of the world's forests (Myers, 1989; Repetto, 1990) and wetlands (Steinhart, 1990; Wallace, 1985); and an accompanying decrease in biodiversity (Soule, 1991; Wilson, 1989) and worldwide loss of arable land (Crossen and Rosenberg, 1989; National Research Council, 1990; Schlesinger et al., 1990).

Human activities are believed to be among the major causes of these threats. The burning of fossil fuels contributes to the atmospheric accumulation of greenhouse gases, air pollution, and the production of acid rain. The use of chlorofluorocarbons as coolants and aerosols is believed to be a cause of the thinning of stratospheric ozone. Water contamination results from runoff from agricultural use of fertilizers and pesticides, from improper disposal of toxic wastes, and from salt used for roadway deicing. The list of ways in which human activities affect the environment is easily extended (Stern et al., 1992).

Because these activities are aimed at satisfying human needs and desires, it is reasonable to assume that they will increase as world population grows. Many of these activities have been much more common in the industrialized world than in underdeveloped countries, because they are associated with industrialization and the products of technology. These activities are expected to grow even more rapidly as countries in many of the underdeveloped regions of the world try to accelerate their rate of technological development.

Previous Research

Psychologists have done much research on how to get people to modify environmentally harmful behavior to make it more environmentally benign (Holahan, 1986; Russell and Ward, 1982; Saegert and Winkel, 1990; Stern, 1992). This work includes studies of the use of incentives, rewards, education and information campaigns, persuasion, and other techniques to motivate people to conserve energy or water, participate in recycling programs, generate less waste and decrease littering, and make other changes in their behavior that would be desirable for environmental preservation (Baum and Singer, 1981; Coach et al., 1979; Cone and Hayes, 1980; Geller, 1986; Geller et al., 1982).

These studies have demonstrated that the desired types of behavior change can be effected through the techniques that have been used. Unfortunately, studies that have checked for the persistence of the new behavior much beyond the end of the intervention period have generally not reported positive results (Geller et al., 1982).

A complementary approach is to try to change technology so as to maintain its effectiveness in meeting human needs while decreasing the opportunities it affords for harming the environment. Except for work aimed at identifying causes of industrial accidents, which often have substantial environmental impact (Reason, 1990; Senders and Moray, 1991), this approach has not received much attention from researchers. It is, however, one the human factors community should be well suited to take. We believe there is a need both for continuation of research on ways to motivate people to behave in environmentally benign ways and for much greater attention to the possibility of shaping technology so that the natural consequences of its use will be less environmentally damaging.

Research Needs and Opportunities

Inasmuch as many of the most significant threats to the environment are direct consequences of the production and use of energy, high-priority objectives must include improving the efficiency of energy use and substituting forms of energy that are not harmful to the environment for those that are. Demands for energy can be decreased whenever goals traditionally served by transporting people and material can be served equally well by transmitting information. Computer and telecommunications technologies have the potential, through such media as electronic mail, teleconferencing, and computer-supported cooperative work facilities, to reduce the need for travel for certain purposes considerably. The extent to which the potential of these technologies is realized in this regard will depend on their acceptability to industry and potential users. A challenge to human factors researchers is to help ensure that such facilities are well designed from the users' point of view; there is, however, also a need for research that will lead to a better understanding of what factors, in addition to interface design, make these systems more acceptable to potential users and of how to facilitate the transition to their use.

Making mass public transportation more attractive, so that it is more often used as an alternative to private automobiles in urban areas, is also in the interest of environmental protection because it will reduce the energy expenditure per passenger mile traveled. This could also help reduce noise pollution, traffic congestion, traffic accidents, and other problems that attend excessive traffic in urban areas. Greater use of carpooling could have similar beneficial effects.

Paper and paper products account for about one-third of all the solid waste produced in the United States. Much of the paper that becomes a waste disposal problem is used to store and distribute information via newspapers, magazines, books, and other print media. The technology to store and distribute much of the information electronically exists and is rapidly becoming available to the public. A challenge to research is to determine how to design the display and input devices that provide people with access to electronic information so that they are acceptable, if not preferred, substitutes for paper.

"Virtual reality" technology—which has been receiving a lot of attention in the press recently, in part because of its potential use in entertainment and education—could have implications for environmental issues. For example, to the extent that virtual systems can be effectively substituted for the sometimes material- and energy-heavy systems that have been required in the past for training purposes, the environment should benefit. Determining how real a virtual reality must be to be effective for specific training purposes is largely a human factors question.

A variety of other challenges for human factors research relate to the problems of waste reduction and waste management. One need is for the development of design criteria for consumer goods that include—in addition to functionality, usability, and user safety—such considerations as longevity, maintainability, and recyclability or disposability. Another is for improvement, from a human factors perspective, of recycling and waste management. The handling of radioactive and toxic wastes poses some especially difficult problems that human factors researchers should be able to help solve.

Other human factors challenges that would help the environment include making the information in environmental databases more accessible to users and potential users who would use that information for research on, and the monitoring and predicting of, environmental change; continued study of the problems of risk assessment, management, and communication; and continued emphasis on designing industrial systems to reduce the probability of human error, especially when it can significantly harm the environment.

The problems of environmental change has not received a lot of attention from the human factors research community in the past, and there has been relatively little debate or discussion about what the community has to offer toward solutions in this area. We believe that human factors research could have an impact on various aspects of these problems. The suggestions made here are but a few of the possibilities; they are offered in the hope of stimulating more thought and discussion by which others would be identified.

COMMUNICATION TECHNOLOGY AND TELENETWORKING

The Problem

Since the experimental establishment of the first networks linking computers from different geographical locations in the mid-1960s, network technology has advanced very rapidly. The ARPANET, which became the largest operational network in the world and remained so for many years, was started as a four-node system by the Advanced Research Projects Agency of the U.S. Department of Defense in 1969 (Heart, 1975; Heart et al., 1978). According to a recent *Science* report, its successor, the Internet, now connects about 1.7 million host computers and between 5 and 15 million users, and the numbers are doubling annually (Pool, 1993). It now appears that this technology will continue to advance rapidly and that its applications will become increasingly pervasive over the foreseeable future.

The establishment and proliferation of computer networks have been accompanied by—indeed made possible by—an ever-increasing blurring of the distinction between computer and communication technologies. For instance, computing resources are heavily used in the operation of communication networks, and the capabilities and services to which these networks provide access include electronic mail and bulletin boards, computer-mediated teleconferencing, and information utilities of many types.

Networks are almost certain to continue to increase in number, in complexity, and in the rate at which they can move information from place to place. If their bandwidth, or carrying capacity, continues to increase at anything like its recent rate, the transmission of enormous amounts of data (including digital voice and high-quality video) from almost anywhere to almost anywhere will be possible at relatively low cost. Concurrently with advances in the technology, the uses to which it is being put are multiplying as well.

The evolving macrosystem of interlinked networks can be thought of as one enormous global nexus that has the potential to increase by many orders of magnitude the extent to which individuals and information resources all over the world are interconnected and therefore accessible to each other. Such capabilities should give people unprecedented access to both information (in libraries, museums, news databases, and other repositories) and people. They have the potential to greatly facilitate not only the distribution and exchange of information in the conventional sense, but also long-distance collaboration involving a real-time sharing of workspaces, tools, and resources (National Research Council, 1993; Wulf, 1993).

Previous Research

A great deal of human factors research has focused on certain aspects of computer and communication technology and especially how the interfaces through which people use this technology for various purposes should be designed. Indeed the general area of human-computer interaction has perhaps been the fastest-growing area of research done by human factors researchers and others in closely allied fields over the last few decades. This work has been reported in several journals that focus on the subject, most of which have come into existence during the last 20 years.

Much less work has focused more directly on the human factors of telecommunications and the resources that are accessible through telecommunications systems. Teleconferencing (Carlisle, 1975; Kerr and Hiltz, 1982) was perhaps the first service provided by telecommunications networks that received some human factors research attention. More recently, cooperative work supported by telecommunication systems has been the focus of some research (special 1992 issues of *Human-Computer Interaction* and *Interacting with Computers*), as have electronic mail and electronic bulletin boards (Sproull and Kiesler, 1991).

Computer-based communication should become a high priority for human factors research for many reasons, not the least of which is the expectation that eventually nearly everyone is likely to be a user of this technology. A growing percentage of the workforce will find it essential in the performance of their jobs; many people will use it for personal reasons—information acquisition and exchange, personal business transactions, entertainment, interpersonal communication. The user community will be as diverse as the general population, and the range of uses will be great. As this technology continues to evolve, challenges to human factors research that we cannot now anticipate are bound to emerge; here we mention only a few that are already apparent.

Research Needs and Opportunities

Human-computer interaction, the focus of much research in the past, deserves continued attention from human factors researchers. Issues of interface design, information finding and utilization, personal information management, and users' comprehension of the systems with which they interact continue to be important research challenges. Most terminals today depend primarily on two-dimensional visual displays and on typewriter-like keyboards, typically complemented with a pointing and drawing instrument such as a mouse or trackball. Speech will become an increasingly feasible option for both input and output (Makhoul et al., 1990; Weischedel et al., 1990), as will three-dimensional, "virtual-reality" representations of objects one can "walk around" and environments with which users can interact (Durlach and Mavor, 1995). Refining these and other innovative input-output techniques requires that a variety of human factors issues be addressed.

Realization of the potential benefits of computer networks requires the development of a variety of tools that facilitate interaction with complex databases by both specialists and the public. Tools are also needed to help people apply information technology effectively to manage their personal data stores and to cope with the information overload that a greatly increased connectivity to information resources and to people can create.

With computer-based systems playing increasingly important roles in people's lives, there is a need to learn more about people's attitudes and beliefs about these systems. We also need to understand how to increase the likelihood that people's conceptualizations of what computer-based systems can do and how they do it are reasonably accurate—or at least not inaccurate in counterproductive ways.

Many important challenges to human factors research arise because computer-based communication technology has the potential to greatly increase the ways in which people can communicate with each other. More and more people are using electronic mail, electronic bulletin boards, electronic forums and discussion groups, and computer-based teleconferencing facilities, but these provide only a hint of what is likely to be possible in the not-distant future. Much human factors research needs to be done on the design and utilization of such facilities to ensure that they develop in directions consistent with the needs and preferences of their prospective users and in ways that really enhance and enrich interpersonal communication. Questions involve how interpersonal communication through computer-based systems compares, favorably or unfavorably, with communication through more traditional media and how the emerging media might be shaped so as to increase the opportunities for interpersonal communication as well as to enhance the quality of the communication that occurs.

Communication technology and telenetworking should have important implications for the problem of increasing the employment of people with disabilities. This technology should be able to greatly increase the access of people with mobility problems to many resources that once were available only to people who could travel to them. The full realization of this potential will require that questions of a human factors nature be addressed regarding how best to design the interfaces and operating procedures that will ensure the usability of the technology by people with various types of disabilities.

Even though the use of network technology to hold conferences among "attendees" located at different places has been of interest since the mid-1960s, teleconferencing has not yet become a widely used alternative to face-to-face meetings, even when the meetings involve the expense and inconvenience of considerable travel. Some of the technical limitations of teleconferencing systems are being eased as networks acquire sufficient bandwidth to support the real-time transmission of sufficiently high-fidelity video representations of participants' images to create a realistic impression of an actual gathering. This should help make teleconferencing more attractive, but there is still much to be learned about the human factors aspects of this technology before it becomes the communication asset its originators intended it to be.

The term *telecommuting* captures the idea of using telenetworks to provide people with the resources to enable them to work at home or in locations other than their traditional workplaces. Substituting the transmission of information via computer networks for the transportation of people to and from places of work is attractive for several reasons, not the least of which is the possibility of conserving energy and easing some of the load on urban areas (traffic congestion, parking problems, air pollution) to which commuting traffic contributes. A significant fraction of the workforce is already doing telecommuting, and increasing this fraction manyfold is believed to be technically feasible. Impediments to greater use of telecommuting include the need for additional user-oriented tools to ensure that workers can perform the desired tasks; other impediments involve issues of worker satisfaction and related psychological variables. Human factors studies are needed to address both types of issues.

With the help of computer networks, colleagues can cooperate at a distance in ways that were impossible until fairly recently. Teams of experts, all in different places but drawing on the same resources and sharing a common "virtual" work space, can collaborate on problems that require their collective knowledge and skills. The success with which this scenario can be played out in specific instances will depend, to no small extent, on how well the many human factors issues relating to the design of the underlying systems are resolved.

As the bandwidth of computer networks continues to increase, it will be possible to transmit an increasingly detailed and veridical representation of a physical situation to a remotely located individual. It seems unlikely, however, that it will be possible, at least anytime soon, to represent most nontrivial situations in sufficient detail that one could not tell the virtual reality from the real thing. Fortunately, such verisimilitude is not necessary for most applications, but the question of how real (in appearance) is real enough is open and probably must be answered on a case-by-case basis. Representations of reality that would be more than adequate for some applications might be inadequate for others. There is a need for work on how to determine the degree of fidelity required in specific instances.

INFORMATION ACCESS AND USABILITY

The Problem

Among the more striking characteristics of modern society is the dependence of its governmental, economic, and social institutions on constant accessibility to accurate, up-to-date information of many types. The primacy of information and knowledge as resources in today's world is reflected in the growing tendency to refer to our society as the information society (e.g., Salvaggio, 1989) or the postindustrial society (Bell, 1973).

An associated development is the greatly increased quantity of information available to people in all walks of life. The material published for professionals in specific fields has long exceeded the capacity of individuals to stay current except in very narrow subspecialties. The amount of information that is available to the public is also enormous and growing. Being available, however, does not necessarily mean being readily accessible, and users, or potential users, of information that is available often experience frustration in locating, accessing, and interpreting the information they want.

Technological developments—including ever more powerful information-manipulation and display facilities, "intelligent" interfaces, and information-finding and utilization aids—may help to address this problem. An essential aspect of any effort to make information more accessible to people who need or want to use it, however, must be attention to a variety of human factors issues relating to the ways in which people might interface with information repositories and tools that are intended to facilitate finding, using, and conveying information. Moreover, this attention will need to extend beyond the technical user of information systems because the proportion of the labor force that engages in information handling has been increasing rapidly and is expected to continue to do so (Koenig, 1990); many people tap into information resources for purposes other than the requirements of their jobs. Given the unlimited diversity of the community of users of information systems, designers cannot rely on user perseverance or high technical skills to overcome poor interface design.

Previous Research

Considerable research has been done on the design of interfaces for databases representing restricted domains. Menu organization, for example, has been the focus of numerous studies (Card, 1982; Giroux and Belleau, 1986; Kiger, 1984; Miller, 1981; Norman, 1991; Shneiderman, 1987; Snowberry et al., 1983). Most of these studies have used relatively small, homogeneous, and often abstract databases (Fisher et al., 1990) unlike those typically found in operational settings, and search questions have generally been goal-directed, which means that the results may not help us understand browsing behavior.

Work has also been done on the design of query languages for providing database access (Belkin and Croft, 1987; Shneiderman, 1987), and studies have addressed the effectiveness of various possible search techniques (Blair, 1984; Muckler, 1987; Vigil, 1985). Investigators have also studied the strategies that people use spontaneously to search for information in a large database (Brooks et al., 1986; Chen and Dhar, 1991; Harter and Peters, 1985). Attempts have been made to enhance database search through artificial intelligence techniques (Croft and Thompson, 1987; Hawkins, 1988).

The versatility of the computer for generating graphical representations has sparked some interest in how databases might be represented so as to make the information in them easier to find. Among the approaches being explored are some that represent data sets as three-dimensional structures that users can inspect from different angles and at adjustable viewing distances, thus permitting views of the data from relatively global as well as narrowly focused perspectives (Card et al., 1990; Clarkson, 1991; Newby, 1992). Little is known, however, about the relative effectiveness of different forms of data representation as aids to understanding complex data sets (Jensen and Anderson, 1987; Liu and Wickens, 1992; Merwin and Wickens, 1991).

Research Needs and Opportunities

A continuing human factors challenge relating to complex information systems is interface design. This applies even to systems that represent highly restricted areas of knowledge, for example, an on-line help system for a word-processing program or an airline flight reservation system. An effective interface should allow users to readily locate and access items of information as they are needed and to make a rapid transition from one set of related items to another. These capabilities relate to the *organizational structure* of the database (e.g., hierarchical, matrix, network; Durding et al., 1977) and the *navigational tools* for moving from entry to entry (Seidler and Wickens, 1992). The organizational structure may or may not be independent of the navigational tools. For example, a database may be organized in a strict hierarchical fashion, but navigational tools may allow the user to access any node in the database from any other node with a single command, obviating the need to travel up and down the paths in the hierarchy. Whatever the relationship between the database structure and the navigation aids, the important requirement from the user's point of view is that it facilitate and not impede access.

There is some evidence that the effectiveness of a menu design for an information system may partly depend on how closely the structure represented by the menu corresponds to the user's mental model of the structure of the database (Roske-Hofstrand and Paap, 1986). In some cases, however, there are many "natural" ways to organize a database, and there is little or no guidance as to whether one way is preferable to another, whether one corresponds more closely than another to a mental model, how homogeneous these mental models are across different users, or whether the organizational structure that is appropriate for one task (e.g., normal operation) differs from what is needed for another (e.g., fault diagnosis and troubleshooting). Many human factors issues must be addressed if interface designs are to be optimized for particular combinations of information systems and the users they are intended to serve. The use of electronic database "maps" has proved to be a promising method of making the organization of a database visible to the user (Vincente and Williges, 1988), but major design challenges confront the human factors community if such maps are to represent the vast number of nodes and entries in very large databases in a truly useful way (Mackinlay et al., 1991; Shneiderman, 1987).

Some databases are fluid, in that the information in them is continually changing and they may not have a permanent structure. The database of human factors research is a case in point; not only is new information being added to it all the time, but people do not agree on where its boundaries should lie. Menu-based interfaces have serious limitations when used with such systems. Query languages that rely on keyword searches have some advantages, but also their own limitations.

A relatively neglected, but important, question pertaining to evaluating the effectiveness of any information system or information search procedure is how to determine the value of information to its prospective users. This issue will have to be considered in the design of automated search and retrieval techniques. Not all the information that is relevant to a given topic of interest is of equal value to a user, but little is known about how to convey to a search procedure the distinction between what would be considered important and what would not. More generally, evaluating the effectiveness of information systems is a continuing challenge. Simply counting "hits" or computing the ratio of hits to false alarms does not suffice, because it begs the question of what should be considered a hit, and it does not distinguish among hits and misses of different degrees of importance. Taking into account the use to which information is put would probably give a more accurate picture of system effectiveness, but often this may not be practical. We also do not know how to assess effectiveness for systems that are intended to facilitate browsing as opposed to structured search, and this will be a difficult problem.

A special challenge for human factors research relates to the unprecedented capability of computer-based systems to represent information in graphical or pictorial form. The rapid development of capabilities for dynamic three-dimensional displays and of tools that permit users to manipulate and interact with such displays has outdistanced our ability to exploit such capabilities to full advantage. There is a need both to analyze the tasks for which sophisticated visualization tools could be especially helpful and to find better ways to evaluate how well visualization-aiding systems help users perform their tasks. The considerable current interest in better understanding the use of visualization in science and the application of computer-based displays to the development of more powerful techniques for visual data representation emphasize the importance of human-factors research in this area.

For systems that are intended to have some built-in intelligence, designers have often stated that one of their goals is to give the system the ability to adapt flexibly to the needs of individual users, and even perhaps to adapt to changes in those needs over time. Too much adaptability, however, could prove to be counterproductive. The research done to date provides little guidance as to the conditions under which intelligent adaptability may stop being beneficial and may actually be harmful. In general, there is a need to explore the trade-off between flexibility (adaptability to users' needs or preferences) and consistency in the characteristics and operational features of information systems.

People already have extensive experience navigating in space; this creates a strong argument for attempting to exploit spatial metaphors in the design of tools for helping people interact effectively with large complex databases and information networks. But effective use of spatial metaphors will require addressing a number of unresolved issues: determining how best to represent data that have a dimensionality greater than two or three, developing effective methods for helping users maintain their orientation (keep from getting lost) when navigating a database, providing easy means of conveying to the system desired control actions (zooming, panning, relocating), and numerous issues involving other aspects of user-information system interaction.

EMERGING TECHNOLOGIES IN WORK DESIGN

The Problem

Remaining competitive in the global marketplace has been an increasingly difficult challenge to many U.S. industries, especially in the manufacturing sector of the economy. The U.S. lead in several industrial areas has been eroded in recent years, and the prospects for competing effectively in these and other areas will depend on the country's ability to use material and human resources more effectively.

It appears that a critical aspect of industrial competitiveness will be the ability to adapt quickly to rapid technological developments and constantly changing market conditions. This means, among other things, being sufficiently flexible to shift production quickly from one item to another and being able to produce items economically in relatively small quantities (Piore and Sabel, 1984). Production flexibility requires a very different job design from that of conventional twentieth-century assembly lines. Other innovative ways to increase competitiveness, such as just-in-time manufacturing, which is intended to save costs by minimizing the need for large inventories, also have implications for work design (National Research Council, 1986).

Emerging technologies—many forms of automation, robotics, and information technologies—will figure prominently in shaping the workplaces of the future. In some cases, workers will be displaced as technological advances make their tasks obsolete or performable by machine. Other jobs will be transformed by the availability of new tools that workers will have to learn how to use. And jobs that do not now exist will be created. Change will be the norm.

Human factors research has something useful to contribute to maintaining the industrial competitiveness of the United States as these changes take place, but the goal of the research must be more than the design of better displays and controls (which is not to discount the importance of this objective). Improving manufacturing, for example, will require attention to human factors issues conceived sufficiently broadly to include the social psychology of the functioning of work groups as well as classical ergonomics. The realization of the need for a relatively broad perspective has led to the current interest in macro-ergonomics.

Previous Research

Job design has been a focus of human factors research from the earliest days of the discipline. Originally attention was focused primarily on the individual worker and the demands of specific tasks. "Taylorism," as a theory of job design, was concerned with the optimization of physical effort in order to increase speed of production. Jobs were subdivided into repetitive tasks to minimize the amount of learning required by the worker and to maximize the speed with which a task could be performed (Knights et al., 1985). "Fordism" introduced the assembly line, which organized Taylorized tasks into patterns that would yield automobiles and other complex consumer products.

The assembly line approach was very effective in mass-producing consumer goods that could be sold at affordable prices, but it left much to be desired from the worker's perspective and often led to stress, low worker motivation, absenteeism, and labor conflict (Kornhauser, 1965; Walker and Guest, 1952). Researchers interested in job design began to see worker satisfaction and quality of life in the work setting as important complements to worker efficiency. In time, the importance of taking into account the interactions among individuals in work settings was recognized, as was the need to apply a systems perspective to the study of complexes involving both people and machines.

Despite the attention that job design has received from human factors researchers, most jobs still are not designed in any rigorous sense, but they evolve—often being forced to change in order to accommodate the introduction of new technology, but not necessarily in optimal or even satisfactory ways. What was already a complex problem has been exacerbated in some instances by the extraordinary rapidity with which information-based technologies have permeated the workplace.

The need for more attention from human factors researchers to the introduction and use of information-based technology in the workplace is seen in such disparate problems as the biochemical and physiological difficulties that are sometimes associated with the use of information-technology devices (e.g., eyestrain and carpal tunnel syndrome) and the claims of some investigators that these technologies do not increase productivity as they were expected to do. There have also been calls for a better understanding of the ways in which the new technologies are changing the skills required of workers (National Commission on Employment Policy, 1986) and of why these technologies appear to be underused or inefficiently used in some contexts, especially manufacturing (National Academy of Engineering/National Research Council, 1991).

Research Needs and Opportunities

If U.S. industry is to compete effectively in world markets, it must give a high priority to product quality. The importance of quality increases with the amount of competition that an industry faces, and many of the industries that are important to the economic health of the nation have much competition on the world scene. Quality flaws in products typically derive either from worker error or from suboptimal work processes and procedures. There are many opportunities for human factors research on both decreasing the probability of human error in the workplace and enhancing product quality by improving the designs of work processes and procedures. The study of human error in certain types of information work—programming or software design, for example—poses a special challenge because errors may be very difficult to identify and it may be almost impossible to trace back to them from their eventual effects.

Workplace health and safety will continue to be major foci of human factors research. Lighting, air quality, potentially dangerous machinery, noise, fatigue, biomechanical stress, and other traditionally important human factors issues relating to work and the workplace will still require attention, but new problems will also surface as jobs and workplace technology continue to change. Certain types of musculoskeletal disorders, sometimes referred to as repetitive strain injury, appear to have been becoming increasingly common among people who make constant use of computer terminals or similar keyboard devices. So also have certain ailments that are probably attributable to psychosocial properties of some jobs that seem to create undue psychological stress. Effectively addressing workplace health and safety will require attention both to the traditional ergonomic concerns of proper equipment design and to psychosocial factors that can directly and indirectly affect workers' health and safety.

The integration of human and machine labor in semiautomated industrial operations continues to be a challenge, and so far it has not been met effectively. The idea that machine intelligence might substitute for human intelligence in many manufacturing processes appears to have been giving way to the idea that what is needed, for the foreseeable future at least, is a better understanding of how to combine human and machine capabilities in truly symbiotic ways (Brodner, 1986; De Greene, 1991; Kellso, 1989; Kuo and Hsu, 1990). If such systems are to be realized, many human factors issues will have to be addressed; they range from the traditional problems of interface design to planning, scheduling, and coordinating activities to top-level policy setting (Sanderson, 1989).

Many industrial jobs of the future will involve people as supervisors of highly automated operations. When things are operating smoothly as intended, "supervisory controllers" may have relatively little to do; however, when systems fail or malfunction, they may have to take quick and decisive action to avoid serious consequences. Such situations pose problems of boredom and skill maintenance. How are supervisory controllers to be kept attentive and interested in their jobs when everything is proceeding as intended without their involvement, and how are their skills to be kept honed so they can do what is required in moments when they are needed to avert a crisis? More generally, there is a need for systematic and detailed descriptive research about the skills, especially the higher-level cognitive skills, that workers need in the various job situations that high technology is creating and that will be even more common in the future.

One effect of the new information technologies on many jobs has been to put some distance, real or symbolic, between workers and whatever they are producing. This is illustrated most strikingly in the use of tele-operators and other remotely controlled devices in some production processes, and it is apparent in other contexts as well. Many machinists no longer do any machining by hand with traditional machine tools; instead they feed their specifications to a programmable machine tool. Even people who generate paper products (reports, drawings, blueprints) typically do so via a computer-based system that yields the paper only after the work has been done. There has been some speculation about the effects of this change on workers and on the skill requirements of their jobs (Zuboff, 1988), but there has been little empirical investigation into the matter.

The emphasis in the foregoing has been on the need to find ways to ensure the competitiveness of U.S. industry in a rapidly changing world; this is indeed a serious challenge for the nation. We believe human factors researchers can significantly contribute to meeting this challenge. It is important, however, that in doing so, this community maintain the complementary objective of ensuring that the work climate, culture, and environment are humane and fulfilling to the human beings who work within it. Job satisfaction will require attention as long as there are jobs, but the answer to the question of how to promote it is likely to change as the characteristics of jobs change, and not always in predictable ways.

TRANSPORTATION

The Problem

Most people interact with transportation systems as operators and as passengers. Millions of us operate motor vehicles, and millions more are passengers on airplanes and railroad trains. In both cases, the goals are to achieve convenient, fast, and safe travel. Automobiles offer individual mobility; air and rail transport offer efficient travel over long distances. As speed and convenience have increased through advances in technology, there has also been increased concern about public safety.

In the United States, motor vehicle accidents are the primary cause of death for people under age 38. Current statistics show that there are also 1.7 million disabling injuries from motor vehicles every year. Because traffic accidents take a disproportionate toll from the younger population, the average number of life years lost is two to three times higher than for heart disease and cancer. A large proportion of these accidents can be attributed to human error. Sources of such error include lack of experience, misperception of events, poor vision, use of drugs and alcohol, fatigue, and high levels of risk tolerance.

Air travel, in contrast, is the safest form of mass transportation in terms of accident occurrence and loss of life. However, when accidents do occur, the toll is often catastrophic. In over two-thirds of these accidents, human error is cited as the primary cause. The sources of human error in air accidents are different from those in automobile accidents. In modern aircraft, pilot errors may stem from poor mental models of the functions being performed by highly complex automated systems, from information or work overload, or from failures in coordination and communication among members of cockpit crews or between cockpit crews and air traffic controllers.

We can anticipate that, as technology advances and greater demands are placed on both motor vehicles and commercial aircraft for convenience and speed, understanding the capabilities and limitations of the human operators will become even more critical. Human factors can make important contributions by developing coherent models of the human as an operator and of the team and organizational environment in which the human must function. In addition, human factors can provide insights into what functions should be automated and how this should be done to ensure efficient and safe operation.

Previous Research

The available knowledge base for understanding proficient driving and traffic safety skills in motor vehicle operation can be characterized as extensive but fragmented. As already mentioned, several sources of human error have been researched; what is missing, however, is a conceptual model that specifies a coherent research agenda for describing driving performance under a variety of conditions with a range of technological alternatives. The main lines of researchable questions have focused on deficits in vision, the influence of alcohol and drugs, attention and automaticity, aging, and the misestimation of risk as potential contributors to reduced driving performance.

Tests for visual acuity currently assess static acuity on one eye at a time. However, recent research (Burg, 1967, 1968, 1971) suggests that static acuity alone is a poor predictor of accident probability and that tests for dynamic acuity are preferable. Moreover, the degree to which both eyes work together to provide the widest possible field of view has also been found to be a crucial factor. One problem is that the current tests used to measure visual acuity for drivers are based on evaluating only the threshold for resolution for high-contrast optotypes whereas, in many situations, the driver must distinguish low-contrast objects under low illumination. The tests of vision that appear to have higher predictive power for safe driving involve peripheral vision, contrast sensitivity, and motion perception (Johnson and Keltner, 1983).

Recent findings on alcohol and drug intoxication suggest that impairment that would affect driving can be detected in individuals with much lower blood-alcohol levels than those now used to determine legal liability. Also, strong interaction effects between individuals and conditions have been recorded; this should lead to a radical rethinking of the legal limits on blood alcohol.

Another line of research on drivers deals with vigilance and the useful field of view. There is evidence that repetitive tasks can result in general drowsiness or a narrowing of the effective field of view for attention, and some studies show that restrictions in this field can pose serious problems for older drivers (Ball and Owsley, 1992). Furthermore, as alertness decreases, reaction time can become significantly longer.

Work is also going forward on driver attitudes toward risk or risk tolerance. Recent findings suggest that some drivers have a risk tolerance threshold that is constant. If such drivers have vehicles with automatic seat belts or air bags that lower risk, they tend to increase risk by driving faster than they otherwise would. The net effect is that the benefits of technological advances are canceled out.

Research on motor vehicle drivers has focused on the characteristics of the individual and how those characteristics interact with performance. Human factors research on aircraft operation, however, has concentrated both on the individual and on teamwork in multiperson operator crews. In recent years programmatic movement in this area has emerged under the rubric of crew resource management (CRM). CRM is based on the recognition that many vehicles—particularly large commercial aircraft—are actually piloted by a team rather than a single individual. The performance of the vehicle as a system thus depends in part on how well the team members accommodate to one another in addition to how well they operate as competent specialists. The approach has, as a first focus, the training of the crew. Current practices put great emphasis on the creation of training situations that exercise the teamwork functions. In a dynamic flight simulator, for example, the whole crew is exercised simultaneously. What takes place is not just skill learning in the sense of interpersonal communication skills but also a solid sense of mutual confidence that can allow team members to compensate for one another's weak points.

Research Needs and Opportunities

There is no shortage of needs for rigorous research in advanced transportation systems. A truly basic need is for a better systemic representation of the human in the system. Ideally, the community of human factors researchers would create a model to identify the human attributes that contribute most to performance. The framework should include the distribution of individual differences for each such attribute and something similar to a multiple regression equation, describing in quantitative terms how each attribute contributes to the quality of performance in challenging work situations. Given such a framework, the analysis of a set of particular driver behaviors would be more straightforward and would be more likely to point clearly to reforms at the practical level, such as the design of tests used to screen drivers for licensure.

Another key opportunity exists in the area of automation and function allocation. The present atmosphere appears to favor the development and evaluation of computer-based aids rather than complete reassignment of operator responsibilities to a computer. For example, it is now possible to use information from satellites to determine one's exact position in three-dimensional space. Such positional information could be used, in turn, to drive map-type displays in a cockpit of an aircraft, bridge of a boat, cabin of a locomotive, or instrument panel in a car. What effect would such displays have on the performance of the operator and on the system as a whole? Would the presumptive improvement in performance make up for the additional cost? Would people drive automobiles more recklessly if they thought that the new display reduced risk?

The time has come when the operator's role could be totally automated in many transportation systems. Would passengers (or, indeed, nonparticipating bystanders) permit the total automation of any common carrier vehicles? What are the predominant attitudes of various cultures toward automation? How do these attitudes influence the willingness of individuals to disengage automated systems and revert to human control when conditions warrant?

A third important issue is the differential reactions of various cultures to team training such as CRM. For example, in collectivist cultures a high value is placed on the group, whereas in individualist cultures, such as the United States, more value is placed on individual performance. These values and their consequences should be explored in detail so that training can be appropriately designed. One important goal of human factors is to define curricula that are relevant to the task, understandable to those being trained, and acceptable in the context of societal values.

Finally, there is a need to develop an improved methodology for understanding human error at the group level and for evaluating group and system performance. Accident investigations have found that a number of factors at the regulatory and organizational level, as well as at the individual level, contribute to the inadequacy of safeguards against fatal decisions. Effort should be directed toward developing a taxonomy of human factors problems at various levels that could be applied to the analysis of accidents and incidents. This would be an invaluable tool for researchers and for those responsible for ensuring safety (Jones, 1993).

The field of transportation provides rich opportunities to expand the scope of human factors in areas in which outcomes can have major consequences. Optimizing the interface between individuals and groups that work with complex technology and systems requires a multidisciplinary approach that embraces the full range of concerns of human factors specialists. Concern with a particular problem, whether vehicular or aviation safety, should not blind researchers to concepts that transcend problem areas and that reflect more broadly on human capabilities and limitations.

COGNITIVE PERFORMANCE UNDER STRESS

The Problem

Stress is a fact of life in modern society and it manifests itself in a variety of ways, sometimes tragically. Incidents of violence in the workplace have sometimes been attributed, at least in part, to stress. Stress has also been implicated both in high-profile accidents, such as the Three Mile Island incident, the downing of an Iranian airliner by the *Vincennes*, and the crash of Air Ontario Flight 363 (Helmreich, 1992; Wickens, 1992). Stress is also involved in the countless less spectacular mishaps that occur everyday in the workplace, in the home, and elsewhere (Druckman and Swets, 1988; Manuso, 1983).

Since Hans Selye (1956) first directed the attention of the scientific community to the problem of stress and articulated his theory of the "general adaptation syndrome," the concept of stress has evolved considerably. In this report the term is taken to mean reactions to perceived significant threats to one's welfare, reactions that often entail heightened emotion (see Keinan, 1987; Yates, 1990). Stress, according to this conception, is dependent on both the person and the situation. What is threatening to one person at a particular time may not be so to another person or to the same person at a different time. A type of threat that is especially germane to understanding how stress affects cognitive performance is the threat of task demands, especially time-limitation demands, that appear to be beyond one's capabilities to meet.

Some occupations are generally viewed as especially stressful—air traffic controller, for example. Generally, any occupation that puts one in the position of being able to cause great harm to other people or to oneself as a consequence of making an error is likely to be considered stressful. Other jobs can be stressful for much more subtle reasons, and the stress they cause may be at a lower level but relatively continuous. Jobs at which people must work continuously at a near-maximum pace to maintain production quotas, or in which inadequate personal space or privacy create social tension, or in which one has little job security are examples of jobs that can cause chronic stress.

Previous Research

The psychological literature on stress is huge, and a considerable amount of work has been done on the effects of stress on cognition (Hamilton and Warburton, 1979; Hockey, 1983, 1986; Hockey and Hamilton, 1983). We believe that the need for research on this topic will increase over the foreseeable future for several reasons: threats to job security that will come from increasing competitive pressures in many industries, the need to accommodate to increasingly rapid change in the workplace because of technological advances and other factors, the greater cognitive demands of many jobs and the possibility that errors will have far-reaching implications, and the stress evoked by new methods of electronic monitoring and surveillance.

There seems to be general agreement among researchers that stress adversely affects short-term memory, although it is not clear whether stress reduces capacity or imposes a greater load on short-term memory (Cohen, 1978). Another consensus conclusion is that stress reduces the scope of perceptual attention (Baddeley, 1972), a phenomenon sometimes referred to as perceptual narrowing or "tunneling," but there is some doubt about which stimuli do and do not receive attention when this narrowing occurs (Yates, 1990). Resolving these and similar theoretical questions that have been prompted by previous research would facilitate the development of adequate approaches to training people to function well under stress.

One difficulty that has plagued laboratory studies of stress and its effects on performance is the impossibility, for ethical reasons, of exposing subjects to situations that are threatening in a nontrivial way. Informed consent requirements also preclude the element of surprise in experimentation, inasmuch as subjects must be told the details of the experimental procedure in advance. For these reasons, much of the laboratory work on the effects of stress on performance has limited applicability to the real-world situations that seriously, and often unexpectedly, threaten people's safety or well-being.

Another limitation of most laboratory studies of stress is that even the mild stressors used are applied for relatively short periods of time. It is very difficult to imagine practically feasible laboratory situations for studying the effects of long-term or chronic stress. But such stress is as much a problem as is short-term or acute stress. As suggested by the work of Cohen (1980) and Cohen and Spacapan (1978), the effects of chronic stress hold surprises that seem inaccessible via standard laboratory techniques.

Research Needs and Opportunities

A priority for future research on the effects of stress on cognitive performance must be a search for feasible solutions to the limitations of laboratory methods. If laboratory results are to be generalizable to the real-world situations of interest, ethical but effective ways must be found to study the influences of stressors that more closely approximate those in real-world human-technology systems. We mention here two strategies that might be applied to this end: (a) analyses of naturally occurring incidents and (b) simulations, including competitive games. Both of these strategies are used currently; we recommend that they be exploited more fully and creatively.

Incident analysis could be considerably facilitated by the development of formal protocols for use in specific incidents, such as those involving nuclear power plant operation, aviation, and surgical procedures. The availability of such protocols, appropriately evaluated, would make it practical for incident analyses to be commissioned regularly (e.g., by the Nuclear Regulatory Commission, the Federal Aviation Administration, and various surgery review boards) to better understand their causes. Statistical analyses of incident-recording databases, such as the Aviation Safety Reporting System that is maintained by the National Aeronautics and Space Administration, is one means of investigating the effects of stress on performance; nonintrusive study of people in stressful situations in which they have voluntarily placed themselves is another.

The simulation of stressful situations is a possible approach if simulators can be built that create a convincing illusion of stress without actually putting subjects at risk. How to build such simulators is itself a research challenge because of the limitations of our present knowledge of the determinants of stress, but common experience with frightening films, amusement park rides, and similar emotion-arousing stimuli attest to the possibility of feeling stressed even while not believing one's self to be in actual danger. Virtual reality should offer the possibility of simulating stressful situations with a high degree of realism without exposing people to physical threats. How best to exploit these possibilities in the study of the effects of stress on performance is a question that will require some research to answer.

In addition to learning more about the effects of stress on cognitive performance, a major challenge for human factors research is the development of effective ways to counteract the more detrimental of those effects. Approaches to handling stress include attempting to eliminate stressors or to lessen their strength, reducing the stress reactions of individuals, selecting people with high tolerance for stress for stressful jobs, training people to function effectively even when stressed, and "stress-proofing" systems so they will function smoothly even when their operators are stressed. All of these approaches are worth pursuing.

We know that people differ both in the situations they find stressful and in their ability to function under a given level of stress (see Hockey et al., 1986, especially Section IV). And considerable attention has been given to the question of how to distinguish people who are more and less stress-resistant (Allred and Smith, 1989; Kobasa, 1979; Parkes, 1986). We need to know more about such individual differences in response to stress, but perhaps even more importantly, we need to be able to predict how individuals are likely to be affected by specific stressors in specific situations.

Some research has been done on how to train people so that they will be able to perform their tasks well even under highly stressful conditions, but more is needed. It is not clear from the results obtained to date whether it is better to train people primarily under nonstressful conditions or to give them a lot of practice under conditions that approximate the stressfulness under which they may have to function. It is clear that high degrees of stress usually impede learning, but it does not follow that, given the need to be able to function under stress eventually, the training methods that are most successful in the short run will prove the most efficient from a sufficiently long-range perspective. There is a need for some systematic research on this problem to explore the range of possible approaches that will best prepare people to perform under the full spectrum of situations they could encounter on the job. These might include hybrid approaches that would, for example, provide training under nonstressful conditions and "overtraining" under stress.

Complementing the need to determine how best to train people to function effectively under stress is the need to make systems as stress-resistant as possible. The research objective in this case is to determine how to design human-technology systems so their operational goals will be realized even when their operators are stressed and more prone to make errors than normally. Considerably less is known about effects of stress on the performance of groups or person-machine complexes than about the effects of stress on the performance of individuals (Davis et al., 1992), so this challenge involves the need to acquire knowledge at a relatively fundamental level. In addition, however, there is a current practical need to identify features that could be designed into systems to safeguard against known effects of stress on human performance. To illustrate the point: because stress often results in restricting one's range of attention, a system might be given the capability to force operators, during emergency situations, to explicitly acknowledge their attention to every item on a checklist designed for that contingency.

Eliminating or reducing high levels of stress in work situations should be a continuing objective of research. However, prudence dictates the assumption that, no matter how successful we are in this objective, there will arise from time to time in many modern work contexts stressful situations that either cannot be anticipated in detail or cannot be precluded. It is therefore essential that research seek effective ways to deal with those situations through personnel selection, training, and system design.

AIDING INTELLECTUAL WORK

The Problem

Almost all work is done with the assistance of tools of one sort or another. This is true of intellectual as well as physical work. As computer technology has become widely available in the workplace, many software tools have been developed to help people perform intellectual tasks, such as writing, designing, decision making, and analyzing. These tools include electronic dictionaries, thesauruses, spelling verifiers, spreadsheets, design aids, conferencing systems for group work, models for weather forecasting, medical diagnostic aids, and systems to help people visualize structures such as complex molecules or airflow patterns on an aircraft wing.

It seems certain that efforts to develop additional and more powerful aids to intellectual work will continue for the foreseeable future. This effort will be driven in part by continuing rapid advances in the enabling technologies and in part by the promise that such aids hold for increasing productivity and creativity in the workplace and for enriching people's intellectual experience in many other contexts.

The human factors issue of overriding importance is to help ensure that the technology really has the effect of enhancing the quality of life on the whole. Human factors efforts are needed (a) to evaluate the effectiveness of such aids, (b) to provide—through task analyses—a better understanding of what further aids would be useful, and (c) to participate in the design, implementation, and iterative improvement of future aids.

Previous Research

Although many electronic aids to intellectual work have been developed, some of which are used daily in work settings, surprisingly few systematic studies have been done to evaluate their effectiveness. The evaluative data that do exist suggest that some aids do not increase productivity (Attewell, 1994).

Studies conducted in the 1970s on the use of dictating equipment illustrate the importance of checking prevailing beliefs about the difficulty or effects of introducing new technologies in the workplace against objective data. In this case, common assumptions about the difficulty of learning to dictate effectively, and about the efficiency of dictation as compared with manual writing for experienced users of dictation equipment, proved to be wrong (Gould, 1980). Experimental studies also showed that people found it easier to compose "voice documents" for listening than to dictate documents to be read. Such findings were instrumental in shifting the focus of some developers from dictation systems to voice-message systems (Gould and Boies, 1983).

Other experiments have assessed the effects on productivity of the increasing trend among office professionals, including managers, to use text processing systems to compose and edit text themselves instead of using the services of a secretary or clerical staff (Card et al., 1984; Gould, 1982). Again, major savings in time or improvements in product quality were not always found.

Field studies of organizations that have introduced innovative technological aids to intellectual tasks in the workplace have also shown that it is possible for these aides to increase productivity while lowering employee morale and job satisfaction (Kraut et al., 1989). Negative reactions sometimes occur, for example, because using the technological aids changes the amount of social contact and face-to-face interaction that people have in their jobs.

Recent research has also focused on the development or evaluation of aids to facilitate collaborative work among the members of a group (Turner and Kraut, 1992). Some of this work has been done in the laboratory, some in real work situations, and some in both (Olson et al., 1992). The main conclusion to be drawn from this early work is that much additional research will be needed before firm design guidelines for the development of aids for intellectual work by groups can be articulated.

A few efforts have been made by human factors groups to initiate the development of one or more aids to intellectual work and to see this through to implementation in a work setting (e.g., Gould et al., 1993; Harris, 1980; Landauer et al., 1993). These are atypical, however; more commonly human factors researchers have become involved in development efforts only after they are under way, usually in support rather than leadership roles, and often for the sole purpose of validation testing at the end of a developmental cycle.

Research Needs and Opportunities

Evaluative studies are needed, not so much to determine the relative merits of existing aids but to provide the insights required to improve upon them and to ensure that they accomplish what is intended. Evaluative research should focus not only on aids, but also on the processes that are used to develop them. We need to know what developmental approaches are most effective and how they might be improved.

There is also a need to study the detailed demands of intellectual work in specific settings with a view to identifying how the development of new tools could aid that work. It would be useful, for example, to study successful physicians diagnosing, nurses providing nursing care, musicians composing, architects designing, and teachers teaching; the aim would be to identify the intellectual skills involved and develop a theory about the task demands that would guide the search for new tools to aid performance. This effort should be conditioned, however, by the recognition that frequently providing new tools results not so much in making the current job easier to do but in changing the nature of the job by making it possible to do things that could not be done before.

Several studies, using either observational techniques or questionnaires, have been done of how white-collar workers spend their time (Klemmer and Snyder, 1972; Kraut and Streeter, 1995; Mintzberg, 1973; Panko, 1992). Although these studies have been valuable, they have not led to the development of new intellectual work aids, nor were they intended to do so. Insights that could lead to such development are most likely to come from studies designed for that purpose, and this probably means studies that relate specific activities to personal and organizational goals and that attempt to determine how well the activities serve those goals. One general conclusion that all of these studies support is that interpersonal communication is a major component of most white-collar jobs, so it would seem to follow that a major opportunity for performance enhancement should lie in the development of more effective aids to that communication.

Human factors researchers can also contribute to the design, implementation, and evaluation of new aids for intellectual work. They bring to the process an iterative design philosophy and a unique focus on users' needs that has proved to be important in the development of systems whose ultimate characteristics were impossible to specify in detail in advance.

Some evaluation studies must focus on people in actual work settings, as opposed to laboratory simulations of these settings. Studies should also focus on the effects of technological aids on both individuals and organizations. There is a need to determine the extent to which changes in individual productivity translate into similar changes at organizational levels, and there is a need too to ensure that productivity gains at an organizational level are not realized at the cost of individual job satisfaction. Acquiring reliable information on the long-term effects of the use of specific technological aids to intellectual tasks in the workplace, especially those intended to facilitate group work, is likely to require studies of considerably longer duration than those that are typically conducted (Cool et al., 1992).

When a work activity that has been performed one way is suddenly performed in a very different way because new technology was introduced, this can have important but quite subtle effects. Malone (1983), for example, has shown that the arrangement of papers on people's desks can serve to remind them of what has to be done and in what priority. When the desktop is replaced by an electronic file and a video monitor, the traditional cues are lost, and if the system has not been designed to provide an effective substitute for them, work may suffer. One human factors challenge associated with the design of intellectual work aids is to identify nonobvious but important aspects of current ways of performing tasks for which some provision should be made in an electronic aid.

For some time, there has been compelling evidence that diagnostic judgments based on actuarial data are, in many contexts, typically more accurate than judgments made by human diagnosticians without the aid of the actuarial data (Dawes et al., 1989). A number of decision aiding systems have been developed that exploit this fact, but the diagnosticians have generally resisted using them. Although many reasons have been hypothesized for the reluctance of experts to use these aids, the situation is still not well understood, and the question of how to get diagnosticians and decision makers to avail themselves of aids that would improve their performance is a continuing challenge.

Successful development and use of new tools to help people perform intellectually demanding work will require collaboration among scientists and engineers from several disciplines. Human factors researchers have a critical role to play in this effort. We believe that they can and should take the initiative in identifying tasks whose performance could be enhanced through appropriately designed aids and that human factors researchers have much to contribute to both the design and the evaluation of these aids. The need for interdisciplinary collaboration is itself a challenge, however, to all the disciplines involved and may require the human factors community to broaden its conceptualization of what constitutes legitimate research. In particular, human factors researchers must be willing, at least in some instances, to take responsibility for initiating and successfully completing interdisciplinary projects to produce significant aids, with all the burdens and long-term commitment this entails. Iterative design with constant ongoing evaluation involving intended users of an aid is an approach that the human factors community has advocated and with which it has had some success.

CONCLUSION

This summary report identifies important areas of need and opportunity for human factors research during the next few decades. It is the committee's hope that this volume will be of use both to sponsors of human factors research and to people who do human factors research.

The committee has not attempted to lay out a research agenda. It recognizes that the priorities of different research-sponsoring agencies will differ depending on the agencies' specific missions and goals and that individual scientists and scientists-in-training will be drawn to different problem areas depending both on funding opportunities and on personal interests, knowledge, and expertise. We believe, however, that there are many opportunities for human factors research to address important national and global problems, some of which have not been the focus of such research in the past. This report and the papers in Part II indicate what, in the committee's view, some of those opportunities are.

REFERENCES

Allred, K.D., and T.W. Smith
  1989   The hardy personality: cognitive and physiological responses to evaluative threat. *Journal of Personality and Social Psychology* 56:257-266.

Attewell, P.
  1994   Information technology and the productivity paradox. Pp. 13-53 in D.H. Harris, ed., *Organizational Linkages: Understanding the Productivity Paradox*. Panel on Organizational Linkages, Committee on Human Factors, National Research Council. Washington, D.C.: National Academy Press.

Baddeley, A.D.
  1972   Selective attention and performance in dangerous environments. *British Journal of Psychology* 63:537-546.

Baker, L.A., A.T. Herlihy, P.R. Kaufmann, and J.M. Eilers
  1991   Acidic lakes and streams in the United States: the role of acidic deposition. *Science* 252:1151-1154.

Ball, K., and C. Owsley
  1992   The useful field of view: a new technique for evaluating age-related declines in visual function. *Journal of the American Optometric Association* 63:71-79.

Baron, S., D.S. Kruser, and B.M. Huey, eds.
  1990   *Quantitative Modeling of Human Performance in Complex, Dynamic Systems*. Panel on Human Performance Modeling, Committee on Human Factors, National Research Council. Washington, D.C.: National Academy Press.

Baum, A., and J.E. Singer, eds.
1981 *Conservation: Psychological Perspectives*, Vol. 3. *Advances in Environmental Psychology* series. Hillsdale, N.J.: Erlbaum.

Belkin, N.J., and W.B. Croft
1987 Retrieval techniques. *Annual Review of Information Science and Technology* 22:109-145.

Bell, D.
1973 *The Coming of the Post-Industrial Society*. New York: Basic Books.

Blair, D.C.
1984 The management of information: basic distinctions. *Sloan Management Review* 26(1):13-23.

Bogner, S.
1991 Human factors and medicine. P. 682 in *Proceedings of the Human Factors Society 35th Annual Meeting*. Santa Monica, Calif.: Human Factors Society.

Bowe, F.G.
1984 *Personal Computers and Special Needs*. Berkeley, Calif.: Sybex Computer Books.

Brodner, P.
1986 Skill-based manufacturing vs "Unmanned Factory": which is superior? *International Journal of Industrial Ergonomics* 1:145-153.

Brooks, H.M., P.J. Daniels, and N.J. Belkin
1986 Research on information interaction and intelligent information provision mechanisms. *Journal of Information Science: Principles and Practice* 12(1):37-44.

Bureau of the Census
1987 *Statistical Abstract of the United States*. Bureau of the Census. Washington, D.C.: U.S. Department of Commerce.
1989 *Statistical Abstract of the United States*. Washington, D.C.: U.S. Department of Commerce.

Bureau of Labor Statistics
1988a *Comparative Real Gross Domestic Product, Real GDP per Capita, and Real GDP per Employed Person, Thirteen Countries, 1950-1987*. Washington, D.C.: U.S. Department of Labor.
1988b *International Comparisons of Manufacturing Productivity and Labor Cost Trends*. Washington, D.C.: U.S. Department of Labor.

Burg, A.
1967 *The Relationship Between Vision Test Scores and Driving Record: General Findings*. Department of Engineering Report No. 67-24. Los Angeles, Calif.: University of California.
1968 *The Relationship Between Vision Test Scores and Driving Record: Additional Findings*. Department of Engineering Report No. 68-27. Los Angeles, Calif.: University of California.
1971 Vision and driving: a report on research. *Human Factors* 13(1):79-87.

Card, S.K.
1982 User perceptual mechanisms in search of computer command menus. Pp. 190-196 in *CHI '82 Conference on Human Factors in Computer Science*. New York: Association for Computing Machinery.

Card, S.K., J.M. Robert, and L.N. Keenan
1984 On-line composition of text. Pp. 231-236 in *Proceedings of Interact'84, First IFIP Conference on Human-Computer Interaction*, Vol. 1. London, England: Elsevier North Holland.

Card, S.K., J.D. Mackinlay, and G.G. Robertson
  1990   The design space of input devices. Pp. 117-124 in *CHI Proceedings*. New York: Association for Computing Machinery.
Carlisle, J.H.
  1975   A Tutorial for Use of the TENEX Electronic Notebook-Conference (TEN-C) System on the ARPANET. Report No. ISI/RR-75-38. Arlington, Va.: Defense Advanced Research Projects Agency.
Chen, H., and V. Dhar
  1991   Cognitive process as a basic for intelligent retrieval systems design. *Information Processing and Management* 27(5):405-432.
Clarkson, M.A.
  1991   An easier interface. *BYTE* February:277-282.
Coach, J.V., T. Garber, and L. Karpus
  1979   Response maintenance and paper recycling. *Journal of Environmental Systems* 8:127-137.
Cohen, S.
  1978   Environmental load and the allocation of attention. Pp. 1-29 in A. Baum, J.E. Singer, and S. Vallins, eds., *Advances in Environmental Psychology*. Hillsdale, N.J.: Erlbaum.
  1980   Aftereffects of stress on human performance and social behavior: a review of research and theory. *Psychological Bulletin* 88:82-108.
Cohen, S., and S. Spacapan
  1978   The aftereffects of stress: an attentional interpretation. *Environmental Psychology and Nonverbal Behavior* 3:43-57.
Commission on the Skills of the American Workforce
  1990   *America's Choice: High Skills or Low Wages*. Rochester, N.Y.: National Center on Education and the Economy.
Commission on Workforce Quality and Labor Market Efficiency
  1989   *Investing in People: A Strategy to Address America's Workforce Quality*. Washington, D.C.: U.S. Department of Labor.
Committee on Human Factors
  1983   *Research Needs for Human Factors*. National Research Council. Washington, D.C.: National Academy Press.
Cone, J.D., and S.C. Hayes
  1980   *Environmental Problems/Behavioral Solutions*. Monterey, Calif.: Brooks/Cole Publishing.
Cool, C., R.S. Fish, R.E. Kraut, and C.M. Lowery
  1992   Iterative design of video communication systems. Pp. 25-32 in J. Turner and R. Kraut, eds., *Proceedings of the Conference on Computer-Supported Cooperative Work*. New York: ACM Press.
Croft, W.B., and R.H. Thompson
  1987 I^3R: a new approach to the design of document retrieval systems. *Journal of the American Society for Information Science* 36(6):389-404.
Crossen, P.R., and N.J. Rosenberg
  1989   Strategies for agriculture. *Scientific American* 261(3):128-135.
Davis, J.H., T. Kameda, and M.F. Stasson
  1992   Group risk taking: selected topics. Pp. 163-199 in J.F. Yates, ed., *Risk-Taking Behavior*. Chichester, England: Wiley.
Dawes, R.M., D. Faust, and P.E. Meehl
  1989   Clinical versus actuarial judgment. *Science* 243:1668-1673.

De Greene, K.B.
1991   Emergent complexity and person-machine systems. *International Journal of Man-Machine Studies* 35(2):219-234.

Druckman, D., and R. Bjork, eds.
1991   *In the Mind's Eye: Enhancing Human Performance.* Committee on Techniques for the Enhancement of Human Performance, National Research Council. Washington, D.C.: National Academy Press.

Druckman, D., and J.A. Swets, eds.
1988   *Enhancing Human Performance: Issues, Theories, and Techniques.* Committee on Techniques for the Enhancement of Human Performance, National Research Council. Washington, D.C.: National Academy Press.

Durding, B.M., C.A. Becker, and J.D. Gould
1977   Data organization. *Human Factors* 19:1-14.

Durlach, N.I., and A.S. Mavor, eds.
1995   *Virtual Reality: Scientific and Technological Challenges.* Committee on Virtual Reality Research and Development, National Research Council. Washington, D.C.: National Academy Press.

Elkind, J.I.
1990   The incidence of disabilities in the United States. *Human Factors* 32:397-405.

Fisher, D.L., E.J. Yungkurth, and S.M. Moss
1990   Optimal menu hierarchy design: syntax and semantics. *Human Factors* 32:665-683.

Gagnon, M., C. Sicard, and J.P. Sirois
1986   Evaluation of forces on the lumbo-sacral joint and assessment of work and energy transfers in nursing aides lifting patients. *Ergonomics* 29:407-421.

Geller, E.S.
1986   Prevention of environmental problems. Pp. 361-383 in *Handbook of Prevention.* New York: Plenum.

Geller, E.S., R.R. Winett, and P.B. Everett
1982   *Preserving the Environment: New Strategies for Behavior Change.* Elmsford, N.Y.: Pergamon Press.

Giroux, L., and R. Belleau
1986   What's on the menu? The influence of menu content on the selection process. *Behavior and Information Technology* 5:169-172.

Goldstein, I.L.
1988   Tomorrow's workforce today. *Industry Week* 41-43.
1992   *Training in Organizations: Needs Assessment, Development, and Evaluation*, 3rd ed. Pacific Grove, Calif.: Brooks/Cole.

Goodman, P.S.
1979   *Assessing Organizational Change: The Rushton Quality of Work Experiment.* New York: John Wiley & Sons.

Goodman, P.S., and T.L. Griffith
1991   Process approach to implementation of new technology. *Journal of Engineering and Technology Management* 8:161-185.

Goodman, P.S., R. Devadas, and T.L. Griffith
1988   Groups and productivity: analyzing the effectiveness of self-managing teams. Pp. 295-327 in J.P. Campbell and R.J. Campbell, eds., *Productivity in Organizations.* San Francisco: Jossey-Bass.

Gould, J.D.
1980   Experiments on composing letters: some facts, some myths, and some observations. Pp. 97-118 in L. Gregg and E.R. Steinberg, eds., *Cognitive Processes in Writing.* Hillsdale, N.J.: Erlbaum and Associates.

1982 Writing and speaking letters and messages. *International Journal of Man-Machine Studies* 16:147-171.

Gould, J.D., and S.J. Boies
1983 Human factors challenges in creating a principal support office system: the speech filing system approach. *ACM Transactions of Office Information Systems* 4(1):273-298.

Gould, J.D., J.P. Ukelson, and S.J. Boies
1993 Improving application development productivity by using ITS. *International Journal of Man-Machine Studies* 39:113-146. Also available as *IBM Research Report, 1991*, RC-17496.

Gray, C.L., Jr., and J.A. Alson
1989 The case for methonal. *Scientific American* 261(5):108-114.

Guzzo, R.A.
1988 Productivity research: reviewing psychological and economic perspectives. Pp. 63-82 in J.P. Campbell and R.J. Campbell, eds., *Productivity in Organizations*. San Francisco: Jossey-Bass.

Hackman, J.R., ed.
1990 *Groups That Work (and Those That Don't): Creating Conditions for Teamwork*. San Francisco: Jossey-Bass.

Hamilton, P., and D. Warburton, eds.
1979 *Human Stress and Cognition*. Chichester, England: Wiley.

Harris, D.H.
1980 Visual detection of driving while intoxicated. *Human Factors* 22(6):725-732.

Harter, S.P., and A.R. Peters
1985 Heuristics for online information retrieval: a typology and preliminary listing. *Online Review* 9(5):407-424.

Hawkins, F.
1988 *Human Factors in Flight*. Brookfield, Vt.: Gower.

Heart, F.
1975 The ARPANET network. In R.L. Grimsdale and F.F. Kuo, eds., *Computer Communication Networks: 1973 Proceedings of the NATO Advanced Study Institute*. Leyden, Netherlands: Noordhoff International Publishing.

Heart, F., A. McKenzie, J. McQuillan, and D. Walden
1978 *ARPANET Completion Report*. Cambridge, Mass.: Bolt Beranek and Newman Inc.

Helmreich, R.L.
1992 Human factors aspects of the Air Ontario crash at Dryden, Ontario. In V.P. Moshansky, ed., *Commission of Inquiry into the Air Ontario Accident at Dryden, Ontario: Final Report*. Ottawa, Canada: Minister of Supply and Services.

Hockey, G.R.J., ed.
1983 *Stress and Fatigue in Human Performance*. Chichester, England: Wiley.
1986 Changes in operator efficiency. In K. Boff, L. Kaufman, and J. Thomas, eds., *Handbook of Perception and Performance*, Vol. II. New York: Wiley.

Hockey, R., and P. Hamilton
1983 The cognitive patterning of stress states. Pp. 331-362 in G.R.J. Hockey, ed., *Stress and Fatigue in Human Performance*. Chichester, England: Wiley.

Hockey, G.R.J., A.W.K. Gaillard, and M.G.H. Coles, eds.
1986 *Energetics and Human Information Processing*. Dordrecht, Netherlands: Martinus Nijhoff.

Holahan, C.
1986 Environmental psychology. *Annual Review of Psychology* 37:381-407.

Houghton, R.A., and G.M. Woodwell
1989   Global climatic change. *Scientific American* 260(4):36-44.
Human-Computer Interaction, special 1992 issue.
Interacting with Computers, special 1992 issue.
Jensen, C.R., and L.A. Anderson
1987   Comparing three dimensional representation of data to scatterplots. Pp. 1174-1178 in *Proceedings of the 31st Annual Meeting of the Human Factors Society*. Santa Monica, Calif.: Human Factors Society.
Johnson, C.A., and J.L. Keltner
1983   Incidence of visual field loss and its relation to driving performance. *Archives of Ophthalmology* 101:371-375.
Johnston, W., and A. Packer
1987   *Workforce 2000: Work and Workers for the 21st Century*. Indianapolis, Ind.: Hudson Institute.
Jones, S.G.
1993   Human factors in incident reporting. Pp. 567-572 in *Proceedings of the Seventh International Symposium on Aviation Psychology*. Columbus: Ohio State University.
Keinan, G.
1987   Decision making under stress: scanning of alternatives under controllable and uncontrollable threats. *Journal of Personality and Social Psychology* 52:639-644.
Kellso, J.R.
1989   CIM in action: microelectronics, manufacturer charts course towards true systems integration. *Industrial Engineering* 21:18-22.
Kelly, R.T., J.R. Callan, T.A. Kozlowski, and E. Menngola
1990   *Human Factors in Self-Monitoring of Blood Glucose*. Task 4 Final Report. FDA/CDRH-90/60. Springfield, Va.: NTIS.
Kerr, E.B., and S.R. Hiltz
1982   *Computer-Mediated Communication Systems: Status and Evaluation*. New York: Academic Press.
Khalil, T.M., E. Abdel-Moty, and T.M. Asfour
1990   Ergonomics in the management of occupational injuries. Pp. 41-53 in B.M. Pulat and D.C. Alexander, eds., *Industrial Ergonomics: Case Studies*. Norcross, Ga.: Industrial Engineering and Management Press.
Kiger, J.L.
1984   The depth/breadth trade-off in the design of menu-driven user interfaces. *International Journal of Man-Machine Studies* 20:201-213.
Klemmer, E.T., and F.W. Snyder
1972   Measurement of time spent communicating. *Journal of Communication* 22:148-158.
Knights, D., H. Wilmott, and D. Collinson
1985   *Job Redesign: Critical Perspectives on the Labor Process*. Brookfield, Vt.: Gower Publishers.
Kobasa, S.C.
1979   Stressful life events, personality, and health: an inquiry into hardiness. *Journal of Personality and Social Psychology* 37:1-11.
Koenig, M.E.D.
1990   Information services and downstream productivity. *Annual Review of Information Science and Technology* 25:55-86.
Kornhauser, A.
1965   *The Mental Health of the Industrial Worker*. New York: Wiley.

Kraus, L.E., and S. Stoddard
1989 *Chart Book on Disability in the United States: An InfoUse Report*. National Institute on Disability and Rehabilitation Research. Washington, D.C.: U.S. Department of Education.

Kraut, R.E.
1987 *Technology and the Transformation of White Collar Work*. Hillsdale, N.J.: Lawrence Erlbaum Associates.

Kraut, R.E., S. Dumais, and S. Koch
1989 Computerization, productivity, and quality of work-life. *Communications of the ACM* 32(2):220-238.

Kraut, R.E., and L.A. Streeter
1995 Coordination in software development. *Communications of the ACM* 38(3):69-81.

Kuo W., and J.P. Hsu
1990 Update: simultaneous engineering design in Japan. *Industrial Engineering* 22:23-28.

Landauer, T., D. Egan, J. Remde, M. Lesk, C. Lochbaum, and D. Ketchum
1993 Enhancing the usability of text through computer delivery and formative evaluation: the SuperBook project. In C.M. McKnight, A. Dillon, and J. Richardson, eds., *Hypertext: A Psychological Perspective*. New York: Horwood.

la Riviere, J.W.M.
1989 Threats to the world's water. *Scientific American* 261(3):80-94.

Latham, G.
1988 Human resource training and development. *Annual Review of Psychology* 39:545-582.

Levi, L.
1990 Occupational stress: spice of life or kiss of death? *American Psychologist* 45:1142-1145.

Liu, Y., and C.D. Wickens
1992 Visual scanning with or without spatial uncertainty and divided and selective attention. *Acta Psychologica* 79:131-153.

Locke, E.A., and G.P. Latham
1990 *A Task Theory of Goal Setting and Task Performance*, Englewood Cliffs, N.J.: Prentice Hall.

MacDuffie, J.P., and J.F. Krafcik
1990 Integrating Technology and Human Resources for High Performance Manufacturing: Evidence from the International Auto Industry. Paper presented at the Transforming Organizations Conference, Massachusetts Institute of Technology, Boston.

Mackinlay, J.D., G.G. Robertson, and S.K. Card
1991 The perspective wall: detail and context smoothly integrated. Pp. 173-179 in S.P. Robertson, G.M. Olson, and J.S. Olson, eds., *Human Factors in Computing Systems: Reaching Through Technology. CHI '91*. New York: Association for Computing Machinery.

Makhoul, J., F. Jelinek, L. Rabiner, C. Weinstein, and V. Zue
1990 Spoken language systems. *Annual Review of Computer Science* 4:481-501.

Malone, T.W.
1983 How do people organize their desks: implications for designing office automation systems. *ACM Transactions on Office Automation Systems* 1:99-112.

Manuso, J.
1983 The Equitable Life Assurance Society program. *Preventive Medicine* 12:658-662.

McDonald, W.I.
  1984   Quality control of home monitoring of blood glucose concentrations. *British Medical Journal* 288:1915.
Merwin, D.H., and C.D. Wickens
  1991   Comparison of 2D planar and 3D perspective display formats in multidimensional data visualization. *Proceedings of the International Society for Optical Engineering.* Bellingham, Wash.: SPIE.
Miller, D.P.
  1981   The depth/breadth tradeoff in hierarchical computer menus. Pp. 296-300 in *Proceedings of the 25th Annual Meeting of the Human Factors Society.* Santa Monica, Calif.: Human Factors Society.
Mintzberg, H.
  1973   *The Nature of Managerial Work.* New York: Harper and Row.
Mohnen, V.A.
  1988   The challenge of acid rain. *Scientific American* 259(2):30-38.
Moss, J.P., and D.E. Delawter
  1986   Self-monitoring of blood glucose. *American Family Physician* 33:225-228.
Muckler, F.A.
  1987   The human-computer interface: the past 35 years and the next 35 years. In G. Salvendy, ed., *Cognitive Engineering in the Design of Human-Computer Interaction and Expert Systems,* Proceedings of the Second International Conference on Human-Computer Interaction, Honolulu, Hawaii. Amsterdam, Netherlands: Elsevier.
Myers, N.
  1989   *Deforestation Rates in Tropical Forests and Their Climatic Implications.* London, England: Friends of the Earth.
National Academy of Engineering/National Research Council
  1991   *People and Technology in the Workplace.* Washington, D.C.: National Academy Press.
National Commission on Employment Policy
  1986   *Computers in the Workplace: Selected Issues.* Washington, D.C.: National Commission on Employment Policy.
National Commission on Excellence in Education
  1983   *A Nation at Risk: The Imperative for Educational Reform.* Washington, D.C.: U.S. Government Printing Office.
National Council on Education Standards and Testing
  1992   *Raising Standards for American Education.* Washington, D.C.: U.S. Government Printing Office.
National Research Council
  1977   *Drinking Water and Health.* Safe Drinking Water Committee, Commission on Life Sciences. Washington, D.C.: National Academy of Sciences.
  1983   *Changing Climate: Report of the Carbon Dioxide Assessment Committee.* Washington, D.C.: National Academy Press.
  1986   *Human Resource Practices for Implementing Advanced Manufacturing Technology.* Manufacturing Studies Board. Washington D.C.: National Academy Press.
  1990   *The Improvement of Tropical and Subtropical Rangelands.* Board on Science and Technology for International Development, Office of International Affairs. Washington, D.C.: National Academy Press.
  1991   *Rethinking the Ozone Problem in Urban and Regional Air Pollution.* Committee on Tropospheric Ozone Formation and Measurement, Board on Environmental Sciences and Toxicology. Washington, D.C.: National Academy Press.
  1993   *National Collaboratories: Applying Information Technology for Scientific Re-*

search. Committee on a National Collaboratory: Establishing the User-Developer Partnership, Computer Science and Telecommunications Board. Washington, D.C.: National Academy Press.

Newby, G.B.
1992   An Investigation of the Role of Navigation for Information Retrieval. ASIS Meeting.

Newell, G.R., and V.G. Vogel
1988   Personal risk factors: what do they mean? *Cancer* 62:1695-1701.

Norman, K.
1991   *The Psychology of Menu Selection.* Hillsdale, N.J.: Erlbaum.

Office of Technology Assessment
1988   *Urban Ozone and the Clean Air Act: Problems and Proposals for Change.* Washington, D.C.: U.S. Government Printing Office.
1990   *Worker Training: Competing in the New International Economy.* Washington, D.C.: U.S. Government Printing Office.

Olson, J.S., G.M. Olson, M. Storrosten, and M. Carter
1992   How a group-editor changes the character of a design meeting as well as its outcome. Pp. 91-98 in J. Turner and R. Kraut, eds., *Proceedings of the Conference on Computer-Supported Cooperative Work.* New York: ACM Press.

O'Malley, K.G., and K.G. Ricca
1990   Optimization of a PACS display workstation for diagnostic reading. Pp. 940-946 in *Medical Imaging IV: PACS System Design and Evaluation.* Proceedings of SPIE, the International Society for Optical Engineering in cooperation with the American Association of Physicists in Medicine, Vol. 1234. Bellingham, Wash.: SPIE.

Panko, R.R.
1992   Managerial communication patterns. *Journal of Organizational Computing* 2(1):95-122.

Parkes, K.R.
1986   Coping in stressful episodes: the role of individual differences, environmental factors, and situational characteristics. *Journal of Personality and Social Psychology* 51:1277-1292.

Pickett, R.M., and T.J. Triggs, eds.
1974   *Human Factors in Health Care.* Lexington, Mass.: D.C. Heath.

Piore, M., and C. Sabel
1984   *The Second Industrial Divide.* New York: Basic Books.

Pool, R.
1993   Beyond databases and e-mail. *Science* 261(August 13):841-843.

Postel, S.
1985   Thirsty in a water-rich world. *International Wildlife* 15(6):32-37.

Reason, J.
1990   *Human Error.* New York: Cambridge University Press.

Repetto, R.
1990   Deforestation in the tropics. *Scientific American* 262(4):36-42.

Roske-Hofstrand, R.J., and K.R. Paap
1986   Cognitive networks as a guide to menu organization: an application in the automated cockpit. *Ergonomics* 29(11):1301-1311.

Rosomoff, H.L., C. Green, M. Silbert, and R. Steele
1981   Pain and low back rehabilitation program at the University of Miami School of Medicine. In K.Y. Lorenzo, ed., *New Approaches to Treatment of Chronic Pain.* NIDA Research Monograph 36. Washington, D.C.: U.S. Department of Health and Human Services.

Russell, J.A., and L.M. Ward
  1982   Environmental psychology. *Annual Review of Psychology* 33:651-688.
Saegert, S., and G.H. Winkel
  1990   Environmental psychology. *Annual Review of Psychology* 41:441-477.
Salvaggio, J.L.
  1989   *The Information Society: Economic, Social, and Structural Issues.* Hillsdale, N.J.: Lawrence Erlbaum.
Sanderson, P.M.
  1989   The human planning and scheduling role in advanced manufacturing systems: an emerging human factors domain. *Human Factors* 31(6):635-666.
Schlesinger, W.H., J.F. Reynolds, G.L. Cunningham, L.F. Huenneke, W.M. Gerrell, R.A. Virginia, and W.G. Whitford
  1990   Biological feedbacks in global desertification. *Science* 247:1043-1048.
Schwartz, S.E.
  1989   Acid deposition: unraveling a regional phenomenon. *Science* 243:753-763.
Seidler, K.S., and C.D. Wickens
  1992   Distance and organization in multifunction displays. *Human Factors* 34:555-569.
Selye, H.
  1956   *The Stress of Life.* New York: McGraw-Hill.
Senders, J., and N. Moray, eds.
  1991   *Human Error: Cause, Prediction, and Reduction.* Hillsdale, N.J.: Erlbaum.
Shneiderman, B.
  1987   *Designing the User Interface: Strategies for Effective Human-Computer Interaction.* Reading, Mass.: Addison-Wesley.
Smith, M.J.
  1987   Occupational stress. Pp. 844-860 in G. Salvendy, ed., *Handbook of Human Factors.* New York: Wiley.
Snowberry, K., S.R. Parkinson, and N. Sisson
  1983   Computer display menus. *Ergonomics* 26(7):699-712.
Soule, M.E.
  1991   Conservation: tactics for a constant crisis. *Science* 253:744-750.
Sproull, L., and S. Kiesler
  1991   Computers, networks and work. *Scientific American* 265(3):116-123.
Steinhart, P.
  1990   No net loss. *Audubon* July:18-21.
Stern, P.C.
  1992   Psychological dimensions of global environmental change. *Annual Review of Psychology* 43:269-302.
Stern, P.C., O.R. Young, and D. Druckman, eds.
  1992   *Global Environmental Change: Understanding the Human Dimensions.* Committee on the Human Dimensions of Global Change, National Research Council. Washington, D.C.: National Academy Press.
Stolarski, R.S.
  1988   The Antarctic ozone hole. *Scientific American* 258(1):30-36.
Stolarski, R.S., R. Bojkov, L. Bishop, C. Zerefos, J. Staehelin, and J. Zawodny
  1992   Measured trends in stratospheric ozone. *Science* 256:342-349.
Strassman, P.A.
  1985   *Information Payoff: The Transformation of Workers in the Electronic Age.* New York: Free Press.

Stubbs, D.A., P.W. Buckle, M.P. Hudson, and P.M. Rivers
    1983   Back pain in the nursing profession, II. The effectiveness of training. *Ergonomics* 26:767-779.
Swanson, G.M.
    1988   Cancer prevention in the workplace and natural environment: a review of etiology, research design, and methods of risk reduction. *Cancer* 62:1725-1746.
Swets, J.A.
    1988   Measuring the accuracy of diagnostic systems. *Science* 240:1285-1293.
Swets, J.A., R.M. Pickett, S.F. Whitehead, D.J. Getty, J.B. Schnur, J.B. Swets, and B.A. Freeman
    1979   Assessment of diagnostic technologies. *Science* 205:753-759.
Tannenbaum, S., and G. Yukl
    1991   Training and development in work organizations. *Annual Review of Psychology* 41:399-441.
Turner, J., and R. Kraut, eds.
    1992   *Proceedings of the Conference on Computer-Supported Cooperative Work.* New York: ACM Press.
Vachon, R.A.
    1990   Employing the disabled. *Issues in Science and Technology* 6(2):44-50.
Vanderheiden, G.C.
    1990   Thirty-something million: should they be exceptions? *Human Factors* 32:383-396.
Vigil, P.
    1985   Computer literacy and the two cultures revisited. Pp. 240-242 in C.A. Parkhurst, ed., *Proceedings of the American Society for Information Science (ASIS) 48th Annual Meeting.* White Plains, N.Y.: Knowledge Industry Publications Inc.
Vincente, K.J., and R.C. Williges
    1988   Accommodating individual differences in searching a hierarchical file system. *International Journal of Man-Machine Studies* 29:647-668.
Walker, C., and R. Guest
    1952   *Man on the Assembly Line.* Cambridge, Mass.: Harvard University Press.
Wallace, D.R.
    1985   Wetlands in America: labyrinth and temple. *Wilderness* 49:12-27.
Weischedel, R., J. Carbonell, B. Grosz, W. Lehnert, M. Marcus, R. Perrault, and R. Wilensky
    1990   Natural language processing. *Annual Review of Computer Science* 4:435-452.
Wexley, K.
    1984   Personnel training. *Annual Review of Psychology* 31:519-551.
Wickens, C.D.
    1992   *Engineering Psychology and Human Performance*, 2nd. ed. New York: Harper Collins.
William T. Grant Foundation
    1988   *The Forgotten Half, Pathways to Success for America's Youth and Young Families.* Commission on Work, Family, and Citizenship. Washington, D.C.: William T. Grant Foundation.
Williams, G.M.
    1991   Causes and prevention of cancer. *Statistical Bulletin of Metropolitan Insurance Companies* 72:6-10.
Wilson, E.O.
    1989   Threats to biodiversity. *Scientific American* 261(3):108-116.
Wulf, W.
    1993   The collaboratory opportunity. *Science* 261(August 13):854-855.

Yates, J.F.
1990   *Judgment and Decision Making.* Englewood Cliffs, N.J.: Prentice Hall.
Zuboff, S.
1988   *In the Age of the Smart Machine: The Future of Work and Power.* New York: Basic Books.

PART II

Background Papers

1

Productivity in Organizations

Paul S. Goodman and Douglas H. Harris

INTRODUCTION

This chapter identifies specific opportunities for research on productivity in organizations in the following four areas: implementation of effective change within the organization; integration of individual productivity into organizational productivity; congruence of technology, people, and organizations; and integration of the enterprise. Our focus is on new research to be conducted by human factors specialists. We also recognize the need for coordination and cooperation with researchers in other disciplines who are involved in addressing issues in organizational productivity. Consequently, this chapter is for all researchers interested in gaining a better understanding of productivity in organizations. We believe it is also relevant to policy makers and research planners with needs and interests in this area.

The Problem Area

An enterprise consists of technology and people organized to accomplish some purpose. The success of an enterprise can be assessed on the basis of its output and/or the processes and inputs that produce this output. Productivity, in general terms, is the ratio of the output of the enterprise to the inputs. We consider both total factor productivity and labor productivity (see Mahoney, 1988, for a more intensive discussion of productivity). In some of the research we review in this chapter, the criterion variable is performance, not productivity. Other measures, such as quality, timeliness, and profitability, will also be considered in this analysis. To simplify the exposition, however, we will focus primarily on productivity.

The important questions we must now answer are the following: How do we identify the major conditions and processes that cause variation in organizational productivity? How do we address these factors to increase organizational productivity?

The main focus of human factors research has been on improving individual task performance. Researchers have obtained measures of individual performance, such as speed, accuracy, and time needed to learn, and have used these to estimate individual productivity. In some cases they have measured productivity directly. The implicit or explicit assumption underlying these efforts has been that increased individual productivity will increase organizational productivity. However, very little research evidence is available to support this assumption. Improvements in individual productivity may not add up to improvements in organizational productivity because a variety of variables may moderate the effects of the individual improvements and because the various productivity increases that occur on an individual level can interact in a complex way.

We need to understand the factors that drive organizational productivity. If the human factors researcher intervenes at the individual or group level to increase productivity, we need to learn whether these changes affect productivity at the organizational level. If researchers intervene at the organizational level with new technology, decision-making aids, training, and so on, we still must identify the conditions under which these changes lead to improvements in organizational productivity.

The Importance of Research on Productivity

We live in an increasingly competitive global economy. The ability of companies and countries to enhance the productivity of their resources is critical for remaining competitive in this environment and, on a national level, for enhancing the standard of living.

The United States has been introducing advanced technologies to enhance productivity and, hence, our competitive position (National Academy of Engineering and National Research Council, 1991; National Research Council, 1986). The returns from some of these investments, however, appear to be relatively small. For example, the data-processing budgets for corporations in the United States have reportedly been increasing about 12 percent per year. Productivity increases, however, have been averaging no more than 2 percent per year (Weiner and Brown, 1989). Other investigations of the impact of information technology on productivity have failed to find any increase at all from the more than $100 billion per year that is spent on hardware, software, and computer services (Attewell, 1994).

Enhancing productivity is clearly a major national challenge. As competition increases in the global economy, we need to find new ways to improve organizational productivity. The role of researchers is to understand more about the inhibitors and facilitators of organizational productivity. As we develop new knowledge in this area, we will be better able to link new innovations, technologies, organizational structures, and capital to enhance organizational productivity.

We begin our discussion with a review of past research on organizational productivity. We then describe the 4 areas that need further research and identify 15 research opportunities within these areas.

A STRATEGY FOR IDENTIFYING RESEARCH OPPORTUNITIES

The purpose of this section is to give the reader a selective review of productivity research in organizations. We draw primarily from the behavioral science literatures (e.g., human factors, industrial psychology, organizational psychology). Research that has been generated from the fields of economics and technology are outside this review. We have also omitted the methodological issues of productivity measurement and analysis (see Mahoney, 1988, for a discussion of these issues).

Studies of Individuals and Groups

The bulk of the research completed by industrial/organizational psychologists and human factors specialists has focused on interventions designed to increase individual performance or productivity (Campbell et al., 1988; Pritchard, 1991). Examples of interventions include training, measurement and feedback of performance, goal setting, work design, and human-equipment interface design. Guzzo (1988) assessed what we know from psychological research about productivity and its improvement and identified training as the most powerful way to increase individual productivity. He further stated that the effect of training was strongest on output measures of productivity. Locke and Latham (1990) have provided some convincing evidence of the positive effects of goal setting on individual performance. Research (Guzzo, 1988) has indicated that changes in work design have improved individual productivity. What we do not know, however, is whether or not these individual increases are linked to productivity increases of the organization.

Human factors specialists and industrial engineers have reported numerous instances in which the application of appropriate behavioral principles to the design of workplaces and human-equipment interfaces resulted in increases in individual productivity. For example, Harris (1984), in his presidential address to the Human Factors and Ergonomics Society, described 30 human factors projects, many of which involved increases in individual productivity in organizations. Examples included production of microelectronic devices, maintenance of army tanks, investigation of criminal activity, purification of water, on-the-road detection of drunk drivers, and military mission planning. In most cases, however, there was no attempt to study whether the positive effects of these increases in individual productivity affected group or organizational productivity.

Over the last decade, there has been a renewed research interest in groups. Particular attention has been paid to the role of autonomous or self-managing groups. The findings emerging from this research (Goodman et al., 1988; Hackman, 1990) are that self-managing teams can increase productivity, quality, and worker satisfaction. However, most of this research has not conceptually or empirically considered whether increases in group or team productivity produce increases in organizational productivity.

Studies of Systems and Organizations

Another source of studies on productivity in organizations comes from researchers interested in how the interaction of environment, technology, organization, and people variables affects organizational productivity (Lawrence and Lorsch, 1969; Scott, 1987; Thompson, 1967). Socio-technical analysis (Trist et al., 1963) provided the early intellectual underpinnings of this work.

One study (MacDuffie and Krafcik, 1990) sought to explain productivity differences in a worldwide sample of automobile assembly plants. The critical findings of this study were that the plants with the highest level of automation were not the most productive. The most productive plants introduced congruent organizational changes as well as technological changes. In general, congruency refers to the fit between the type of technology, specified in this case by levels of interdependence, and the nature of the organizational arrangements, specified by the flexibility in organizational arrangements. Better fit means higher scores in productivity or effectiveness.

Studies in the shipping industry (Walton, 1987) and coal industry (Goodman, 1979) also show that high organizational productivity is associated with congruency among innovation in technology, organizational arrangements, and people factors.

The Gaps in What We Know

We know something about how to increase individual and group productivity. However, there is no compelling research evidence that these changes lead to changes in organizational productivity. We also know that congruency among technological, organizational, and people factors can contribute to organizational productivity.

Yet we do not know how changes at the organizational level are related to changes in individual or group productivity. Also, the systems perspective does not inform us ahead of time how conditions of congruency will be determined. There are many gaps in our knowledge of how new forms of technology, new organizational forms, and changes in a competitive environment will impact on rates of changes in a organizational productivity. Although we have made progress in understanding organizational productivity, we need to address some of these gaps with new theories and new methods.

FOUR RESEARCH OPPORTUNITY AREAS

We will outline the research that is needed to define the processes and conditions that will facilitate this transformation. Our focus, then, is not on productivity per se, but on how human factors and other organizational researchers can develop theories and methodologies to understand the link between changes in individual, group, and organizational productivity.

We organize our discussion of research in terms of four opportunity areas. We describe the problem and rationale for each of these areas, along with specific research needs. Throughout this analysis we consider productivity a multilevel and multidimensional concept.

Implementation of Effective Change Within the Organization

Billions of dollars have been spent introducing new forms of organizational and technological change in organizations (Bikson et al., 1987; Leonard-Barton, 1988; National Academy of Engineering/National Research Council, 1991). Implementation deals with the process of effectively introducing changes in organization. There is a growing body of evidence (Goodman and Griffith, 1991) to show that many organizational and technological interventions, which should inherently improve organizational productivity, are not successfully implemented. This means the interventions are rejected, resisted, only partially accepted, or only temporarily adopted and then rejected.

If technological and organizational interventions have clear objectives or inherent benefits but cannot be successfully implemented, then they will not improve individual or organizational productivity. Indeed, successful implementation is key to our analysis of organizational productivity. For example, we recently studied the introduction of a vision system into a sophisticated computer-integrated manufacturing environment (Goodman et al., 1990). The purpose of this vision system was to enhance quality through new forms of monitoring. By most objective measures, the new technology dominated current technology for monitoring and improving quality. However, the vision systems remained basically unused two years after their introduction.

If we cannot establish improvements through successful implementation at the individual, group, or organizational level we cannot expect to see productivity gains at the organizational level. Understanding the process of successful implementation is therefore the starting point in our analysis of research opportunities.

New Research Topics

Over the last decade there has been an increasing body of research on the process of implementation. In much of that work, researchers have examined how technological-, organizational-, and individual-level variables affect the implementation of new technologies. They have paid less attention to identifying underlying explanatory mechanisms. Thus, there are frameworks but no well-defined theories. Despite the large number of possible predictor variables, there have been few multivariate studies and few longitudinal designs to capture implementation over time.

Research needs include the following:

• Developing better conceptualizations and measures of implementation success. Many studies focus on user satisfaction and frequency of use. There has been little attempt to link individual and organizational dimensions of implementation success.

• Developing more parsimonious theory. Much of the research has used a contingency theory approach. That is, researchers have tried to identify a contingent set of variables or conditions (e.g., top management support or participation) under which technologies will be more effectively implemented and conditions under which implementations will be less effective. Because of the large number of variables that might affect implementation success, such an approach has great limitations. Recently, there has been movement toward a process-oriented theory (Goodman and Griffith, 1991). The key idea is to develop a better understanding of the mechanisms that lead to successful implementation.

• Assessing the value of alternative forms of training and learning. Training is an important part of the implementation process. It is clear that formal classroom training plays an important role in the successful implementation of new technology. It is also true that other forms of learning, such as observation, modeling, and apprenticeship, bear on successful implementation of new technology or organizational arrangements. We have little systematic knowledge of how and under what conditions these alternative forms of learning affect implementation success.

• Studying the effects of alternative forms of commitment. Participation is a common prescription in implementing new changes successfully. However, the empirical evidence on the effectiveness of participation is quite mixed (Goodman and Griffith, 1991). Perhaps we should explore other ways to enhance commitment: for example, fostering recognition of the inherent benefits of the new technology or a normative consensus about the new technology. Alternatively, we could explore the specific mechanisms evoked by different forms of participation and how they bear on changing levels of commitment.

• Studying adaptation and redesign. A significant finding (Leonard-Barton, 1987) in recent implementation studies is the importance of a redesign function. That is, as a piece of technology is introduced into an organization, one needs a mechanism to adapt the technology to changing user needs and organizational contexts. We know little about how to design alternative forms of adaptation mechanisms or the effectiveness of these mechanisms in different contexts. This is important if we are going to understand the successful implementation of new technology and organizational arrangement.

Integration of Individual Productivity into Organizational Productivity

Although there has been considerable research on productivity at the individual, group, and organizational levels, we know little about how changes in productivity at one level affect changes in productivity at another level.

New Research Topics

Research opportunities for examining the linkages between individual and organizational productivity are relatively unlimited, given the paucity of studies in this area. We recommend examining the different types of linkages to discover whether increases in individual productivity inhibit or facilitate increases in organizational productivity.

Research needs include the following:

• Studying productivity linkages within jobs. Jobs are bundles of linked activities. Most interventions to improve productivity focus on specific sets of activities. There is some evidence that increasing the productivity of certain sets of activities within a job can reduce the productivity of other job activities. For example, Kraut et al. (1989) examined the impact of a new computerized information system on the productivity of customer service representatives. They learned that productivity increased for routine tasks, but dropped for nonroutine tasks. So the net increase in productivity of that job may be close to zero. Similarly, increases in productivity in certain job activities can simply create additional slack, which does not enhance productivity for the total job or for the organization. Furthermore, some job activities are central to core production activities, while others (e.g., recordkeeping) are more peripheral. Improving productivity in peripheral activities is less likely to affect organizational productivity. We do not understand the linkages between different activities within a job. When will productivity changes in certain activities stimulate positive changes in other activities? When will the opposite occur? We also do not understand when slack will be generated, when slack can have a positive effect on productivity, and when slack will mitigate any greater effects of productivity, given increases in certain job activities.

• Studying productivity linkages between jobs or units. Jobs in organizations are interdependent in varying degrees. Most research has focused on increasing productivity in individual jobs, but not on the consequences of this productivity for jobs horizontally or vertically linked to them. There is some evidence that productivity increases in a particular job may (1) reduce productivity in other jobs, (2) simply create slack that does not lead to productivity in other jobs, or (3) encounter constraints in interdependent jobs that prevent the productivity increases from having an impact on any other part of the organization. The situation is analogous to what occurs among activities within a job. Thus, increases in productivity in one job may decrease productivity in other jobs, leading to a net productivity gain of zero. The basic research question is, how do changes in one job positively or negatively affect other linked jobs? There is a need for a well-developed theory or research paradigm to help us predict when increases in productivity in one job would enhance the productivity in a horizontally, or vertically, linked job.

• Studying linkages and organizational type. Organizations differ in the interdependence and complexity of their linkages. Some organizations, for example, have pooled interdependence: different departments are fairly independent, but the products of all these departments need to be pooled together to achieve an organizational product. A department store is an example of pooled interdependence. Some organizations have other forms of interdependence: an assembly line is a form of serial interdependence; a hospital might illustrate reciprocal interdependence. Even organizations with the same type of interdependence may vary in complexity or simply in the number of different linkages. Linkages between individual, department, and organizational productivity in pooled interdependence differ from the linkages in serial or reciprocal interdependence. For example, because department store units are relatively independent, increasing productivity in one unit should contribute fairly directly to increases in organizational productivity. In contrast, the effect of productivity increases in any unit of an assembly line depends, to a large extent, on what happens in the other interdependent units and, hence, will condition the impact of that unit on organizational productivity (see National Research Council, 1994). We need to understand how the mechanisms that translate changes in individual-level productivity to organizational-level productivity differ by organizational type. These comparative organizational analyses should provide new ideas about and new insights into the relationship between individual and organizational changes in productivity. Although we have defined organizational type in terms of the form of interdependence, other measures such as size may be relevant.

Congruency of Technology, People, and Organizations

A growing body of literature indicates that the fit or congruency among technological, organizational, and people factors is the key to understanding differences in organizational productivity (Goodman, 1979; Walton, 1987). For example, in a study of auto assembly plants, MacDuffie and Krafcik (1990) observe that the most productive plants, identified as "lean production systems," have (1) a technological system characterized by highly interdependent technology, no buffers, a mechanism that immediately stops the system when downtime occurs, and no repair areas and (2) an organizational system characterized by fewer job classifications, workers with multijob skills, intensive training in problem solving, teams, fewer status barriers, and contingent reward systems.

The critical research question is, why does this configuration of technological and organizational arrangements lead to high productivity? What does congruency or fit mean? An increasing number of studies indicate that fit makes a difference. However, we need some before-the-fact explanations for what fit among technology, organization, and people factors means. This becomes particularly important as new technologies proliferate and new organizational arrangements evolve.

The importance of this research area should be clear. The future will bring rapid changes in both technology and organizational arrangements, as well as new demographic characteristics. We need to develop theories and bodies of research findings that will help us understand and predict how different combinations of these factors will affect organizational productivity.

New Research Topics

There are important, exciting research opportunities concerning the congruency among technological, people, and organizational factors as drivers of organizational productivity.

Research needs include the following:

• Identifying the critical processes or mechanisms that explain why certain configurations among technology, organizations, and people lead to higher organizational productivity. Goodman et al. (1994) identify five critical processes underlying the congruency question: coordination, problem solving, focus of attention, organizational evolution or redesign, and motivation. Are there other processes? Should some of these be combined? To what extent are the processes independent explanations and to what extent do the processes reinforce one another? Reward systems at the organizational level both improve motivation and focus attention on organizational-level outcomes. How do these processes function under different technological and organizational arrangements?

• Connecting the issues of linkage and congruency. A central question in this chapter is, how do changes in individual-level productivity lead to changes in organizational productivity? This is an important and relatively unresearched area. Although, as noted above, there is a body of literature supporting the importance of congruency among technology, organizations, and people to enhance productivity, there is little theory or evidence to explain when, how, and why this congruency functions. Research is needed to integrate the theoretical work on linkages and congruency in some empirical context. Are there common processes that enhance both linkages and congruency among technological and organizational systems? How do they function? Where is the synergy? Research in this area might, for example, focus on redesign. The implementation literature points out that redesign is necessary to adapt new technology to changing organizational factors and user needs over time. If a successful redesign process is in place, there is a higher probability that individual productivity can be sustained and, therefore, a greater opportunity to contribute to organizational productivity. Without a redesign process, there is a much lower probability that individual productivity will affect organizational productivity. The redesign process also appears in the literature about highly productive, congruent systems. In the high-productivity, lean production systems discussed above, there was a continuous process of redesigning the organizational and technological systems to ensure better fits and higher productivity. It is important to note that the redesign process is central to both explanations (linkage and congruency) for higher productivity.

• Studying the effects on productivity of dramatic changes in technology, organization, and the workforce. Over the next decade, we expect to see rapid and dramatic changes in all three areas. New forms of technology will proliferate as tools are created for decision making, communication, and working across space and time. On an organizational level, there will probably be a continuation of the movement toward smaller and more autonomous units, the structuring of organizations around processes rather than functions, and increased interest in such processes as organizational learning. The characteristics of the workforce will also continue to change, possibly altering the "contract" between the employees and the organization. We need research on how changes in each of three areas—technology, organization, and the workforce—affect productivity in the other two. It may be helpful to do some of this research in different national contexts.

Integration of the Enterprise

Developments since the late 1980s point to the need to integrate organizational processes better, for example, by computer-integrated manufacturing. At the same time, there has been a push to integrate external constituencies such as customers and suppliers. One important question in the 1990s is how to integrate these external activities to improve organizational productivity. The broader question is how to integrate internal and external activities into a total enterprise to improve overall productivity and effectiveness.

The design process is a good exemplar of this issue. Traditionally, the design process was a serial activity with primarily an internal focus. It went from product conception to the design, manufacturing, and marketing of the product. Now, the importance of integrating customer needs into the design process is well understood, and the necessity of moving away from serial interdependence to a more simultaneous consideration of customer, designer, and manufacturing needs is becoming more apparent (Clark, 1991). Integration, both internally and externally, is what makes simultaneous consideration of these multiple needs possible. The consequences of good design processes are readily apparent. Producers with better externally and internally integrated design processes get goods to the market faster, are more responsive to customer needs, and produce goods with higher quality and productivity.

Another example of enterprise integration concerns the relationship between an organization and its suppliers and vendors. Traditionally, this was an arm's length relationship. Now, there are major forces to integrate suppliers and customers into the daily operation of the organization. "Just-in-time" environments demand a close association between the suppliers and the focal organization.

If our ultimate objective is to better understand organizational productivity, we must pay attention to integrating the enterprise. This is not a well-researched area. In the past, much of the work in this area has focused on specific levels of analysis and on interventions to increase productivity at specific levels within the organization. The challenge in the 1990s is to understand the process of integrating external demands into the internal processes of the organization.

New Research Topics

Increasing our understanding of the process of integration should increase our knowledge of productivity in organizations. However, this problem is challenging simply because of the scope of the problem statement. Enterprise integration focuses on the totality of the organization and all the relevant constituencies of its environment. Given the paucity of research in this area, we have identified research areas that would inform us about the enterprise integration process.

Research needs include the following:

• Investigating the role and consequences of technology. With the costs of buying and using computers declining and with the levels of computer power and networking capability increasing, technology will be a major factor in the integration process. Much of the early work in enterprise integration focused on internal integration of machines. The new challenges are using technologies to integrate operational planning and strategic decision making into a total operating system. Some of the research on electronic data interchange shows how certain types of external integration can improve organizational performance indicators (Kekre and Mukhopadhyay, 1992; Srinivasan et al., 1991). An important research endeavor will be to document the functional and dysfunctional consequences of this and other technologies in achieving internal and external integration. What types of information and transactions are necessary in this integrated enterprise, and how will electronic communication (versus other forms of communication) contribute to the productivity and quality of these information exchanges?

• Studying the process of coordination. Implicit in the concept of integration is the need to coordinate activities across varied constituencies. The design example, discussed above, involved different groups (customers, designer, manufacturer) who had different goals and interests. The tasks were to identify critical forms of information and to coordinate this information so as to optimize all the relevant processes (design, manufacturing, marketing). New theory and research on negotiation and coordination seem to be critical for understanding the broader question of enterprise integration (Malone and Crowston, 1994).

- Studying the process of developing new partnerships. The concept of enterprise integration is not grounded solely in technological solutions. Customers, suppliers, and firms all have different interests. Entering a collaborative arrangement requires some form of organizational change. These new organizational forms require new forms of reward, commitment, and structural relationship. How to define effective collaboration is an important research question that bears on our broader question of enhancing productivity in organizations (Kanter, 1991).
- Studying the effects of linking external and internal productivity. We have focused on the need for research on how productivity increases at the individual level lead to productivity increases at the organizational level. Our analysis, however, has been within the organization. Another approach would be to examine how productivity increases in the customer's or supplier's organization leads to productivity increases in the focal organization. If the focal organization intervenes in the customer organization to increase productivity, how do these productivity increases link back into the focal organization?

CONCLUSION

We have identified four broad research opportunity areas, describing each problem area and possible research topics. We believe that understanding these four areas is critical to understanding the general question of productivity in organizations. We think there is important and challenging research to be done in each area. We derived 15 research needs from these 4 opportunity areas. Each research need can be a point of departure for understanding organizational productivity.

Although we have focused primarily on productivity, we could have switched our focus to other criterion variables, such as quality. We also think the research issues generated in this chapter bear equally on many different organizational forms. The research questions are relevant to both manufacturing and service organizations, and to both profit and nonprofit organizations.

REFERENCES

Attewell, P.
    1994   Information technology and the productivity paradox. Pp. 13-53 in D.H. Harris, ed., *Organizational Linkages: Understanding the Productivity Paradox*. Panel on Organizational Linkages, Committee on Human Factors, National Research Council. Washington, D.C.: National Academy Press.

Bikson, T.K., B. Gutek, and D.A. Mankin
    1987   *Implementing Computerized Procedures in Office Settings*. Santa Monica, Calif.: RAND Corporation.

Campbell, J.P., R.J. Campbell, and associates
  1988   *Productivity in Organizations.* San Francisco: Jossey-Bass.
Clark, K.B.
  1991   *Product Development Performance: Strategy, Organization and Management in the World Auto Industry.* Boston: Harvard Business School Press.
Goodman, P.S.
  1979   *Assessing Organizational Change: The Rushton Quality of Work Experiment.* New York: John Wiley & Sons.
Goodman, P.S., and T.L. Griffith
  1991   Process approach to implementation of new technology. *Journal of Engineering and Technology Management* 8:161-185.
Goodman, P.S., R. Devadas, and T.L. Griffith
  1988   Groups and productivity: analyzing the effectiveness of self-managing teams. Pp. 295-327 in J.P. Campbell, R.J. Campbell, and associates, *Productivity in Organizations.* San Francisco: Jossey-Bass.
Goodman, P.S., T.L. Griffith, and D.B. Fenner
  1990   Understanding technology and the individual in an organizational context. Pp. 45-86 in P.S. Goodman and L.S. Sproull, eds., *Technology and Organizations.* San Francisco: Jossey-Bass.
Goodman, P.S., F.J. Lerch, and T. Mukhopadhyay
  1994   Individual and organizational productivity: linkages and processes. Pp. 54-80 in D.H. Harris, ed., *Organizational Linkages: Understanding the Productivity Paradox.* Panel on Organizational Linkages, Committee on Human Factors, National Research Council. Washington, D.C.: National Academy Press.
Guzzo, R.A.
  1988   Productivity research: reviewing psychological and economic perspectives. Pp. 63-82 in J.P. Campbell, R.J. Campbell, and associates, *Productivity in Organizations.* San Francisco: Jossey-Bass.
Hackman, J.R., ed.
  1990   *Groups That Work (and Those That Don't): Creating Conditions for Teamwork.* San Francisco: Jossey-Bass.
Harris, D.H.
  1984   Human factors success stories. Pp. 1-5 in *Proceedings of the Human Factors Society 28th Annual Meeting.* Santa Monica, Calif.: Human Factors Society.
  1994   *Organizational Linkages: Understanding the Productivity Paradox.* Panel on Organizational Linkages, Committee on Human Factors, National Research Council. Washington, D.C.: National Academy Press.
Kanter, R.M.
  1991   Improving the development, acceptance, and use of new technology: organizational and interorganizational challenges. Pp. 15-56 in National Academy of Engineering and National Research Council, *People and Technology in the Workplace.* Washington, D.C.: National Academy Press.
Kekre, S., and T. Mukhopadhyay
  1992   Impact of electronic data interchange technology on quality improvement and inventory reduction programs: a field study. *International Journal of Production Economics* 28:265-282.
Kraut, R.S., S. Dumais, and S. Koch
  1989   Computerization, productivity, and quality of work-life. *Communications of the ACM* 32(2):220-238.

Lawrence, P.R., and J.W. Lorsch
1969   *Organization and Environment: Managing Differentiation and Integration.* Homewood, Ill.: Irwin.

Leonard-Barton, D.
1987   The case for integrative innovation: an expert system at Digital. *Sloan Management Review* Fall:7-19.
1988   Implementation as mutual adaptation of technology and organization. *Research Policy* 17:251-267.

Locke, E.A., and G.P. Latham
1990   *A Task Theory of Goal Setting and Task Performance.* Englewood Cliffs, N.J.: Prentice Hall.

MacDuffie, J.P., and J.F. Krafcik
1990   Integrating Technology and Human Resources for High Performance Manufacturing: Evidence from the International Auto Industry. Paper presented at the Transforming Organizations Conference, Massachusetts Institute of Technology, Boston.

Mahoney, T.A.
1988   Productivity defined: the relativity of efficiency, effectiveness, and change. Pp. 13-39 in J.P. Campbell, R.J. Campbell, and associates, *Productivity in Organizations.* San Francisco: Jossey-Bass.

Malone, T.W., and K. Crowston
1994   The interdisciplinary study of coordination. *ACM Computing Surveys* 26(1):87-119.

National Academy of Engineering and National Research Council
1991   *People and Technology in the Workplace.* Washington, D.C.: National Academy Press.

National Research Council
1986   *Human Resource Practices for Implementing Advanced Manufacturing Technology.* Manufacturing Studies Board. Washington, D.C.: National Academy Press.

Pritchard, R.D.
1991   Organizational productivity. In M.D. Dunnette and L.M. Hough, eds., *The Handbook of Industrial and Organizational Psychology,* 2nd ed. Palo Alto, Calif.: Consulting Psychologists Press.

Scott, W.R.
1987   *Organizations: Rational, Natural, and Open Systems.* Englewood Cliffs, N.J.: Prentice-Hall.

Srinivasan, K., S. Kekre, and T. Mukhopadhyay
1991   Impact of electronic data interchange technology on JIT. *Management Science* 40(10):1291-1304.

Thompson, J.D.
1967   *Organizations in Action.* New York: McGraw-Hill.

Trist, E., G. Higgins, H. Murray, and A. Pollock
1963   *Organizational Choice.* London, England: Tavistock.

Walton, R.E.
1987   *Innovating to Compete: Lessons for Diffusing and Managing Change in the Workplace.* San Francisco: Jossey-Bass.

Weiner, E., and A. Brown
1989   Human factors: the gap between humans and machines. *The Futurist* May-June:9-11.

2

Training and Education

Joyce L. Shields, Joseph B. Cavallaro, Beverly M. Huey, and Harold P. Van Cott

INTRODUCTION

This chapter focuses on the relationship between job-oriented training and human factors research. The underlying premise is that human factors research can help solve some of the difficult problems confronted by the people who design, develop, and manage training programs and who deliver training services to trainees.

Education is included as a topic because there are strong links between general education and training for the workplace. Education prepares for job-oriented training and influences it. To a large extent, the educational attainment of prospective trainees determines what the training provider can do and how the training can or should be presented. In this sense, the level of educational attainment of the trainee is a parameter in the assembly of any given training program.

In job-oriented training, training and human factors are to some extent reciprocal. When a new system is being created, the goal is to design the system so that training requirements are as light as possible. If there are design deficiencies, the operators of the system must be trained to overcome or compensate for them. Thus, the absence of effective human factors inputs into the system design process generates technical problems for the people responsible for the training of operator and maintenance personnel.

Human-centered systems design and development also come into the process of providing the tools used for training. The complete ensemble of such tools is itself a system and is, therefore, a reasonable domain for the kinds of systems research carried out by human factors scientists.

There are many ways in which human factors research is and continues to be useful in addressing some of the problems that now are being confronted by the people engaged in providing job-oriented training. In the rest of this chapter, we will explore these problems and the current and potential contribution of human factors to them.

NOTABLE PROBLEMS

Changing Conditions of Work

The very nature of work is changing. Instead of the physical manipulations that formerly characterized most industrial jobs, more and more jobs are based on the manipulation of symbols—words and numbers. New technologies and work processes are shifting skill requirements in some types of jobs from manual and job-specific to generic, from concrete to abstract. For example, in many modern factories, the production process is so computerized that many workers no longer physically execute tasks but are now responsible mainly for monitoring automated processes. Whereas previously a machinist adjusted and maintained machinery by hands-on manipulation, today a technician works through a keyboard and CRT display and consequently has a more distant relationship to both process and product.

A direct consequence of computerization is that tangible work outcomes are less evident. There is some evidence suggesting that spontaneous learning and skill transfer do not work as well in highly computerized environments as in more traditional workplaces. For example, the difference in performance between the best and the average appears to be greater in computerized work than in traditional jobs. This finding leads to the conjecture that less incidental or informal learning takes place in work that involves symbol manipulation. This conjecture is supported by the fact that much about symbol manipulation is invisible, unlike the visible actions involved in object manipulation. The findings also suggest that John Seely Brown (1988) and others are correct when they call for new forms of training such as a "cognitive apprenticeship," which would somehow make the mental procedures and decision processes of exemplary performers "visible" so that others might learn from them.

Changing Workforce Demographics and Changing National Patterns

Similar concerns are expressed about the changing demographics of the workforce. In this era of vigorous global competition, rapid technological change, and shifting economic conditions, trainers must be able to adjust to such changes. In *Workforce 2000* (Johnston and Packer, 1987), the following "demographic facts" are identified as shaping the workforce of the future:

- There will soon be a slowdown in the growth of the population and the workforce—the greatest since the 1930s.
- The average age of the workforce will rise and the pool of young workers will decline.
- More women will enter the workforce, although the rate of increase will be smaller than in the past.
- Minorities will be a larger share of new labor force entrants.
- Immigrants will represent the largest share of the increase in the population and the workforce since the First World War.

Thus, the "next decade will usher in a workforce unlike anything Corporate America has ever seen—teeming with women, minorities, and the elderly, the antithesis of the workforce present at many companies today" (Goldstein, 1988). Employers will soon have to utilize these nontraditional sources of labor supply in order to sustain output—even if equal opportunity regulations are abolished.

Human resource implications of demographic shifts are numerous. For example, Taylor (1989) lists several training and development issues that will arise from the "age wave" of older workers, including the need to recognize and emphasize the contributions that older employees can make at all levels of the organization. It will be incumbent on the employer organization to provide effective training and retraining programs for these older workers.

Because people will remain in the workforce longer, there will be an increasing need to retrain people who have already reached positions of some seniority and authority. Some of these people will be the victims of the buggywhip scenario—their jobs will just become obsolete. For others, the job will still be there to be done, but it will be done by a robot or a computer.

Older workers might need to be retrained through substantially different techniques than those used to train true novices. In some instances, a highly proficient worker may be required to unlearn some of the skills that gave status in the original job. Such skills may actually be counterproductive in the latest version of some jobs.

Discussing the greater numbers of women in the workforce, Johnston and Packer (1987) predict that the distinctions between males and females in wage rates and in terms of concentrations in particular jobs will decline in response to market pressures. They also predict that the proportion of "part-time, flexible, and stay-at-home jobs will increase, and [that] total work hours per employee are likely to drop in response to the needs of women to integrate work and child rearing." On the plus side, training requirements may become more uniform if both men and women are doing the same work. Negatives in terms of increased costs could result from having more people to train if more jobs are shared by two individuals.

The Office of Technology Assessment (1990) discusses the number of new workforce entrants who will be coming from minority groups and who are often disadvantaged because of inadequate basic educational opportunities. Furthermore, many young workforce entrants, regardless of their ethnic heritage, will confront a daunting challenge if the predicted shift to more highly skilled jobs within the manufacturing sector takes place (Personick, 1989). If, as seems likely, a similar shift is taking place in the service sector as it continues to expand, the country might find itself in the ironic condition of having large numbers of workers unemployed while there are labor shortages in relatively well-paying job categories.

At the national level, the growing service sector is expected to account for over 90 percent of the 18 million new jobs forecasted for the next 10 years. Much of the growth is expected to occur in the health, education, business, and food services. In both the service and manufacturing sectors, systems analyst and computer programmer jobs will be among the fastest-growing occupations, and workers in other occupations will need to be increasingly computer literate. Improved office technology will continue to limit the growth of administrative support occupations, which will be among the slower-growing groups of occupations (Bureau of Labor Statistics, 1991).

Changing Organizational Arrangements

Numerous changes are also occurring in the nature and structure of work at the organizational level. Changes instituted by organizations include decentralizing and flattening organizational structures, establishing work teams, pushing decision-making responsibilities downward throughout the organization, and streamlining the management ranks through the elimination of middle- and lower-level managers.

Flattened organizational structures reduce the hierarchical flow of information and should accelerate decision making. Job classifications are broadening as work teams assume some or all of the responsibilities of the first-line supervisor such as inspection, quality control, production scheduling, work allocation, and coordination with other departments (Office of Technology Assessment, 1990).

Decentralizing the organizational structure and establishing work teams increase the worker's autonomy and responsibility. In a decentralized organizational structure, fewer layers of management result in managers' supervising greater numbers of workers. Managers are unable to manage at the same level of detail as before, and some of the responsibility and authority previously held by middle managers is delegated to workers in lower levels of the organization (Kravitz, 1988).

The development of cross-functional work teams has also increased the responsibilities of workers. Employees no longer perform fragmented tasks, but have broad responsibility for knowing how to perform all aspects of a job, a situation that draws greater initiative from workers and increases their involvement in decision making (Bailey, 1990). Consequently, social skills will also become more important as organizational structures change and as more service jobs are created. Interpersonal skills, teamwork, and communication skills will become critical for every employee, from the chief executive to the line worker (Carnevale et al., 1988).

The nature of work is also likely to change owing to developments in areas such as telecommunications (e.g., teleconferencing, voice mail, picture phones, and facsimile machines) and personal computers (e.g., wide-area networks, modems, communications software, laptop and notebook computers). These technologies are making it possible for more and more workers to work out of the home. As more people work at home, a variety of management and organizational questions are raised, including how to structure teamwork, how to train, and how to measure performance.

HUMAN FACTORS CHALLENGES

Conceptual Framework

It seems wise to have some sort of conceptual model to guide the process of laying out a research program at the strategic level. Human factors researchers are fortunate to have several such models from which to choose. One was generated in the 1980s when the Army developed the policies, procedures, and tools to more fully integrate the inclusion of manpower, personnel, training, systems safety, health hazards, and human factors engineering trade-offs into the design of systems (Booher, 1990). This program, referred to as MANPRINT (Manpower and Personnel Integration), successfully refocused the design process on the performance of the total system in the hands of the user. The program has led to the design and redesign of weapon systems with greatly reduced training demands and enhanced performance capabilities. Lessons learned in this program have potential application to the acquisition and design of a wide range of complex systems.

Another approach was introduced by Rouse (1991), who provided a systematic framework for ensuring that the concerns, values, and perceptions of all participants in a design effort, as well as the ultimate users, are considered and balanced. This approach seeks to enhance human abilities, overcome human limitations, and foster user acceptance throughout the design process. Rouse's method recommends asking the following questions in the design and evaluation of systems:

- Viability—are the benefits of using a system sufficiently greater than its costs?
- Acceptance—do organizations/individuals use the system?
- Validation—does the system solve the problem?
- Evaluation—does the system meet the requirements?
- Demonstration—how do observers react to the system?
- Verification—is the system put together as planned?
- Testing—does the system run, compute, and so forth?

From a designer's perspective, the process begins with measures of effectiveness and then addresses understandability and, finally, compatibility. From a user's perspective, the evaluation begins with compatibility, and goes on to understandability and then effectiveness (top-down design vs. bottom-up evaluation).

Both of these macro-ergonomic approaches can be applied to the design and development of training systems. However, the field would benefit from additional competition. Researchers should be encouraged to develop additional concepts for this area of discourse.

Building Bridges Between Theory and Practice

A major research issue related to designing and implementing effective training is closing the gap between (1) theory and research and (2) applications. In an article in *Human Factors*, Cannon-Bowers et al. (1991) addressed the challenge of linking the theory and practice of training. This is particularly difficult because research and practitioner communities exist separately and are not well coordinated. The authors highlight linkages between training-related theory and training-related techniques in the areas of training analysis, design, and evaluation. This represents a good start in illustrating the possibilities; however, as the authors note, the current body of training research does not provide the practitioner with a full complement of training techniques that are fully grounded in logical principles. Further, there remains the question of how to enhance the acceptance and translation of research into practice.

Boff and Rouse (1987) emphasized the issue of ensuring an adequate supply of useful and usable data resources in formats that are specifically intelligible to design engineers. Although there is a substantial volume of research information that has potential value and relevance to systems design, Boff and Lincoln (1988) found that the data had low usability in terms of accessibility, interpretability, and applicability. If these data were more usable, they could greatly enhance the effectiveness of new training systems; therefore, ways to exploit the existing investment in research findings need to be developed.

This theme is also present in a report by the Office of Technology Assessment (1990). The report indicates that research that could lead to improvement in the efficiency and quality of training often fails to be integrated into training practices. The report recommends that federal agencies (1) develop and disseminate information about best-practice approaches and technologies and (2) work with industry to develop and implement operating standards for training technologies and related software.

An example of such an approach in the human factors area is the Crew System Ergonomics Information Analysis Center (CSERIAC). The objective of CSERIAC is to acquire, analyze, and disseminate timely information on crew system ergonomics. CSERIAC's principal products and services include technical advice and assistance; customized bibliographies; written reviews, analyses, and assessments; workshops, conferences, symposia, and short courses; and special studies.

The challenge to the human factors community is to devise, evaluate, and perfect ways to transfer the methods, data, and principles of human-centered design into the procedures used by training system designers and managers. Human factors scientists should also ensure that feedback from applications experiences closes the loop between research and practice.

Strategic Methodologies

A central methodological problem is the lack of a comprehensive array of performance measures that are acceptable to both researchers and practitioners. The basic premise is that one cannot know whether a training system or training practice is better than an alternative unless one can (1) measure system performance reliably and accurately and (2) link such bottom-line measures with specific attributes of the training system.

One cause of the problem is that there are so many different tasks in the world of work, and each task has its own mix of desired outcomes. Consequently, research that is intended to provide general findings is often not convincing to the practitioner, who sees a distinction between the supposedly generic laboratory task and the task for which his or her students are being trained.

A closely related methodological problem is the lack of criteria for doing an overall evaluation of training programs. Therefore, comprehensive evaluations are rarely conducted, even though their importance is recognized. When evaluations are done, about half of all companies and the vast majority of large, successful corporations use crude measures such as student ratings or inferences drawn from informal follow-up interviews. This may be due in part to the difficulty and expense of specifying better measures as well as to the concerns of trainers and educators that negative results may jeopardize future efforts (Goldstein, 1992). The human factors community can contribute to the resolution of these issues.

An example of such a contribution comes from the work of Kraiger et al. (1992), who presented a theoretically based model of training evaluation in an attempt to link evaluation to learning objectives and provide a more conceptual framework for evaluation. The authors conceive of learning as a multidimensional process of cognitive, behavioral, and affective change and believe it is important to move away from the influence of behaviorism in evaluating training. They argue that the emphasis on behaviorism has stifled the development of training as a scientific discipline. The framework they present leads to a large number of suggested measures and provides a stimulus for further research on the development and testing of measures that could lead to a better understanding of training effectiveness.

Toward a similar end, Goldstein (1992) points out that training can be characterized as attempting to achieve one of four types of validity: (1) training validity, (2) transfer validity, (3) intraorganizational validity, and (4) interorganizational validity. The first two validities are measured by the performance of the individual trainee at the time of training or on the job. Both training and transfer validity begin with needs assessment. But transfer validity is greatly affected by the organizational environment of the trainee on the job. Therefore, special consideration has to be given to the collection and interpretation of the results.

The last two types of validity refer to the effectiveness of the training program for a new group of employees in the same organization or in different organizations. In both cases, data should be collected from new populations of trainees to determine the effectiveness of the training program for different trainees and organizations. Validities can change because of (1) changes in tasks, people, or the organization; (2) the rigor of the data collection/evaluation; and (3) changes in the training program. Carefully planned and executed evaluation programs are clearly required to assess and demonstrate the value of training.

A significant development in evaluation has been the estimation of the dollar value or utility of training interventions (Clegg, 1987). A variety of equations have been derived for a variety of situations, from estimations of return on investment to cost-benefit analyses using organizational levels of measures of performance (Wexley and Baldwin, 1986; Paquet et al., 1987). However, only a third of the large, successful companies evaluate bottom-line improvement to assess training or educational success.

It is clear that conducting evaluations of the relevance and effectiveness of education and training programs has been complex and expensive. The questions remain: How can information be collected in an affordable manner to aid decision makers in determining the value of training? And how can organizations be encouraged to employ "bottom-line" methods for assessing their training programs?

Evaluation studies would be more logical and convincing if they were tied to a formal assessment of the real needs of the worker and the requirements of the job. A common complaint about existing training programs is that they are not responsive to these areas.

Tannenbaum and Yukl (1991) tend to confirm this allegation. They report that comprehensive needs analyses are rarely done in organizations. Therefore, training program content is not always linked to organizational strategy or job performance requirements.

When needs analysis *is* attempted, the predominant approach is likely to be to use the procedures outlined by McGehee and Thayer (1961). These link organization, task, and personal attributes in a basic stimulus-response type of logic structure. Given the major changes in workplace training in the last 30 years and the requirements of the modern workplace, the validity of this needs assessment framework is now problematic.

The starting point for specifying the requirements of a job is often a task analysis in which a human factors practitioner focuses on the technical and procedural portion of the task. Task analysis is a general-purpose method for describing how tasks are performed. Devised by Robert B. Miller in 1954, it was first used to specify the characteristics of the displays and controls needed to enable good operator performance. The method was later expanded to identify and document the skills, knowledge, and abilities (SKAs) that a person must have to perform a task and job effectively. These SKAs formed a database called qualitative and quantitative personnel requirements information, which was used to forecast manpower, human factors engineering, and training needs for new systems. Task analysis was subsequently adopted by educators and test construction specialists.

The methods of task analysis in general use today assume that the content of jobs can be defined by their stimulus and response components. This assumption is understandable considering that task analysis had its origin in behavioristic psychology. Although the stimulus-response approach may be acceptable for characterizing the psychomotor tasks that have historically been dominant components of most jobs, it cannot be used to identify the cognitive content of jobs. Our ability to analyze the growing cognitive elements in everyday tasks has not kept up with events in the real world. Consequently, we need methods to handle tasks that may involve more complex cognitive and social skills. Such forward-looking analysis procedures have been recommended (Tannenbaum and Yukl, 1991) but have not been tested.

One way to broaden the scope of task analysis is to view cognitive functions as similar to computer programs. A computer cannot function without a protocol, a program that tells the computer how to perform a given application. Similarly, a protocol is needed to describe the overt and cognitive tasks that make up a particular job. Without such a protocol, no rational basis, other than tradition or intuition, exists for systematically defining and describing the content of a job training program or the procedures to be followed in performing job tasks.

This approach has yet to be fully implemented, and alternative approaches do not exist. Although several attempts have been made to overcome the limitations of conventional task analysis (Glaser, 1992), no one has yet devised a method that is easily learned and acceptable to practitioners as well as researchers. To be most useful, such a method should allow for the description of the cognitive, sensory, and motor components of a task as part of a comprehensive systems and functions analysis.

Specific Research Projects and Programs

Two major studies conducted recently document the increasing importance accorded to fundamental workplace skills. One comprehensive study was a joint two-year project of the American Society for Training and Development and the U.S. Department of Labor. Their report, *Workplace Basics: The Essential Skills Employers Want* (Carnevale et al., 1990), identified 16 basic skills within 7 broad skill groups:

- the foundation: learning to learn;
- competence: reading, writing, and computation;
- communication: listening and oral communication;
- adaptability: creative thinking and problem solving;
- personal management: self-esteem, goal-setting/motivation, and personal/career development;
- group effectiveness: interpersonal skills, negotiation, and teamwork; and
- influence: organizational effectiveness and leadership.

Another widely cited study, by the Department of Labor (Secretary's Commission on Achieving Necessary Skills, 1991), proposed a set of 14 fundamental skills within 3 categories that they regard as the foundation for effective job performance:

- basic skills—reading, writing, arithmetic/mathematics, listening, and speaking;
- thinking skills—creative thinking, decision making, problem solving, knowing how to learn, and reasoning; and

- personal qualities—individual responsibility, self-esteem and self-management, sociability, and integrity.

This study is part of a major effort by the U.S. Department of Labor to examine worker skill requirements and other issues related to work-based training and the transition from school to work. The challenge for the human factors research community is to understand how fundamental skills relate to the performance of technical tasks, how to design comprehensive training programs incorporating knowledge of worker skills, and how to evaluate these skills in the workplace.

Training individuals to perform effectively as teams is another area of increasing importance and requires greater research attention (Salas et al., 1992). Work teams are increasing in frequency in American organizations as "total quality management" programs are introduced. The performance of these teams is critical to organizational success. Research is needed in order to develop a greater understanding of work teams and of when and how they are likely to be most effective. In addition to training, other interventions to improve team effectiveness include team building, job re-design, and group incentive programs. In designing training programs, team-building interventions, and other programs for complex systems, human factors researchers are in a unique position to assist in developing information on the relationship between team processes and team performance over time and on the relationship between task type, stage of team development, and team performance.

The design of educational and training programs to meet these new demands offers special challenges. Researchers who have focused on the more traditional areas of how to design, deliver, and evaluate training and educational programs will have to integrate research from a number of fields in order to give trainees the resources they need to develop the skills required by the changing work environment. Furthermore, technical training must increasingly be integrated with a broader set of skills. Bringing all this material together in the design and development of training systems is difficult. The typical human factors specialist may not be very well versed in such subjects as group dynamics, which are rarely taught in system design curricula. As indicated in a recent publication by the National Research Council (Van Cott and Huey, 1992), although human factors practitioners may learn much about training design and development, they receive little formal education in these "softer" areas. Therefore, the human factors research community faces the challenge of developing the research base to adequately educate and prepare practitioners to develop training systems for the workplace that meet the demands of improving group rather than individual performance.

Flexible, Individualized Training

Thomas Bailey (1990), working with the National Center for Research in Vocational Education, reports that the approximately half of U.S. youth who do not go on to college receive little help in making the transition from school to work. The challenge is how to design educational and training programs to create an active learning environment that can ease this transition. Research is demonstrating that individuals have different learning styles and that the effectiveness of alternative teaching and training methods corresponds to how well it matches an individual's preferred learning style (Martel, 1991). The importance of individual differences in learning styles also has implications for education and training design. For example, Snow and Lohman (1984) found that students of low ability benefited more from high-structure/low-complexity programs and that for students of high ability the reverse was true. Much of this research is for grades K-12; more research is needed to determine the full parameters of aptitude treatment interactions and also to determine how to train/educate students so they can continue to learn in natural settings (Goldstein, 1992). Goldstein believes that "the processes involved in designing instructional systems and the revolution first described by both Gagne and Brenner in the 1960s has only begun to be understood." Research is needed on methodologies to support the design process, so that training developers can effectively incorporate *what* is to be learned with *who* the learners are and *what* characteristics they possess.

Research in cognitive psychology has also highlighted shortcomings in traditional approaches to teaching. Cognitive psychology focuses on how learning occurs by addressing ways the human mind processes and uses information. It has shifted the emphasis in learning to the mental operations and attributes of the learner/trainee and has provided a new theoretical basis for training and instructional technology. Cognitive research suggests that teaching/training should emphasize the development of thinking and problem-solving skills rather than the learning of facts through expository lectures (Molnar, 1989). In addition, much current research is addressing the effect of context on training and the transfer of information obtained during training to the situation in which it must be used. (See Druckman and Bjork, 1994, for a full discussion of the effect of the training context on effective transfer to other situations and systems.) These lines of research have significant implications for the design of programs that enhance learning and retention of knowledge and skills, that increase the probability of transfer of learning from one situation to another, and that support lifelong learning as well as the roles of teachers and trainers.

Cognitive science research has also now embraced the mental models concept (Stevens and Gentner, 1983; Johnson-Laird, 1983), which focuses on the ways in which humans understand systems. According to Rouse and Morris (1986), mental models allow a person to understand and to anticipate the behavior of a system, which implies that mental models must be capable of prediction. These models are domain-, context-, and situation-specific. "With respect to training, a number of theorists have hypothesized that training that fosters development of accurate mental models of a system will improve performance" (Cannon-Bowers and Salas, 1990:4). However, training in only general principles of system design and function is of limited utility (e.g., Kieras and Bovair, 1984; Morris and Rouse, 1985). To obtain optimal training, guidance is required in how to apply the knowledge in order to accomplish the task (Rouse and Morris, 1986).

Changing Roles of Teachers and Trainers

Just as new job-specific and managerial skills are required to meet today's demands, the roles of the teacher and trainer are changing. Rather than functioning strictly as lecturers, teachers and trainers will increasingly need to become facilitators. The major reason for this change is the exponential growth in information resources. Teachers can no longer be viewed as all-knowing sources of information. Because of the huge and expanding body of knowledge, teachers cannot possibly remain abreast of information in all areas (Molnar, 1989).

The focus of teachers and trainers is also changing. They must help others learn how to access information as needed and be able to do this themselves. Research is required on learning to learn and on how to teach people both how to determine what information is required for specific tasks and how to access it quickly. At the same time, research is needed on the design of job aids to assist teachers and workers in acquiring information on demand, as needed.

Instructor aids form another area in which technological advances should be promoted and in which potential innovations should be evaluated from a human factors perspective. For example, some progress has been made over the past 15 years or so in the construction of software packages for desktop computers that aid instructors in composing lesson plans, exercises, simulation scenarios, and examinations. These systems appear to vary markedly in their utility. Consistently high quality in such aids for training program managers might be obtained if a properly authorized organization were to sponsor rigorous evaluations of new products before these products were released to the market.

Engineering Applications

A key human factors research issue concerning the use of simulation technology in training programs is fidelity. Technology is available to make simulators realistic, comprehensive representations of a system; however, there is limited knowledge on the relationship of physical and functional fidelity and training effectiveness. In 1985, a National Research Council working group on simulation commented (Jones et al., 1985):

> A comprehensive body of behavioral principles and methods relevant to simulation does not exist. Integrative assessments of the research and operating experience are almost completely lacking for important topics such as fidelity . . . . A result is that there is no history of simulation research and utilization technology in a readily accessible form that serves as a guide for the future.

That need was partially addressed by Hays and Singer (1989), who provided an extensive review of the literature on simulation fidelity. However, the need for a comprehensive body of data on the behavioral principles and methods for determining the effectiveness of fidelity persists.

Despite assertions by human factors specialists that physical fidelity and psychological fidelity may not be equivalent—and in the absence of any empirical data to support their view—simulator manufacturers have increased simulation fidelity as fast as they were paid to do it and the developing new technology allowed. This practice will continue until studies test this assumption empirically.

There is a need for laboratory and field research to address the issue of how much and what kind of fidelity is required for cost-effective simulator training. This research will become increasingly important as simulation technology begins to provide some form of augmented reality that can be used to create training environments. The relationship between fidelity and training effectiveness remains unknown. In truth, the attributes of psychological fidelity need to be defined first. A metric is needed that can then be used to measure and compare psychological fidelity with physical fidelity. These areas all need attention from human factors research.

On a more practical level, studies have compared the cost-effectiveness of alternative training techniques and technologies (Orlansky and String, 1977). The results showed that simple, low-cost, part-task, and procedural trainers can be more effective for certain types of training than more costly full-scale, high-fidelity simulators. In the 1960s, task analysis and the systems approach to training—two methods for the qualitative definition of training objectives and the skill, knowledge, and abilities required for any job training course to be valid—were developed. In the 1970s, new research on team training was initiated.

In the 1980s, researchers built on that knowledge to develop crew resource management training; this technique is now used throughout the airline industry to train crews in team skills and has recently been extended to other systems as well. Also in the 1980s, human factors researchers, faced with the high costs of conventional training devices and classroom instruction, devised the concept of "embedded training."

Embedded training, a training capability built into or added onto operational systems, presents an opportunity to practice tasks and skills, especially those that should be performed automatically under high-stress conditions. It uses the computer and display capabilities of operational equipment to provide opportunities for task training, practice, and performance feedback on the same equipment that is used in operations. This approach is better suited for refresher training than for training in new skills (Oberlin, 1988) and can provide high-quality, standardized, individualized training. Today, the technique is used to train Army tank gunners, the control room crews of commercial process control facilities, and people in many other occupations.

Computer networking and telecommunications will also continue to open up new methods of team training. Simulation networking (SIMNET), the simulator network developed under the sponsorship of the Defense Advanced Research Projects Agency for training teams in tank warfare, uses workstations that simulate Army tank workstations to train soldiers in different geographic locations in battle tactics (Thorpe, 1993). The workstations are interactive so that the responses of one tank or tank unit alter the engagement environment for training participants, allowing trainees to practice a range of communication and decision-making tasks and roles in a realistic environment. The SIMNET concept appears to be generalizable to other education and training domains that have geographically dispersed workers.

Virtual reality, sometimes called cyberspace or augmented reality, is the experience of perceiving and interacting through sensors and effectors in a computer-modeled and computer-generated environment. A person who experiences a virtual world is immersed in and surrounded by simulated sight, sound, touch, and feedback of force from simulated objects. Eventually, as the technology develops, the experience of virtual reality will become increasingly similar to experience of the real world. This technology was recently addressed by the National Research Council's Committee on Virtual Reality Research and Development (Durlach and Mavor, 1995).

The proper use and exploitation of the potential of virtual reality is fundamentally a psychological rather than an engineering or computer science problem. Engineering and computer science provide the basic tools, but psychology and human factors will be essential in shaping and using these tools in productive ways.

Since virtual reality is still in its infancy, it is impossible to predict all of the research that will be needed to make proper use of it. However, some needs are apparent. Two of the most basic are (1) the need to study the transfer of training from a virtual to a real-world environment and (2) the need to understand the psychological as well as the physical basis for transfer of training. The high costs of modeling virtual environments makes research on transfer of training and fidelity especially important.

Other research on virtual reality in training will be required to better understand such questions as the following:

- Do individual differences affect the ability to adapt to and use virtual reality?
- What methods will be needed to identify the instructional elements that will be sufficient to meet stated learning objectives in a learning environment and at the same time exclude "virtual noise"?
- Can virtual reality systems be used to train teams?
- Does having been immersed in and having adapted to a virtual training environment have any negative effects on cognitive or psychomotor performance in the real world?

CONCLUSIONS

Training is sometimes regarded as an alternative to human factors and vice versa. The role of the human factors specialist has been to design systems that minimize the need for training. Success from either perspective is measured by improved performance, productivity, and efficiency of individuals and organizations. The human factors/systems perspective should not replace or usurp the perspective of the training specialist but can be constructively supplemental.

Among the issues that could benefit from human factors research, the highest priorities should be given to the design and evaluation of complex training systems and the design and application of learning technologies. Human factors scientists can add to the understanding of these issues by analyzing existing data, synthesizing such data across multiple studies, and conducting new empirical studies. Examples of the type of research that is required include the development of systematic design principles for training systems, empirical tests of alternative delivery practices, development and tests of theories to support the understanding and evaluation of training effectiveness from the systems perspective to the perspective of the individual learner, and development and validation of cost-effective means and technologies to provide training on demand.

The issue of performance measurement runs through every portion of the literature on training, from basic or scholarly articles to articles for the teachers in the trenches of vocational education (Fitz-enz, 1994). The problem of developing good measurement methods and evaluating these methods exists on at least two levels. The easier of the two levels is the measurement of performance gains in a specific task setting. In such a situation, the job provides its own criteria. The main problem with these rather obvious criteria is that variances between individuals and within the same person during the learning sequence tend to be small unless the trainee is put under serious task stresses. In other words, these obvious measures of performance are relatively insensitive and may not differentiate between alternative training procedures that are otherwise very unlike one another.

The second level of measurement is at the far end of the outcome sequence—that is, indices of the net impact of a training program on system or organizational effectiveness. The obvious difficulty with such measures is that the connection back to particular manipulations of the training regimen are tenuous at best. The real-world systems tend to be too noisy; too many factors other than variations in training techniques change during the course of the training process. Furthermore, the usual cure for such a noisy research environment, multiple trials, is out of bounds in the real-world settings in which impact research must be conducted.

A closely related problem that must be pursued at a basic level of research is how to specify skill requirements for specific jobs. One need only to contemplate how useful it would be to be able to list a set of measurable skills for each occupational specialty in the military jobs system or for each occupational description in the list of occupations used by the U.S. Department of Labor. We already have most of the tools for this in the skill taxonomies developed by human factors specialists.

We should also be able to match trainees with training strategies. In spite of such programs as MANPRINT, there are still significant gaps in our ability to fit the training to the learner (Clark and Taylor, 1994). Individual variability in learning styles may be a relatively fragile construct, but individuals do differ in their capabilities and limitations. Research is badly needed on how to develop descriptive profiles that can in turn be used for prescriptive actions in the delivery of the training service to the learner.

REFERENCES

Bailey, T.
 1990 *Changes in the Nature and Structure of Work: Implications for Skill Requirements and Skill Formation.* Berkeley, Calif.: National Center for Research in Vocational Education.
Boff, K., and W.B. Rouse
 1987 *System Design: Behavioral Perspectives on Designers, Tools, and Organizations.* New York: North Holland.
Boff, K.R., and J.E. Lincoln, eds.
 1988 *Engineering Data Compendium.* Armstrong Medical Research Laboratory. Dayton, Ohio: Wright-Patterson Air Force Base.
Booher, H., ed.
 1990 *MANPRINT: An Approach to Systems Integration.* New York: Van Nostrand Reinhold.

Brown, J.S., A. Collins, and P. Duguid
1988   *Situated Cognition and the Culture of Learning*. Technical Report No. IRL88-0008. Palo Alto, Calif.: Institute for Research on Learning.
Bureau of Labor Statistics
1991   *Occupational Outlook Quarterly* Fall. Washington, D.C.: U.S. Department of Labor.
Cannon-Bowers, J.A., and E. Salas
1990   Cognitive Psychology and Team Training: Shared Mental Models in Complex Systems. Paper presented at the annual meeting of the Society for Industrial and Organizational Psychology, Miami, Fla.
Cannon-Bowers, J.A., S. Tannenbaum, E. Salas, and S. Converse
1991   Toward an integration of training theory and technique. *Human Factors* 33(3):281-292.
Carnevale, A.P., L.J. Gainer, A.S. Meltzer, and S.L. Holland
1988   Skills employers want. *Training and Development Journal* 42(10/October):22-30.
Carnevale, A.P., L.J. Gainer, and A.S. Meltzer
1990   *Workplace Basics: The Essential Skills Employers Want*. American Society for Training and Development/the U.S. Department of Labor. San Francisco: Jossey-Bass.
Clark, R.C., and Taylor, D.
1994   The causes and cures of learner overload. *Training* 31(7):40-43.
Clegg, W.
1987   Management training evaluation: an update. *Training and Development Journal* February:65-71.
Druckman, D., and R.A. Bjork, eds.
1994   *Learning, Remembering, Believing: Enhancing Human Performance*. Committee on Techniques for the Enhancement of Human Performance, National Research Council. Washington, D.C.: National Academy Press.
Durlach, N.I., and A.S. Mavor, eds.
1995   *Virtual Reality: Scientific and Technological Challenges*. Committee on Virtual Reality Research and Development, National Research Council. Washington, D.C.: National Academy Press.
Fitz-enz, J.
1994   Yes . . . . you can weigh training's value. *Training* 31(7):54-58.
Glaser, R.
1992   Learning, cognition, and education: then and now. Pp. 239-265 in H.L. Pick, P.W. Van den Broek, and D.C. Knill, eds., *Cognition: Conceptual and Methodological Issues*. Washington, D.C.: American Psychological Association.
Goldstein, I.
1988   Tomorrow's workforce today. *Industry Week* 41-43.
1992   *Training in Organizations: Needs Assessment, Development, and Evaluation*, 3rd ed. Pacific Grove, Calif.: Brooks/Cole.
Hays, R., and M. Singer
1989   *Simulation Fidelity in Training System Design: Bridging the Gap Between Reality and Training*. New York: Springer-Verlag.
Johnson-Laird, P.N.
1983   *Mental Models: Towards a Cognitive Science of Language, Inference, and Consciousness*. Cambridge, Mass.: Harvard University Press.
Johnston, W., and A. Packer
1987   *Workforce 2000: Work and Workers for the 21st Century*. Indianapolis, Ind.: Hudson Institute.

Jones, E., R. Hennessy, and S. Deutsch, eds.
1985   Human Factors Aspects of Simulation. Working Group on Simulation, Committee on Human Factors, National Research Council. Washington, D.C.: National Academy Press.

Kieras, D.E., and S. Bovair
1984   The role of a mental model in learning to control a device. Cognitive Science 8:255-273.

Kraiger, K., J. Ford, and E. Salas
1992   Integration of Cognitive, Behavioral, and Affective Theories of Learning Into New Methods of Training Evaluation. Unpublished paper. U.S. Naval Warfare Center, Orlando, Fla.

Kravitz, D.
1988   The Human Resources Revolution: Implementing Progressive Management Practices for Bottom Line Success. San Francisco: Jossey-Bass.

Martel, L.
1991   The Integrative Learning System: A Review of Theoretical Foundations. Hilton Head Island, S.C.: National Academy of Integrated Learning.

McGehee, W., and P.W. Thayer, eds.
1961   Training in Business and Industry. New York: Wiley.

Molnar, A.
1989   Information and communications technology: today and in the future. In Lifelong Engineering Education, Proceedings from a Symposium of the Royal Swedish Academy of Engineering Sciences, IVA Rapport 365, Stockholm, Sweden.

Morris, N.M., and W.B. Rouse
1985   The effects of type of knowledge upon human problem solving in a process control task. IEEE Transactions on Systems, Man, and Cybernetics SMC-15:698-707.

Oberlin, M.
1988   Some considerations in implementing embedded training. Human Factors Society Bulletin 31(4):1-4.

Office of Technology Assessment
1990   Worker Training: Competing in the New International Economy. Washington, D.C.: U.S. Government Printing Office.

Orlansky, J., and J. String
1977   Cost-Effectiveness of Flight Simulators for Military Training. Paper No. 1275, August 1977. Arlington, Va.: The Institute for Defense Analyses.

Paquet, B., E. Kasl, L. Weinstein, and W. Waite
1987   The bottom line: here's how one company proved the business impact of management training. Training and Development Journal May:27-33.

Personick, V.
1989   Industry output and employment: a slower trend for the nineties. Monthly Labor Review 11(November 2):24-41.

Rouse, W.
1991   Designing for Success: A Human Centered Approach to Designing Successful Products and Systems. New York: Wiley.

Rouse, W.B., and N.M. Morris
1986   On looking into the black box: prospects and limits in the search for mental models. Psychological Bulletin 100(3):349-363.

Salas, E., T. Dickinson, S. Converse, and S. Tannenbaum
1992   Toward an understanding of team performance and training. In R. Swezey and E. Salas, eds., Teams: Their Training and Performance. Norwood, N.J.: Ablex.

Secretary's Commission on Achieving Necessary Skills
  1991   *What Work Requires of Schools: A SCANS Report for America 2000.* Washington, D.C.: U.S. Department of Labor.
Snow, R., and D. Lohman
  1984   Toward a theory of cognitive aptitude for teaming from instruction. *Journal of Educational Psychology* 71:347-376.
Stevens, A.L., and D. Gentner, eds.
  1983   *Mental Models.* Hillsdale, N.J.: Erlbaum.
Tannenbaum, S., and G. Yukl
  1991   Training and development in work organizations. *Annual Review of Psychology* 41:399-441.
Taylor, S.
  1989   The aging of America. *Training and Development Journal* October:44-50.
Thorpe, J.
  1993   Synthetic Environments Strategic Plan. Draft 3B. Defense Advanced Research Projects Agency, Alexandria, Va.
Van Cott, H.P., and B.M. Huey, eds.
  1992   *Human Factors Specialists' Education and Utilization: Results of a Survey.* Committee on Human Factors, National Research Council. Washington, D.C.: National Academy Press.
Wexley, K.N., and P.T. Baldwin
  1986   Management development. *Journal of Management* 12(2):277-294.

3

Employment and Disabilities

Jerome I. Elkind, Raymond S. Nickerson, Harold P. Van Cott, and Robert C. Williges

INTRODUCTION

In all aspects of daily life, people with disabilities encounter a multitude of problems that limit their ability to live independently, to travel, to use recreation facilities, and to obtain and perform jobs. The effect of these limitations on the individual and their cost to society are great; government, through various legislative initiatives, has made it possible for people with disabilities to participate more fully in the activities and opportunities enjoyed by the population at large. In addition to prohibiting discrimination against people with disabilities in the workplace and educational institutions, these initiatives require that equipment, facilities, and even jobs be designed to be reasonably accessible to disabled people. The human factors community can make important contributions to these designs and thereby enhance the ability of disabled people to fulfill their potential. This chapter discusses the nature of these contributions and the research that is needed to enable them.

The disabled population is very large. According to Census Bureau data, approximately 30 percent of the U.S. population, about 75 million people, report having sensory, physical, or cognitive disabilities (Bureau of the Census, 1987, 1989). These figures do not correct for the incidence of multiple disabilities in individuals, but, even with such correction, the numbers would be very large. Table 3.1 shows the incidence of the different types of disability. The number of people whose disabilities are sufficiently

TABLE 3.1 Incidence of Major Types of Disability (1990 estimates)

Disability	Persons (000s)	Percentage of Population
Sensory		
Visually handicapped	8,600	3.4
Hearing handicapped	22,000	8.8
Motor		
Orthopedic impairments	23,400	9.6
Cognitive		
Specific learning disabilities	18,700	7.5
Speech impaired	2,000	0.9
Mentally retarded	3,000	1.2
Total	77,700	31.4

severe to interfere with work and other activities is, of course, substantially less, but still very large. Approximately 30 million people have severe disabilities; their distribution is shown in Table 3.2 (Elkind, 1990).

People with disabilities find it much more difficult to be gainfully employed than do people who are able-bodied, and their earning power is thereby much diminished. The rate of unemployment among disabled people who would like to work is several times higher than the national average (Kraus and Stoddard, 1989). The percentage of people who live below the poverty line in the United States is between two and three times greater for people with a disability that interferes with their ability to work than for the total working-age population (Vachon, 1990). As a result, 50 percent of the disabled population have an annual household income at or below $15,000, principally derived from social security, public assistance, and other transfer programs. Almost half of the disabled population live at or near the poverty level. Maintenance support programs cost approximately $100 billion per year, and this has doubled in the last 10 years (Vanderheiden, 1990). Other economic costs of disability, such as lost productivity, are even greater.

The population of people with disabilities can, however, also be viewed as an underutilized resource for the country. Although nationwide employment is relatively high, the workforce is growing much more slowly than it has in the past (Rauch, 1989; Vaughan and Berryman, 1989). Some economists are concerned that, as a consequence of the slowdown in the number of people entering the workforce and the alarming statistics on school dropout rates and academic achievement, the country's workforce could be inad-

TABLE 3.2 Summary of Severe Impairments (1990 estimates)

Disability	Persons (000s)	Percentage of Population
Sensory		
Visual impairment		
Legally blind	580	0.2
Severely impaired vision	1,500	0.6
Hearing impairments		
Severe to profound bilateral hearing loss	2,400	1.0
Motor		
Disability interfering with work	13,300	5.3
Orthopedic impairments		
Back or spine	3,300	1.3
Hips and lower limbs	4,100	1.6
Severed spinal cord	580	0.2
Cognitive		
Specific learning disabilities	6,200	2.5
Impaired speech interfering with work	200	0.1
Mental retardation	3,000	1.2
Total	33,160	14.0

equate to meet the demands that are likely to be placed on it over the next decade or so. The disabled population, if better utilized, could help make up for the expected shortfall.

Thus, the need to increase employment opportunities for people with disabilities can be compellingly argued on both equity and economic grounds. These concerns are responsible for the legislation on accommodations for the disabled.

The great advances made in recent years in computer and communications technologies are making it possible for disabled people to perform the essential functions of an increasing number of jobs. Personal computers have been especially important in making these technologies affordable by individuals and are fostering a flowering of assistive devices and systems for people with disabilities (Bowe, 1984). The combination of computer and communications technologies—telenetworking—makes it possible to take jobs to people who find it difficult or impossible to take themselves to jobs (see Chapter 6). It has great potential for enhancing the possibility of employment for disabled people and for generally increasing their access to other people and resources (Nickerson, 1986).

Related technologies that have considerable potential for enhancing the employment opportunities of people with disabilities are robotics and teleoperator systems. These should be especially germane to the design and operation of prosthetic and orthotic devices (Sheridan, 1980).

PROBLEM DEFINITION

The term *disability* encompasses a wide range of conditions that can affect individuals as a result of a birth defect, an accident, a disease, or as an accompaniment of aging. The Americans With Disabilities Act of 1990 (ADA) indicates the scope of the term as it is now being used by those concerned with public policy. The act defines a person with a *disability* as someone with a *physical or mental impairment* that substantially limits one or more *major life activities* (or a person with a record of such an impairment, or a person regarded as having such an impairment). A *physical impairment* is defined as any physiological disorder or cosmetic disfigurement or an anatomical loss affecting one or more of the following body systems: neurological, musculoskeletal, special sense organs, respiratory (including speech organs), cardiovascular, reproductive, digestive, genito-urinary, hemic and lymphatic, and skin and endocrine. *Mental impairment* includes mental retardation, organic brain syndrome, emotional and mental illness, and specific learning disabilities. *Major life activities* include such things as walking, talking, seeing, hearing, caring for oneself, and working (Texas Young Lawyers Association, 1990).

Thus, the ADA focuses attention on the interaction between activities and disability. We represent this interaction by the matrix in Figure 3.1. One axis shows the different kinds of impairments grouped under the general categories of sensory, motor, and cognitive. The other axis shows the major activities of living grouped under the general categories of work,

Task/Activity	Functional Ability		
	Sensory	Motor	Cognitive
Workplace			
Home			
Transportation			
Recreation			
Safety/security			

FIGURE 3.1 Task/activity: functional ability matrix.

home, transportation, recreation, and safety/security. This structure is consistent with the definitions in the ADA and is similar to the structure used in a recent Committee on Human Factors report on aging (Czaja, 1990).

The matrix of Figure 3.1 defines a vast problem domain that covers virtually all the activities and functions that people perform in daily living and all sensory, motor, and cognitive systems (since any of them might be impaired). As is apparent from the title of this chapter, we do not attempt to discuss human factors contributions to this entire domain. The committee decided that focusing on a single activity/function subdomain would make the scope of the discussion more manageable. It chose to focus on employment in recognition of the pivotal role of employment in an individual's independence, self-esteem, and access to a fuller and richer life. Many of the techniques, data, and even designs that are used to enhance employment potential will also be appropriate for life activities in other areas, such as the home and recreation. Yet even the topic of employment is huge.

Increased employment for disabled people will have to come largely in the information-oriented segment of our society. Already more than 55 percent of the workforce is engaged in information-oriented jobs; this percentage has been increasing annually (Strassmann, 1985) and is expected to continue to increase for the foreseeable future (Kraut, 1987). Given that so many jobs and so much growth are in the information sector, an important challenge is to find ways to make information jobs more accessible to disabled people and to increase their ability to compete for these jobs. Greater accessibility can come from making the tools that information workers use and the environment in which they work less restrictive in terms of the capabilities of the user. Competitiveness can be enhanced through training and rehabilitation, but also through augmentation—often computer-based—of the capabilities of disabled persons in order to improve job performance. For these reasons, our discussion will focus on increasing access to computer-based equipment by disabled people and on the use of computer technology to improve access in general.

Let us now consider the problems that people with disabilities encounter in performing jobs for which they are qualified. The ADA is also helpful here in providing us with some useful definitions. The ADA prohibits discrimination against a *qualified individual* in matters relating to employment. It defines a *qualified individual* as one who with or without a *reasonable accommodation* is able to perform the essential functions of a particular job. *Accommodation* includes making existing facilities accessible, restructuring jobs, acquiring or modifying equipment or devices, and modifying work policies. For the accommodation to be considered *reasonable*, it must not be unduly burdensome because of the difficulty or expense involved. Accommodation can be made by changing the nature of the job or the equipment used to perform it or by providing individuals who have a disability with assistive devices so that they can perform their jobs adequately. The definition used in the ADA allows for a growing set of reasonable accommodations. As both technology and our knowledge of disabilities advance, it will be possible to provide more capable assistive devices, a wider range of equipment modifications, and better access to facilities as well as to restructure more jobs to fit individual capabilities at acceptable costs.

The language of the ADA leads us to think in terms of modifying existing work environments, equipment, and jobs. It would be far better if modification were not necessary because the needs of people with disabilities had been taken into account when the environments, equipment, and jobs were designed initially. Given that it is probably not always possible to accommodate every type of disability at the outset, a second objective of initial design should be to provide access facilities that allow assistive equipment to be readily added so that disabled persons can work effectively. In fact, this kind of access was the objective of earlier federal legislation, the Rehabilitation Act of 1973 as amended in 1986 by Public Law 99-506, which required equal access for disabled people to office equipment, transportation, and buildings. Newell (1995), Vanderheiden (1990), and others argue that the extraordinary needs of people with disabilities are often only the exaggerated needs of the able-bodied population, and that taking these extraordinary needs into account produces better and more widely useful design solutions for everyone. They argue for designs that both able-bodied and disabled people can use effectively with no additional modification or, at worst, with augmentations that are easy to add.

We encounter some especially challenging problems when we design to meet the extraordinary requirements of people with disabilities. First, design data on the capabilities and performance of the disabled population are often not available, and, when available, they are much more limited in scope than data on the able-bodied population. Most research over the years has focused on the able-bodied population and on establishing average performance characteristics. Studies of individual differences in performance are the exception (Egan and Gomez, 1985). Further, it is common practice to exclude disabled people from studies of population averages, so that they are not even statistically represented in the results. Since the capabilities and performance of disabled people in the areas of their disabilities are, by definition, well outside the range exhibited by the able-bodied population, the data about the able-bodied and the design assumptions used for them do not provide much guidance on how to accommodate disabilities in design. This makes designing for disabled people difficult and the outcome uncertain. Clearly obtaining a better database about the capabilities of the disabled population is an important need.

Second, the disabled population is not homogeneous but exhibits a wide range of individual differences, both within a disability subgroup and across disability subgroups. The variability in performance among individuals with disabilities is great, and the needs for augmentative support are quite diverse. For example, a computer workstation for word processing that is to be used by people with different types of disabilities should be able to provide displays in large type for people with visual impairments, speech output for those who are blind, somewhat differently designed speech output for people with dyslexia, and special keyboards and speech input for those with motor impairments. And these are just some of possibilities. The wide range of individual capabilities has led to a tendency to design to individual requirements. While individual design may lead to good results for the individuals, it is expensive and its cost will constrain the number of people who can be served. There is a need to develop generic approaches to design that are aimed at the major disability subgroups and that allow tailoring to individual needs.

Third, employment is a large and complex domain. There are many different kinds of jobs requiring different combinations of skills. For most jobs, we do not have either a good characterization of the component skills required or a good specification of the level of competence required in each skill area. Thus, the design problem is not well specified from a requirements side. Furthermore, each work situation must be considered as a system that involves many different work-related tasks, a set of personal care and mobility activities, and a variety of interactions with other people inside and outside the firm. All of these must be taken into account at a satisfactory level for there to be a successful work situation.

Our discussion concentrates on the technology and on other tangible aspects of enabling disabled people to obtain employment. Technology by itself does not solve the problems of unemployment. One must also deal with the sociology of the workplace and with the disincentives to going back to work. For disabled people there can be many disincentives to working, such as lack of social acceptance, difficulty in transportation, and economic penalties. Ability to do the work is just one aspect of the problem, but it is the one to which the human factors community can contribute most directly, so it is the focus of this chapter.

HUMAN FACTORS CONTRIBUTIONS

The unemployment and underemployment of disabled people is a complex problem with technical, psychological, sociological, political, and economic components. Disabled people encounter many disincentives to seeking employment, but a key one is inability to function satisfactorily at the workplace; that is, to perform the tasks required by the job and to take care of personal needs that arise. The human factors community is well equipped

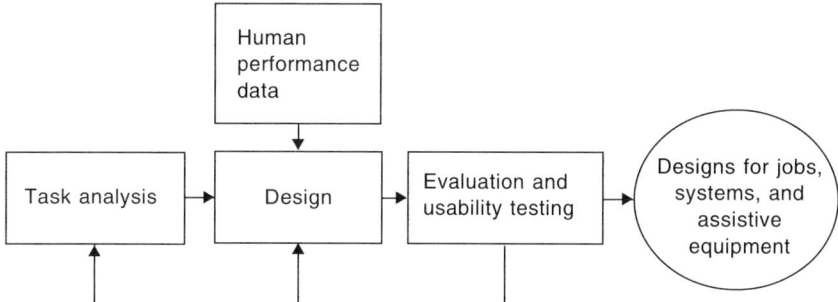

FIGURE 3.2 Human factors design process.

to address this problem by designing accommodations that lower barriers to work and that enable disabled people to perform the jobs for which they are qualified. Moreover, efforts by human factors specialists to accommodate the needs of the disabled population in equipment design will advance human factors as a discipline by directing its attention to designs and methods that will be effective over a much wider range of individual capabilities (Griffith et al., 1989).

This is a classical human factors systems design problem of matching tasks to human capabilities through analysis and redesign of tasks, design of work environments, design of equipment used for performing work-related tasks, and design of assistive devices that enhance individual performance or that provide special interfaces to equipment. The design process proceeds iteratively through the stages illustrated in Figure 3.2 until a satisfactory result is achieved. The stages of the design process are the following:

- gathering data about relevant human performance characteristics;
- conducting an analysis of tasks required for successful performance of the job;
- design of the job, systems for performing work tasks, and assistive equipment; and
- evaluation of the designs through the use of models, simulation tools, and user experiments.

Human Performance Design Data

Human factors design proceeds from a base of knowledge about human capabilities, capacities, and limitations that is relevant to the design problem. A large body of such information is available in databases, design guides, and models to inform human factors design for able-bodied users (see, e.g., Boff and Lincoln, 1988). However, very few data of this kind are available for most disabled populations. Not only does this make informed design for people with disabilities difficult, but it also gives the human factors community an opportunity to extend the knowledge base to include the disabled population (see Kelley and Kroemer, 1990, on the extension of the anthropometric database to the elderly).

Task Analysis

Task analysis is the appropriate starting point for human factors design. Human factors designers draw upon a portfolio of techniques to develop an understanding of the tasks that people must perform in a system and of the perceptual, motor, and cognitive capabilities needed to perform them. These techniques include observations, video analyses, taxonomies, scenario generation, timeline, workload, and input/output analyses (Elkind et al., 1990). In principle, these techniques are directly applicable to designing for people with disabilities. Typically, however, jobs and tasks from common work situations have not often been subject to this kind of detailed examination. Further, it is necessary to probe beneath the surface of how jobs are performed by their present incumbents, who are for the most part able-bodied, and extract the essential requirements for effective performance in these jobs. Disabled workers may have to perform jobs differently, and it is important to characterize the jobs at a level that allows restructuring to accommodate people with specific limitations. These kinds of analyses are properly within the domain of human factors.

Design

We are concerned with three types of design:

- design or redesign of jobs so that disabled people can perform the essential tasks and not be disqualified by inability to perform nonessential tasks;
- design of equipment and systems so that they are accessible for use by disabled people; and
- design of assistive equipment to enhance the performance of people with disabilities or to allow them to use systems and equipment designed for the general able-bodied population.

There are extensive opportunities and an enormous need for human factors contributions to these design efforts. In most cases, these designs will be carried out by teams in which human factors specialists work with specialists from other engineering disciplines. As with other human-machine systems design, human factors focuses on the functions users are asked to perform, user-equipment interfaces, user performance, usability evaluation, and other activities related to the user and the impact of the system or device on the user. The human factors specialist would be expected to bring knowledge about the special characteristics and requirements of the disabled subgroups for whom the design is intended and about how to customize designs to the particular requirements of individuals within a subgroup.

Of the three types of design activities mentioned above, job design for disabled people has probably received the least attention from human factors specialists. Analysis of jobs often reveals opportunities for redistributing tasks among members of a team so that the disabled members are given assignments they can perform or so that the job can be accomplished in some other way that does not require the use of capabilities that are impaired. An example of redistribution would be reassigning the task of writing status reports from a person with severe dyslexia to a team member who does not have this disability. An example of a different way of accomplishing a task would be allowing a mobility-impaired worker to use electronic and computer technology to do office tasks from home (Williges and Williges, 1995).

A key problem in design of accessible equipment and systems is to make them usable by both able-bodied and disabled persons without imposing performance penalties on the able-bodied or excessive cost penalty on anyone. Much of the attention in this area has been on interface designs that are usable by people who do not have normal sensory, manipulative, or communication abilities. If accessibility by both disabled and able-bodied people is not possible, then a backup position is to design so that interfaces are available that allow special assistive devices to be added to make the original equipment or system accessible. This problem has received considerable attention (Vanderheiden, 1988). An example of an accessible system is one that provides both visual and auditory displays of all information that is presented so that persons with either impaired vision or impaired hearing can function as well as persons with normal sight and hearing.

Literally thousands of assistive devices have been designed to enhance the performance of disabled people or to enable them to use systems designed for the able-bodied (ABLE DATA, 1989; Trace Research and Development Center, 1988). Increasingly, these devices are being implemented with computer hardware and software that provide disabled users with capabilities that they are lacking. Examples are speech generation devices for those with severe speech impairments and eye tracking systems as a substitute for keyboard input for those with severe manipulatory impairments. Many of these devices have had the benefit of human factors contributions, but many have not. This is a rich area given the range of disabilities that must be accommodated, as shown in Figure 3.1, and the variety of tasks, equipment, and systems that are encountered in the workplace.

Evaluation Models, Tools, and Experiments

Evaluation of proposed designs and usability experiments are essential to the design process. It is important to conduct user experiments to determine how well disabled individuals can use both existing and proposed designs. Only by careful examination of the shortcomings of existing equipment is it possible to know where to place the effort on making improvements. These studies must take a wide range of disabilities into consideration. Such experimental studies are not easy to conduct and offer an opportunity for a major contribution from the human factors community, which is skilled in this area.

Not all design evaluation studies need be done with human subjects. It is often very effective to use computer models and simulations of human performance to evaluate alternative designs. Using such models for evaluation is less expensive and time-consuming than using human subjects. Selecting and using such models and tools is an area in which human factors specialists have expertise. For example, models and design evaluation tools have become increasingly important to human factors aspects of the design of aircraft (Elkind et al., 1990).

However, most current models describe the normal population and do not encompass disabled populations. Models need to be extended, and this will require using and adding to the body of relevant data about the performance of disabled populations. Human factors specialists can assist in these tasks and help develop models of human performance that represent the performance of people with specified disabilities.

CURRENT STATE OF THE ART AND HUMAN FACTORS NEEDS

Human Performance Design Data

Human factors—with its focus on applying knowledge about human characteristics to the design of tasks, machines, machine systems, and environments—requires a body of systematic and quantitative data about human characteristics. The design references and texts that compose the database for human factors contain little that directly addresses the characteristics of disabled segments of the population. This should not be a surprise since, until recently, the human factors literature has paid little attention to the problems of the disabled. Only a small percentage of the papers in the journal *Human Factors* deals with disabilities. Although this journal has published two special issues on disabilities, in 1978 and 1990, relatively little has appeared between the two issues. Handbooks sometimes include a chapter on disability (Salvendy, 1987) and sometimes have data relevant to the elderly (Boff and Lincoln, 1988), but for the most part these areas are not covered with any degree of completeness. Vanderheiden (1990) comments that disabilities and other functional limitations are only peripherally mentioned in textbooks. Courses on disability in human factors curricula are rare. Little or no attention is paid in the standard curriculum to the needs, characteristics, or design considerations required to accommodate persons with reduced abilities.

In the last few years, human factors researchers have become more interested in the problems that disabled people encounter in gaining employment and in being able to use the facilities that able-bodied people take for granted. Federal legislation has been directly responsible for the increased concern for the disabled population on the part of business and their product designers, and this has influenced the human factors community. A 1990 *Human Factors* special issue, *Assisting People With Functional Impairments*, a similar *Human Factors* special issue on *Aging* also in 1990, and a forthcoming book on disabilities and human computer interaction (Edwards, 1995) are all indications of the growing interest in this area. So there is hope that the body of human factors design data for the disabled will grow more rapidly in the future, but it will take many years before it becomes adequate.

Other disciplines, however, have been active over the years developing large bodies of data about different segments of the disabled population. There is an extensive literature from the medical, rehabilitation, occupational therapy, prosthesis, and other specialized communities on the characteristics of various disabled populations. Not much of this information has found its way into the human factors literature, in part because it is not well organized for use in human factors design.

The field of aging provides an interesting example of the problem, one that is relevant to our purposes inasmuch as elderly people who are disabled compose a major component of the disabled population. Smith (1990) comments that, although 50 years of laboratory research have provided a large amount of information about the physical and psychological changes that occur with aging, there has been little application of this information to human factors design. This is because much of the information is not in the reference works and because the information that is included is largely prescriptive and not quantitative. Smith points out that major effort is required to review the research data on aging, extrapolate from them, fill in important gaps, and structure the data so that they are useful for human factors design.

The situation for the principal areas of disability is much the same. Therefore, an important research need is to build a more adequate human factors database for the major disability subgroups and to get these data included in the standard design references and texts. Much of this database can be obtained from the study of existing literature from other fields, but in many areas new laboratory research will almost certainly be required.

Given the number of different disabilities and the fact that they can affect all areas of performance, it will take considerable effort to assemble a satisfactory body of data and considerable care to organize it into useful form. It will be important to segment the disabled population into subgroups of individuals with reasonably common characteristics of impairment and to identify parameters of performance that allow representation of individual differences. The models and simulations of performance discussed above provide a directly usable and practical framework for organizing the data for many areas of disability. Therefore, it should be kept in mind that the impaired functions of disabled people are usually not sharply segmented from those of the able-bodied population. That is to say, the distribution of performance is not bimodal. Therefore it should be possible to characterize the performance capabilities and limitations of people with disabilities by performance parameters at the tail of the distribution.

Task Analysis

The state of the art is much better in task analysis for disabilities than in human factors design data. In designing for disabled people, human factors specialists can apply a collection of already well-developed and widely used analysis techniques. The human factors designers do not need to develop or extend the techniques themselves but must merely apply them to jobs and tasks that disabled people are capable of performing. The lack of descriptive data may inhibit the use of these techniques, but that is a problem with the data and needs to be addressed separately, as discussed above, not a problem with the analysis techniques themselves.

Skill Requirements for Jobs

Task analysis at the job level is clearly important given our focus on increasing the employment potential of people with disabilities. Much work has already been directed toward establishing the skill requirements of different types of jobs to provide a principled method for selecting personnel to perform these jobs (see Chapter 2). Much of this work has been in the context of personnel selection for jobs in the military; a recent example is the Army Selection and Classification Project (Campbell, 1990; Peterson et al., 1990).

In this approach, existing jobs are examined; the tasks currently composing them are identified; the skills of current incumbents are determined; a set of tests of knowledge, aptitude, and performance is administered to ascertain skill levels of successful incumbents in these jobs; and the results are used to establish criteria for selecting job candidates. The same approach has been applied to jobs in the civilian sector, and in particular by companies that have a large number of people in a single job category, for example, telephone information operators. There is a need to carry out this kind of skills requirements analysis on more jobs in the civilian sector, and it would probably make sense to do so with a representative sample of jobs that provides insight about the mismatch between the skills required and the skills possessed by disabled people in major segments of the economy.

There are two problems with applying this approach to disabled people. The most important is that the method analyzes existing jobs and existing incumbents. Thus, it assumes that the jobs, which have evolved with able-bodied workers, will be done in the same way by disabled workers. Furthermore, it tends to require that disabled workers have the skill levels of the able-bodied. It will be necessary to separate the *essential* functions of a job from the ancillary functions so that it can be determined whether or not people with a particular type and level of impairment are *qualified* for such a job. (We deliberately use the language of the ADA here.) The second problem is that the method takes for granted certain ancillary skills or capabilities, such as mobility, that may be impaired in disabled populations. Thus, it will be necessary, when applying the method, to cast a somewhat wider net than is usual when identifying the tasks that compose a job.

Taxonomies

We would like to avoid having to analyze every job in detail and be able to work in terms of broader job types or categories. To do this, we need to have taxonomies of functions that workers perform in various types of jobs. The notion of taxonomy is simple; it takes effort to develop taxonomies for different types of jobs, and as a result coverage is spotty. Office information work is a particularly important type of job because more than half the workforce is employed in such jobs, and the number is increasing (Kraut, 1987; Strassmann, 1985). There has been some effort to develop taxonomies of office information work activities, but no generally accepted, detailed taxonomy exists. Williges and Williges (1995) provide a preliminary taxonomy of information tasks and task activities performed in professional offices, as does Czaja (1987), who identifies and analyzes tasks such as document preparation, meeting, conferencing, filing and retrieving, communication, decision making, and transacting. Taxonomies of meetings also have been developed (Short et al., 1976). This is an area that requires more work.

Timeline Analysis

Timeline analysis is a detailed unfolding in time of the tasks that a person performs, the information required for the task, the decisions made, and the actions taken. The analysis can be done at increasing levels of detail and, at the limit, examines individual manipulative motions in the spirit of time and motion study. It should not be a surprise to find that Gilbreth and Gilbreth ([1917] 1973) applied this kind of analysis to the study of disabled soldiers and how to return them to gainful employment. Their work provides the intellectual origins for what human factors calls task analysis, and even for our special interest in the employment of people with disabilities.

Timeline analysis provides a detailed look at job and task requirements, with an emphasis on speed of performance and on competition for cognitive, sensory, and motor resources (Corker et al., 1986). Slow performance and competition for reduced resources are common problems encountered by disabled people. Timeline analysis is therefore a very useful tool in determining what kinds of accommodations are required to make jobs accessible. Detailed timeline analyses of office and other jobs, especially when performed by disabled workers, are generally not available. It would be valuable to have an analysis of a representative sample of jobs being performed by people with different kinds of impairments.

Design

Design and Redesign of Jobs

An enormous amount of work has been directed toward design for disabled people, but almost all of it has focused on equipment and systems design. Many assistive devices are described in the technical literature, as well as in catalogs and product brochures. There is little in the technical literature about design or redesign of jobs to accommodate disabled workers. As part of an interest in communications and computer alternatives to travel and commuting, Williges and Williges (1995) address this subject in terms of office information jobs. It would be very useful to have a collection of case histories of jobs that have been designed or redesigned to accommodate specific disabilities.

As is pointed out in other chapters of this book (Chapters 2 and 8), the primary skill requirements of many jobs have been shifting from physical to cognitive. This is true not only in the service sector but also in manufacturing and industrial production. Many jobs have been redesigned to accommodate this shift. To be sure, the redesign has been motivated, for the most part, by new requirements imposed by the introduction of new technologies in the office or plant, and not by the desire to create new work possibilities for people with disabilities. Nonetheless, the fact that redesign is occurring on such an extensive scale provides opportunities for rethinking job requirements with disabled workers in mind.

Another major development is that technology is spawning a growing variety of aids to the performance of intellectual work (see Chapter 11). Again, for the most part, these aids were not designed to increase job opportunities for disabled people. Yet some of the aids have the potential for doing that, and others could, with some adaptation, be used to that end.

As a way of maximizing the effective use of limited resources, there is much to be said for the strategy of looking for ways to turn technological developments to the advantage of people with disabilities. We believe that many of the technological developments around computer and communication technologies have the potential to enhance the job opportunities of disabled people greatly. Realizing this potential will require some attention to a variety of human factors issues relating to the adaptation of devices to special user needs.

Design for Accessibility

It would be enormously beneficial to disabled populations if common devices, systems, and environments were designed to be accessible to people with disabilities as well as to able-bodied people. However, as Vanderheiden (1990) points out, the disabled population, although large in total, is composed of many subgroups, many of which are small. It is clearly impossible to design everything to be accessible to all subgroups, just as it is inefficient to have distinct product designs for each disability subgroup. The sensible path lies somewhere between these extremes. For most types of impairments there are economical ways of designing products so that they are accessible, or at least more accessible, to major segments of the disabled population (Newell, 1995).

There are many examples of successful accessible design (Mueller, 1990; Sorensen, 1979). Public facilities for physically disabled people are the most visible example. It is now routine to have cuts in sidewalks so that wheelchair users and others with mobility impairment can handle street crossings, toilets in public buildings are routinely designed to accommodate wheelchairs, and ramps are now provided as an alternative to stairs. In general, it is now known how to design buildings so that they accommodate people in wheelchairs. This is done in new construction at what is now considered to be an acceptable cost and, in some areas, at negligible cost.

Electronic equipment is not so far along. Some computer systems do offer facilities for those with impaired hand control, for example, "sticky" keys, which allow a command that normally requires simultaneous activation of keys to be invoked by sequential activation. However, computers have not typically been designed with functionally impaired users in mind and are simply not accessible to many people with mobility, sensory, or cognitive limitations. Many of the capabilities that computer systems need in order to be accessible to people with sensory and physical impairments are summarized in a design guide addressing this issue (Vanderheiden, 1988).

When it is not possible to include accessibility facilities in a design, it may still be possible to provide standard interfaces that can be used to integrate special equipment designed for particular disabilities groups. Blind users, for example, need a device that transforms visual images on a computer display into auditory or tactile displays. It is not economical to provide such facilities for all users when only a small number will use them. Yet the data from which the visual display was constructed can be made available to software and hardware makers, which can implement the alternative displays. This is not likely to be a burden on the design (if it is part of the initial specifications). Enabling such special adaptations requires careful design of the interfaces. Such design, in turn, requires good understanding of the kinds of adaptive devices that disabled users will need. Interface requirements and recommendations to enable the use of devices to assist blind and physically handicapped people in using computer systems have also been developed (Vanderheiden, 1988).

Clearly more can be done to improve the built-in accessibility of products and facilities. We need more specific data about how to assist people with different kinds of impairment, and we need to develop new techniques for providing needed assistance. This is especially true for cognitive disabilities. Making fuller provision for the disabled will certainly have an impact on system design.

Design of Assistive Devices and Systems

Many adaptive hardware devices and specially designed software packages have been developed to enable disabled people to use computers (Casali and Williges, 1990; ABLE DATA, 1989; Trace Research and Development Center, 1988). The number of devices available to assist the disabled in using other kinds of equipment (e.g., automobiles) and devices that replace or improve upon an impaired function (e.g., eyeglasses, communication devices) is also large. New assistive devices and systems that perform better, offer new functions, or are cheaper continue to be developed. In particular, we can expect increased attention to and progress in the development of assistive aids to cognitive functions as a result of advances in computer technology and cognitive science. This is important because people with cognitive disabilities are a very large subgroup, one often bypassed in the development of assistive devices because the impairments are difficult to address. The need for better assistive devices remains large and the opportunities for progress are great.

The rich menu of assistive devices and systems raises the problem of how to locate and select a particular device that matches the needs of an individual user. Large databases catalog the devices that are available (ABLE DATA, 1989; Trace Research and Development Center, 1988), but these do not provide detailed information on the characteristics of the devices or the types of users for which they are best suited. Obtaining detailed information about device or system characteristics is not easy, especially for complex software for which the demands on cognitive ability, and even motor and sensory abilities, are difficult to determine and even harder to describe. Without detailed information, the only way to select a device from a set of candidates is by trial and error.

There have been several attempts to provide a systematic method for selecting assistive devices and systems. For example, Behrmann (1989) describes an expert system for the selection of a computer input-output aid that uses subjective ratings by therapists of the suitability of each type of equipment for persons in each of several disability subgroups. Casali (1995) and Rosen and Goodenough-Trapagnier (1989) describe a more comprehensive and quantitative approach to this problem; their approach uses detailed information about particular types of devices together with a set of tests for evaluating the residual capabilities of the disabled user, a tool for eliciting information from the user regarding task needs, a method for integrating this information to select the appropriate combination of device characteristics, and a test of usability. Casali addresses computer cursor control devices; Rosen and Goodenough-Trapagnier address augmentative communication devices. Theirs is a logical approach, but it requires a large body of information and the development of efficient tests of capabilities and methods for integration. Systematic selection remains a major problem for which there are only hints of a solution.

Evaluation Models, Tools, and Experiments

Design is iterative and requires repetitive evaluation of prospective designs that eventually converge on a final solution that satisfies the requirements. Evaluation is facilitated by models of human performance and computer tools to run the models, so that it is not necessary to rely entirely on experiments with human subjects for evaluation. Models and tools covering many areas of human performance have been developed. There are many gaps in coverage, especially in cognitive functions, but this is an active area and progress continues to be made. For example, anthropometric models have been developed to evaluate cockpit design and layout as well as maintenance operations (McDaniel and Askren, 1985; Paquette, 1990). The

National Aeronautics and Space Administration (NASA) and the U.S. Army are developing a human factors computer-aided engineering facility for the design of helicopter cockpits (Army-NASA Aircrew/Aircraft Integration program, Hartzell et al., 1984) that includes models of workload (Corker et al., 1986) and vision (Larimer et al., 1989; see also Elkind et al., 1990 for a summary of vision models being considered). Simple cognitive models have been developed. Very important and widely used are the Model Human Processor and the GOMS model for evaluating human computer interface designs developed by Card et al. (1983). Recently, a more comprehensive model of cognitive function, SOAR cognitive architecture, has been proposed by Newell (1990).

These computer-based tools are potentially directly applicable to the disabled population, but the models they use provide, at best, only a framework for representing the performance of this population. The problem is that these models are derived from psychological research that attempts to be rigorously grounded and relatively general—characteristics that are achieved through experimental controls and statistical analysis and, thus, at the expense of richness and detail. The resulting models, paradoxically, describe the normal case even if there is not even a single exemplar of that case. Individual differences dissolve into variance from the mean, or worse, are considered outside the scope of the model. Thus, almost by definition, the capabilities of people with disabilities often lie outside general models.

A general model has heuristic value so long as we are willing to modify it significantly when it fails to adequately describe a specific person's condition—in other words, when we are willing to bend the model to the person rather than the person to the model. The resulting new model can then be usefully studied to evaluate designs and to produce new designs that are likely to better address the specific needs of persons with disabilities. Elkind and Shrager (1995) applied this method to computer-supported composition by dyslexic individuals. It is important to keep in mind that each person with a disability has specific impairments from a complex class of impairments and must be understood individually. Assistive devices and systems must be adjusted for these individual differences. To adapt general models so that they represent the characteristics of disabled populations and of individuals within these populations will require new data from these populations and probably some model extensions. This will take considerable effort.

SUMMARY OF RESEARCH NEEDS

In the preceding sections, we have mentioned several areas in which research and technology development are needed. In this section we summarize these needs to provide specific recommendations for a human factors program to enhance employment opportunities for people with disabilities.

Human Performance Design Data

An important research need is to build a more adequate human factors design database for the major disability subgroups and to get these data included in the standard design references and texts. Much of the needed database can be obtained from a study of existing literature from other fields, but in some areas new laboratory research will be required. Such data are a prerequisite for a principled methodology for design for disabled people.

Data are required for all major disability subgroups. A structure must be developed and key performance parameters identified so that the data can be applied to disability subgroups that have distinctly different performance characteristics and still allow customization to individual performance, when required. An effort should be made to integrate the data into existing computer models and simulations of human performance in order to guide the data collection process and to make these models useful for design for people with disabilities.

Task Analysis

A set of interesting representative jobs should be selected (a) as a focus for efforts aimed at understanding job requirements and (b) to provide a set of useful case studies. Given the importance of information work as a source of employment, this sector should be given priority in selection of jobs. Standard skills requirements methods should be applied to these jobs to establish their requirements, or a body of skill requirements data should be obtained from the literature. For each job, essential functions should be extracted from the total set of requirements and used to identify the disability subgroups whose impairment does not disqualify them for that job.

Taxonomies of the functions that workers perform in each of the representative set of jobs should be developed. An effort should be made to structure the taxonomies so that a core set of functions is identified that applies across many jobs.

Timeline analyses of the representative set of jobs should be performed with workers from different disability subgroups. These analyses should be used to identify areas where specific disability groups encounter performance limitations and resource constraints in order to guide job redesign and design of assistive devices and systems.

Design

A set of case histories of jobs that have been designed or redesigned to accommodate specific disabilities should be developed from actual experience or collected from the literature. These will serve as examples to disabled people and employers.

A concerted effort should be made to improve the accessibility of computer applications. The major computer systems used in work situations should be identified, their accessibility evaluated, and the changes required to make them accessible to the disabled workforce specified. Principal software vendors should be encouraged to implement these changes. Design guides for achieving accessible designs should be extended to cover a wider range of applications, systems, and disabilities.

These guides should include specification of interfaces required to facilitate integration of assistive devices and software into computer operating systems. Special attention should be given to making systems more accessible to users with cognitive disabilities. New technologies and improvements on existing technologies for assisting disabled users in performing cognitive functions should be investigated and developed.

Continued support should be provided for the development of new and improved assistive device technology that will enable a larger fraction of the disabled population to perform a wider variety of jobs effectively. Systematic methods for characterizing assistive devices and systems and matching them to particular users should be investigated and tested.

Evaluation Models, Tools, and Experiments

Performance data for people with specific disabilities should be incorporated into existing computer-based performance models and simulation tools so that they can be applied to designing for people with disabilities. Where necessary, these models should be extended so that they can represent the performance of disability subgroups. They should also be used to help distinguish between performance aspects of tasks that are essential to getting the tasks done and those that are not and to help identify ways in which tasks can be modified so they are more manageable by people with specific disabilities. Many of the necessary performance data do not now exist and will have to be obtained through experimentation.

Another kind of tool that could be very useful for people with disabilities is an accessible and easy-to-use database that would provide information regarding job-related (and other) resources that they can tap into. Consider the problem that people with specific disabilities currently have in finding out whether there exist assistive devices that could benefit them, and if so, where they are, whether they work as claimed, and how to go about acquiring them. The establishment of systems that could provide this kind of information seems to us to be a worthwhile goal. Clearly, there are human factors issues involved in ensuring the usefulness and usability of such systems.

One can even imagine a nationwide network-accessible information exchange tailored to the needs of people with disabilities. This could provide up-to-date information not only about assistive devices but also about employment opportunities, services, training programs, activities of special-interest groups, recent legislation pertaining to disabilities, and who is doing what in disability-related research and development, as well as suggestions and advice on the use of specific devices. Such an exchange would put disabled people in two-way contact with other individuals and groups with common problems/interests. It would permit the asking of "Does anybody know?" questions of the type one commonly finds on electronic bulletin boards. And it could serve as a gateway to other network-accessible resources.

In our view, one of the attractions of modern telecommunications technology is its potential to enrich the lives of people with disabilities in a variety of ways. It is not safe to assume, however, that this potential will be realized without explicit efforts by technologists, particularly those with a special interest in the human dimensions of technological change.

REFERENCES

ABLE DATA
    1989   *ABLE DATA: The Resource for Disability-Related Consumer Products.* Newington, Conn.: Adaptive Equipment Center (181 E. Cedar St.).

Behrmann, M.
    1989   Development of an Expert System for Assistive Technology Identification. Presentation at Closing the Gap conference, Minneapolis, Minn.

Boff, K.B., and J.E. Lincoln
    1988   *Engineering Data Compendium, Human Perception and Performance.* Wright-Patterson Air Force Base, Ohio: Armstrong Aerospace Medical Research Laboratory.

Bowe, F.G.
    1984   *Personal Computers and Special Needs.* Berkeley, Calif.: Sybex Computer Books.

Bureau of the Census
    1987   *Statistical Abstract of the United States.* Washington, D.C.: U.S. Department of Commerce.
    1989   *Statistical Abstract of the United States.* Washington, D.C.: U.S. Department of Commerce.

Campbell, J.P.
    1990   An overview of the Army Selection and Classification Project (Project A). *Personnel Psychology* 43:231-239.

Card, S.K., T.P. Moran, and A. Newell
    1983   *The Psychology of Human-Computer Interaction.* Hillsdale, N.J.: Erlbaum.

Casali, S.P.
1995 A physical skills-based strategy for choosing an appropriate interface method. In A.D.N. Edwards, ed., *Extra-Ordinary Human-Computer Interaction*. London, England: Cambridge University Press.

Casali, S.P., and R.C. Williges
1990 Data bases of accommodative aids for computer users with disabilities. *Human Factors* 32:407-422.

Corker, K., L. Davis, B. Papazian, and R. Pew
1986 *Development of an Advanced Task Analysis Methodology and Demonstration for Army-NASA Aircrew/Aircraft Integration*. Report 6124. Cambridge, Mass.: Bolt Beranek and Newman.

Czaja, S.J.
1987 Human factors in office automation. Pp. 1587-1616 in G. Salvendy, ed., *Handbook of Human Factors*. New York: Wiley.

Czaja, S.J., ed.
1990 *Human Factors Research Needs for an Aging Population*. Panel on Human Factors Research Issues for an Aging Population, Committee on Human Factors, National Research Council. Washington, D.C.: National Academy Press.

Edwards, A.D.N., ed.
1995 *Extra-Ordinary Human-Computer Interaction*. London, England: Cambridge University Press.

Egan, D.E., and L.M. Gomez
1985 Assaying, isolating, and accommodating individual differences in learning a complex skill. Pp. 173-217 in R.F. Dillon, ed., *Individual Differences in Cognition*, Vol. 2. Orlando, Fla.: Academic Press.

Elkind, J.I.
1990 The incidence of disabilities in the United States. *Human Factors* 32:397-405.

Elkind, J.I., and J. Shrager
1995 Modeling and analysis of dyslexic writing using speech and other modalities. In A.D.N. Edwards, ed., *Extra-Ordinary Human-Computer Interaction*. London, England: Cambridge University Press.

Elkind, J.I., S.K. Card, J. Hochberg, and B.M. Huey, eds.
1990 *Human Performance Models for Computer-Aided Engineering*. Committee on Human Factors, National Research Council. San Diego, Calif.: Academic Press.

Gilbreth, F.B., and L.M. Gilbreth
[1917] 1973 *Applied Motion Study*. Easton, Pa.: Hive Publishing.

Griffith, D., D.J. Garder-Bonneau, A.D.N. Edwards, J.I. Elkind, and R.C. Williges, eds.
1989 Human factors research with special populations will further advance the theory and practice of the human factors discipline. Pp. 565-566 in *Proceedings of the Human Factors Society 33rd Annual Meeting*. Santa Monica, Calif.: Human Factors Society.

Hartzell, E.J., E.W. Aiken, and J.W. Vorhees
1984 Aircrew-aircraft integration issues in future U.S. Army helicopters. In *AGARD Human Factors Considerations in High Performance Aircraft*. Moffett Field, Calif.: NASA Ames Research Center.

Kelley, P.L., and K.H.E. Kroemer
1990 Anthropometry of the elderly: status and recommendations. *Human Factors* 32:571-595.

Kraus, L.E., and S. Stoddard
1989 *Chart Book on Disability in the United States: An InfoUse Report*. National Institute on Disability and Rehabilitation Research. Washington, D.C.: U.S. Department of Education.

Kraut, R.E.
  1987   *Technology and the Transformation of White Collar Work.* Hillsdale, N.J.: Lawrence Erlbaum Associates.
Larimer, J., A. Arditi, J. Bergen, and N. Badler
  1989   *Visibility Modeling Project.* Moffett Field, Calif.: National Aeronautics and Space Administration.
McDaniel, J.W., and W.B. Askren
  1985   *Computer-Aided Design Models to Support Ergonomics.* Wright-Patterson Air Force Base, Ohio: Harry G. Armstrong Aerospace Medical Research Laboratory.
Mueller, J.
  1990   *The Workplace Workbook: An Illustrated Guide to Job Accommodation and Assistive Technology.* Washington, D.C.: Dole Foundation (available from RESNA Press).
Newell, A.F.
  1990   *Unified Theories of Cognition.* Cambridge, Mass.: Harvard University Press.
  1995   Extra-ordinary human-computer interaction. In A.D.N. Edwards, ed., *Extra-Ordinary Human-Computer Interaction.* London, England: Cambridge University Press.
Nickerson, R.S.
  1986   *Using Computers: Human Factors in Information Systems.* Cambridge, Mass.: MIT Press.
Paquette, S.P.
  1990   *Human Analogue Models for Computer-Aided Design and Engineering Applications.* Natick, Mass.: Army Natick Research and Engineeering Center.
Peterson, N.G., L.M. Hough, M.D. Dunnette, R.L. Rosse, J.S. Houston, and J.L. Toquam
  1990   Project A: specification of the predictor domain and development of new selection/classification test. *Personnel Psychology* 43:247-276.
Rauch, J.
  1989   Kids as capital. *Atlantic Monthly* August:56-61.
Rosen, M.J., and C. Goodenough-Trapagnier
  1989   The Tufts-MIT prescription guide: assessment of users to predict the suitability of augmentative communication devices. *Applied Technology* 1(3):51-61.
Salvendy, G., ed.
  1987   *Handbook of Human Factors.* New York: Wiley.
Sheridan, T.B.
  1980   Computers and human alienation. *Technology Review* 83:60-67, 70-73.
Short, J., E. Williams, and B. Christie
  1976   *The Social Psychology of Telecommunications.* New York: Wiley.
Smith, D.B.D.
  1990   Human factors and aging: an overview of research needs and applications opportunities. *Human Factors* 32:509-526.
Sorensen, R.J.
  1979   *Design for Accessibility.* New York: Wiley.
Strassmann, P.A.
  1985   *Information Payoff: The Transformation of Workers in the Electronic Age.* New York: Free Press.
Texas Young Lawyers Association
  1990   *The Americans With Disabilities Act: An Overview.* Austin: Texas Young Lawyers Association (P.O. Box 12487).
Trace Research and Development Center
  1988   *Trace Center Resource Book.* Trace Research and Development Center on Communication, Control, and Computer Access for Handicapped Individuals. Madison: University of Wisconsin.

Vachon, R.A.
   1990   Employing the disabled. *Issues in Science and Technology* 6(2):44-50.

Vanderheiden, G.C.
   1988   *Considerations in the Design of Computers and Operating Systems to Increase Their Accessibility to Persons With Disabilities.* Design Consideration Task Force, Trace Research and Development Center on Communication, Control, and Computer Access for Handicapped Individuals. Madison: University of Wisconsin.
   1990   Thirty-something million: should they be exceptions? *Human Factors* 32:383-396.

Vaughan, R.J., and S.E. Berryman
   1989   Employer-sponsored training: current status, future possibilities. *National Center on Education and Employment NCEE Brief* 4:1-4.

Williges, R.C., and B.H. Williges
   1995   Travel alternatives for the mobility impaired: the surrogate electronic traveler (SET). In A.D.N. Edwards, ed., *Extra-Ordinary Human-Computer Interaction.* London, England: Cambridge University Press.

4

Health Care

Roberta L. Klatzky and M.M. Ayoub

INTRODUCTION

In considering human factors research needs related to health care, we are particularly concerned with new research topics that are likely to reflect societal and technological developments in the coming decade. We also consider more long-standing problems that previous research has failed to address. In keeping with the general goals of this volume, our goal is to highlight problems in which human capabilities and processes play a critical role and thus to identify areas in which human factors researchers can make a significant contribution. Health care consumes a large and growing proportion of the U.S. gross national product; new developments have consequences for the quality and duration of many lives. Even a small change in the effectiveness of workers, medical devices, or care-giving environments can translate into a large impact on costs and human comfort.

Health care has not been a traditional focus for human factors research. This was noted two decades ago in a report on an international NATO symposium (Pickett and Triggs, 1974), whose goal was to provide examples of how human factors techniques could be applied to medical issues. In an introductory overview to that report, Rappaport cited a variety of contributions that human factors specialists could make to the health care field (see also Rappaport, 1970). He foresaw the increased use of computers in medicine and suggested that human factors specialists might interact directly with manufacturers of medical equipment to improve designs.

These seminal efforts notwithstanding, human factors researchers have not been greatly involved in the area of health care. For example, despite the long-standing concern of human factors specialists with sources of error, a symposium on human error in medicine held at the 1991 meeting of the Human Factors Society (Bogner, 1991) appears to have been the first of its kind (Van Cott, 1991). The dearth of attention to the potential contributions of human factors research is unfortunate; the field has much to offer with respect to designing medical devices, human-machine interfaces, and medical environments such as operating rooms and laboratories. Another area of application is training—not only of medical personnel but also of patients and home caregivers who use medical devices. Further, because of rapid technological change, the skills of medical personnel and of others working with medical devices may require frequent updating.

While we emphasize new research needs in this chapter, we should not forget that the health-care arena offers ample opportunities to apply already well-known principles of human factors. Medical errors point to the need for such application; there is a huge number of contexts in which error can and does occur.

Operating rooms are one example. There the medical practitioner might be compared to a member of a flight crew, as many of the same human factors issues apply. There is a necessity for teamwork, often with critical timing constraints (Helmreich and Schaefer, 1994; Regnier, 1993). Good communication among team members is imperative. There can be peaks of high stress, which are likely to coincide with points when timing is particularly important. The workplace is complex, with multiple instruments giving independent readings that must be integrated by the user. And of course, errors resulting from these sources may have dire consequences. Bogner (1991) has cited the substantial number of potentially preventable incidents in which anesthesia has resulted in brain damage or death.

The administration of drugs is another area in which errors frequently occur. Many medical errors within the home can be traced to inadequate instruction in the use of medicines or devices or to the complexities of labeling. Although steps to reduce such errors could be taken on the basis of existing human factors principles, it is also important to evaluate the target context carefully; this may call for further research (Cook, 1991).

The plan of the chapter is as follows. We first discuss some general technological and societal trends that are likely to have an impact on medical practices in the future and consider the human factors implications of these trends. We then describe, in more detail, several broad areas that are likely to raise specific issues for human factors research: enhanced understanding of risk factors and its impact on programs for behavioral change and disease prevention, advances in medical information technology, technological advances in medical instrumentation, and ergonomic issues that arise in health care procedures. Clearly, we cannot hope to introduce every potential area of application for human factors research in medicine. The sample we do present is intended to demonstrate the importance of increased involvement in the medical field by human factors specialists. A recurring theme in this discussion is how technological advances in medical care create new needs for human factors research. Although new health care needs may arise independently from technology—for example, through demographic or social trends or through new understanding of requirements for device design—responses to these needs increasingly reflect today's highly computerized environment.

NEW TRENDS IN HEALTH CARE

A number of trends in medicine indicate ample opportunities for contributions from human factors research in the coming decade. Some of these trends are related to the development of new health care *needs*, whereas others are related to new *tools*, particularly tools made possible by advances in electronics.

Demographic Change

As the relatively numerous post-World War II generation ages, it will inevitably place greater demands on the medical system, and the concomitant increase in demand for health care by that aging population is likely to raise a number of issues that human factors research can address. (Human factors research needs for the aging have recently been reviewed in a publication of the Committee on Human Factors—Czaja, 1990.)

Whereas the human factors industry has been extensively concerned with the design of workplace environments, there is a need for research on the design of appropriate domestic environments for the aged population. One goal should be to minimize the risk of accidents, which involve the elderly disproportionately (Czaja, 1990). A broader question is how environmental design might promote good health, both physical and psychological. The answer should take account of the effects of communal living, which seems to be increasingly likely in the face of high housing and health costs. The goal of constructing a living space that encourages social interaction while maintaining individual privacy is a challenging one. Another problem is how dwelling designs might encourage exercise in an otherwise sedentary population without imposing undue physical stress.

Aging also introduces problems of compliance with health recommendations, for example, difficulty in adhering to an adequate diet or remembering to take pills. These problems may be significantly ameliorated by simple external reminder systems that aid what has been called prospective memory. The human factors area has contributed to research on sensory, biomechanical, and psychomotor effects of aging (see Small, 1987). There are also cognitive changes that can fruitfully be studied from a human factors perspective with the goal of improving the health practices of an aging population.

It is also important to note that the current population of middle-aged adults is different from its predecessors: it is healthier, better informed, and more interested in health care. Many patients today seek to play an active role in prevention and medical treatment. The trend toward increasing patient participation in medical decision making raises many important questions: How can we inform patients about risk factors, preventive measures, and treatment options? How can we aid their decisions? We will consider these issues in more detail below.

Societal Change

An event of substantial societal impact is the emergence of highly infectious and/or untreatable diseases. The prospect of disease transmission in the workplace indicates an important problem for the human factors area, and medical personnel are particularly vulnerable. Interactions between patients and health care workers are coming under increasing scrutiny because of highly publicized cases of disease transmission. Clearly, there is a call for measures to safeguard the welfare of both groups, and human factors research should be fully involved in the development of such safeguards.

The increasing cost of health care is another trend, one outcome of which is decreased hospitalization and, concomitantly, increased home care. More and more frequently, home care incorporates medical devices that may be quite complex to maintain and operate. Human factors concerns should include the design of these devices, the nature of instructions, how individuals are trained to use them, and how stress might alter user competence.

For a substantial number of individuals in the United States, neither home care nor hospitalization is an alternative. Homeless people combine a high propensity for medical needs with a low ability to pay for treatment. They may be shut out of the medical system, and when those who find medical help are given recommendations for treatment, they may be unable to purchase needed drugs or conform to medical regimens. What human factors research can contribute is less clear in this area than in others such as workplace or instrument design. Yet it is in keeping with the goals of this volume that we mention the need to create avenues for diagnosis and treatment among those least able to seek help.

Technological Advances

Technologically advanced tools for health care include not only instruments for treatment of disease and trauma, but also systems for communication among medical researchers, practitioners, and supervisory agencies. Computer technology has created such diverse phenomena as new methods for medical imaging, large-scale databases related to patients and treatments, on-line bibliographies, automated decision aids, and computer-controlled surgical devices. As was noted above, new medical devices also make home treatment possible for patients who would once have required care in a health facility. New technology for assisting and retraining the disabled is discussed in Chapter 3.

As is generally the case, technological advances have produced new tools faster than research can be done on human factors related to their use. We need to understand how to make new techniques and devices as effective and accessible as possible.

RISK ASSESSMENT AND PREVENTIVE PRACTICES

In both the United States and abroad, individuals are increasingly seeking control of health care practices. This reflects a growing knowledge of risk factors and greater interest in taking preventive steps. The computer has had significant impact on people's knowledge of health risks, a development that is in keeping with our general theme of research needs based on technological advances.

Risk Assessment

One role of the computer in medicine has been to provide data about patient populations, giving rise to assessments of risk factors and prognostic factors in disease. For example, a very large database has been compiled by the Surveillance, Epidemiology, and End Results (SEER) program of the National Cancer Institute. Combined with data from the National Center for Health Statistics, which provides mortality rates, the SEER data provide direct assessment of cancer incidence and mortality over time. These data have been used to study a variety of issues related to etiological factors in cancer. Another example of a significant data source is the Breast Cancer Detection Demonstration Project (BCDDP), with over a quarter of a million participants. Data from the BCDDP have led to a technique for absolute risk assessment in breast cancer (Gail et al., 1989). Numerous databases are available from epidemiological surveys and experimental studies in the United States and abroad.

It is increasingly clear that behavioral, demographic, and familial characteristics are predictors of the risk of incurring disease. Among the most significant risk factors that have been identified are behavioral characteristics, including whether an individual smokes and how much dietary fat he or she consumes. For example, it has been estimated that approximately 75 percent of deaths from cancer are attributable to lifestyle (Williams, 1991). Cessation of smoking could reduce lung cancer deaths by over 75 percent; modification of diet could reduce cancer deaths by over 25 percent (Newell and Vogel, 1988).

Health risks arise not only from personal behaviors and attributes; risks due to substances found in the workplace and natural environment are being identified by both experimental and epidemiological techniques (see Swanson, 1988). On-the-job exposure to radiation is a health risk. Vibration may cause damage to the circulatory and nervous systems and to bones and joints. Increasingly, occupational stress is being considered a risk factor (Levi, 1990; Smith, 1987).

Some newly encountered risk factors pose formidable challenges. These include the risk of AIDS transmission to health care workers from needle sticks or handling of bodily fluids. In communities with particularly high rates of HIV infection, it has been estimated that dental workers may have risk levels equivalent to those of homosexuals (Morris and Turgut, 1990). In one survey of medical residents in training during 1989, more than half those surveyed had HIV patients under their care, and 9 percent reported having been exposed to HIV from a needle stick (Hayward and Shapiro, 1991). Hepatitis B virus is an even greater occupational risk for the medical community, whose members have a risk of infection up to 10 times that of the public (Hadler, 1990).

Communication of Risk Factors

An approach known as "health risk appraisal" began some two decades ago in an effort to make use of risk information (for a general review, see *Health Services Research*, 1987). The general idea behind the approach is that risk factors can be measured for a person (objectively or by self-report), and an estimate of that person's risk of disease can then be derived. A health professional would then communicate this personalized estimate to the patient, along with information about what factors are contributing to risk and what possible lifestyle changes he or she can adopt in response. If the person decides to make the suggested changes, the risk factors should be altered and risk should accordingly be reduced.

Each of the basic assumptions underlying health risk appraisal has been criticized. The adequacy of risk-factor assessment has been questioned, relying as it does on the patient's ability to retrieve personal data and his or her inclination to report them accurately. The models used to derive risk have also been criticized for making statistical assumptions that may be simplified or inaccurate. In addition, there is still not a great deal of evidence for the effectiveness of risk assessment in inducing risk-reduction behaviors.

These criticisms have not, however, prevented the public from being intensely interested in risk factors and disease prevention. Public interest in reducing cholesterol, for example, has considerably altered dietary habits in the United States. National agencies have used known risk factors such as familial incidence and age to adjust recommended regimens of screening tests for diseases. It seems clear that risk assessment (population and personalized), along with attendant recommendations for behavioral change, will increasingly be part of the American health scene.

Prevention Programs in the Workplace and Community

Once people are informed of risk factors, preventive practices may take various forms, including controlling toxic substances, designing environments that reduce local exposure, and promoting risk-reducing behaviors. Of these interventions, design of the physical workplace has been a traditional focus of human factors and will continue to be important in reducing exposure to risk factors and occupational hazards. A less obvious contribution to be made by human factors is in helping to change the behaviors of individuals. In particular, an important vehicle for behavioral change is the institution of educational and motivational programs for groups.

Group prevention programs have taken place in educational, community, and work environments. For example, the Pawtucket Heart Health Program attempted to treat obesity in an entire community (Lasater et al., 1991; Carleton et al., 1991). Components of the program included a monthly citywide "weigh in," labeling of items in grocery stores and on menus, blood-cholesterol screening, a student cook-off, and church programs, all facilitated by enlisting a large number of community volunteers. Substantial reductions in serum cholesterol were observed in the initial pilot study. Another example is the "Know Your Body" project, funded by the National Heart, Lung, and Blood Institute and the National Cancer Institute. This attempted to modify risk-inducing behaviors in schoolchildren and was found to be successful in reducing the rate of cigarette smoking several years later (Walter, 1989). Community screening for breast cancer is another success story (Paryani et al., 1990). The movement toward risk-prevention and screening programs in the workplace is growing (Breslow et al., 1990).

One scenario for the creation of a community or workplace prevention program is to begin with controlled trials that demonstrate the effectiveness of a preventive behavior, followed by extensive dissemination of results to enhance public awareness. This has been done in the case of the link between reduced blood cholesterol and decreased incidence of heart disease. Following one well-publicized prevention study, there was a substantial increase in public acceptance of the beneficial effects of cholesterol reduction (Schucker et al., 1987). This undoubtedly created a receptive audience for community interventions like the Pawtucket program.

Human Factors Challenges Related to Risk Assessment and Preventive Practices

There is a clear need for further human factors research on how information about risk and its reduction should be communicated and how these communications are interpreted. Risk communication is complex, as a recent National Research Council publication attests (National Research Council, 1989). Even when the effects of a risk factor can be reasonably estimated, the nature of the risk may be communicated in different ways. Should we be told that a risk is 5 percent for a year or (equivalently) 0.4 percent for a month? Is it more meaningful to know the risk of death or the expected reduction in years of life expectancy? Does it help to have a standard of comparison, for example, to know that the risk of HIV to surgeons performing 25 operations on infected patients is about the same as that of death on a Louisiana oil rig over the course of a year (Orient, 1990)? Once communicated, risk statistics are subject to cognitive reasoning processes that may introduce other sources of error or distortion. For example, whether a risk is framed in terms of the probability of positive or negative outcomes will affect an individual's response (Tversky and Kahneman, 1981). Another problem is that the small probabilities associated with many medical risks may be difficult for people to interpret. Perhaps that is why having personal knowledge about someone who contracts a disease has such a potent effect. For example, the national incidence of tests for breast and colon cancer increased dramatically after Nancy and Ronald Reagan incurred these diseases.

Becker and Janz (1987) have provided a theoretical analysis of health risk appraisal that points to some other relevant issues. One is whether personalized risk estimates are more effective than population estimates, and if so, why. Another is how to motivate risk-reducing behavior, for example, by promoting belief in the alterability of risk factors or by increasing the awareness of risky behavior. The introduction of warnings on cigarette packages is an example of public policy intended to increase risk awareness.

The design of group programs to increase positive health practices is another area in which human factors research can make important and widespread contributions. Programs aimed at reducing smoking, decreasing fat intake and increasing fiber, reducing recreational sun exposure, and inducing compliance with screening programs for such diseases as breast and colon cancer are among those that promise to have very beneficial results (Fink, 1987). Research to increase the effectiveness of such programs would be of considerable societal value.

INFORMATION TECHNOLOGY

The computer can not only store data but also deliver data directly to medical practitioners and patients. The general area concerned with information encoding, representation, and communication in medicine is called medical informatics. Chapter 7 discusses medical informatics in the broader context of information systems. In this section we consider a number of medical contexts in which information delivery is critical and the implications of these contexts for human factors research needs.

On-Line Information Retrieval

Several databases are directly available to medical workers and, in some cases, patients as well. Examples are MEDLINE, which provides titles and abstracts of recently published articles in the medical literature and was developed by the National Library of Medicine, and CANCERLIT, a similar bibliographic system with a narrower range of citations. MEDLINE is available on CD-ROM from several companies and can be accessed by modem.

The National Cancer Institute has developed Physician Data Query (PDQ), a computer database for information about advances in cancer treatment and clinical trials (see Hubbard et al., 1987). Protocols of currently active clinical trials are entered into PDQ and are indexed by disease and eligibility criteria. In addition, PDQ provides state-of-the-art information about disease characteristics and prognosis. There has been considerable effort to make PDQ available to the medical community. The system can be accessed at thousands of medical libraries and centers, and several commercial vendors have been licensed to distribute PDQ via networks and CD-ROM.

Decision Aids and Expert Systems

Information delivery can be extremely helpful in medical decision making. Computer programs variously known as knowledge-based systems, decision aids, and expert systems are meant to facilitate the many medical decisions that must be made by both medical workers and patients. Whether these programs succeed depends in part on how well they are tailored to the humans who use them.

Automated decision systems emerged at least three decades prior to the 1990s (Ledley and Lusted, 1961). These systems generally have two components—a knowledge base of facts, often in the form of rules, and procedures for using the facts, sometimes called the inference engine. When applied to a specific problem, the system also requires data about its characteristics. *Expert systems* are so called because the underlying knowledge is derived from experts. The term *decision aid* suggests a program that provides some of the information that is useful in decision making, often quantified or structured data, but that leaves other elements to the user. In practice, an expert system generally functions as a decision aid in that it contributes to decision making rather than governing it.

One of the best-known expert systems is MYCIN (Shortliffe, 1976), which is designed to identify the source of a bacterial infection and to recommend treatment. MYCIN has been augmented by instructional programs that train potential users. Recently the National Library of Medicine has developed AI/RHEUM, a diagnostic system for rheumatologic diseases (Kingsland et al., 1986). This program features an extensive help system that incorporates video images as well as text.

The range of application of knowledge-based systems is now very wide, including, for example, diagnosis, prescription, causal interpretation of data, medical consulting, and selection of adjuvant therapies. When one considers that each of these areas can be applied to many specific diseases or problem areas, it is clear that the number of potential programs is immense. Commercially available shells, which specify a format for the knowledge base and provide an inference engine, are intended to simplify the development of new systems.

Computerized decision aids are attractive in part because of evidence that human decision making is far from ideal. People have been characterized as subject to heuristics and biases that introduce error (for review in a medical context, see Dowie and Elstein, 1988). People's decisions differ depending on the way problems are presented (framing effects), how much has been expended on a solution in the past (sunk costs), what sample solutions are provided (anchoring and adjustment), and the extent to which they can retrieve past solutions from memory (availability). Medical decisions are complex; the computer can facilitate decision making by providing as much relevant data as possible, along with a content-free objective decision process.

Human Factors Challenges Related to Information Technology

Past human factors research in several areas has intersected with needs in medical informatics. Relevant research topics include modeling of decision making and design of usable interfaces for expert systems. We suggest here, then, not so much new substantive topics as specific applications that present new research needs.

A traditional goal of human factors research has been to improve the accessibility of tools. This goal is critical to the successful application of systems for information delivery, as Chapter 7 discusses at length. A need for increased accessibility of medical information systems is clear in that the explosion of information available to medical practitioners has not necessarily resulted in widespread use. The National Cancer Institute recently conducted a formal evaluation of the PDQ system (Czaja et al., 1989). Despite intensive publicity about the system and efforts to make it maximally available to health care workers, fewer than 50 percent of physicians in cancer specialties were aware of it, and only about half these physicians reported using it. Use by community physicians who were not in oncological specialties was termed "very low." The greatest use was by employees of the Cancer Information Service, who provide information to the public.

The low level of PDQ use was attributed in part to physicians' discomfort with computers. The primary use of computers in a clinical practice tends to be for administration and word processing rather than for access to information or for automated decision facilitation. There are also problems with negative attitudes about computers. For example, only 5 percent of non-oncologists in the PDQ review indicated that computerized data retrieval was very important.

It seems clear that there is a yawning gap between the availability of technological tools—which provide data about disease risk, diagnosis, prognosis, and treatment as well as automated aids for decision making—and the use of these tools. Human factors research can play an important role in bridging this gap. Some of the problems that must be considered are the following:

• The user interface. It is by now a given that the success of a computer program designed for general use depends on its "user friendliness." Because the designers of PDQ were aware that medical practitioners might be unfamiliar with computers, PDQ software offers menus as well as direct commands; some instructions are also given on-line. The evaluation suggests, however, that these measures are insufficient to promote widespread use.

• Medical training. Medical schools do not currently include extensive training in medical informatics in their curricula, although efforts to increase such training are under way (Ball and Douglas, 1990). The National Library of Medicine has created an elective medical informatics course for advanced medical students, which is intended to create a "seed crop" of researchers in this area (National Institutes of Health, 1991). Given the demands of the medical curriculum, any such training must be highly efficient; it is unlikely to receive a substantial portion of overall education time. Postgraduate training in the use of specific programs—conducted, for example, in connection with professional meetings—may be a way to increase computer fluency.

Medical schools also do not currently provide extensive training in epidemiology and statistics (Klatzky et al., 1994); again, it is unclear that such training can be justified in the general curriculum. Yet the evaluation of retrieved information is critical, and this may depend on an understanding of the statistical measures and models that were used to derive it. Klatzky et al. suggested that one approach to this problem is to provide on-line introductions to basic statistical concepts as part of a database retrieval program or decision aid.

- Attitudes about technological aids. Negative attitudes about the usefulness of computerized data retrieval were cited as a potential barrier to the use of PDQ by practicing physicians. More generally, discomfort with decision aids such as computer-guided diagnoses or treatment plans may preclude their use. Efforts to bring technological aids to the practicing clinician should include attitude change as a goal. It should be made clear that computer programs can greatly assist the physician in acquiring relevant data and putting them to use.

It should not be forgotten that patients, too, play a role in decision making. There has been relatively little effort to provide data to patients or to assist them in making decisions about treatment. Research is needed to ascertain the extent to which patients wish to play a decision-making role as well as to determine what resources might aid them. A straightforward approach is to construct patient-oriented databases of general information about diseases and appropriate references. The question then becomes how to maximize the accessibility of the information.

Patients often make decisions under considerable stress. Research needs related to the effects of stress on cognition are reviewed in Chapter 10, but it is important to note here that there has been little research on the stress induced by diagnosis of disease. In some diseases, for example, breast cancer, a range of treatment options are available. Typically, patients make decisions about treatment shortly after diagnosis, when stress is likely to be greatest. Meyerowitz (1980) suggests that the psychological impact of breast cancer diagnosis and treatment may be as threatening as the disease itself. Human factors researchers could address such questions as the effects of stress on decision making in this context and the potential role of automated data retrieval and decision aids. The prevalence of breast cancer suggests that it could provide a useful paradigm for research on stress among medical patients.

MEDICAL INSTRUMENTATION

The continued presence of the familiar blood-pressure cuff notwithstanding, the computer has virtually revolutionized medical instrumentation. *Bioinstrumentation* is a term that has been applied to the combination of electronic and biological technology (Wise, 1990). Bioinstruments include, for example, biomedical imaging systems that take advantage of the computer for gathering, enhancing, and displaying data; electrochemical sensors that incorporate whole living cells in their construction; artificial sensing devices such as tactile sensors based on force-sensitive polymers; and transducers such as cochlear implants.

As is the case with any technological change, advances in medical instrumentation lead to problems in mastering the new technology and create needs for human factors research. The pace of advance means that researchers must "enter the loop" quickly. This has frequently not been the case. Here we briefly describe examples of new developments in instrumentation and consider related human factors needs.

Medical Imaging

Medical imaging is a century old; the X-ray was discovered in 1895. The fundamental requirements for an imaging system remain the same. What is needed is some means of acquiring information about the imaged site and some means of displaying that information. However, technological advances have greatly expanded the ways to meet both of these requirements and have added other capabilities, particularly new means of analyzing, enhancing, and interpreting the data.

A variety of techniques for acquiring information about a targeted site are in widespread use today. Two-dimensional images like that provided by the X-ray are the basis for computed tomography, in which a contiguous sequence of cross-sectional images is used to synthesize a three-dimensional representation. Magnetic resonance imaging relies on the activity of atoms when placed in a magnetic field to construct a two-dimensional image. Positron emission tomography uses the distribution of radioactivity to produce two-dimensional slices that can then be used to construct a three-dimensional model. Still other methods are thermography, which measures the distribution of temperature, and ultrasound.

The resulting data are often subjected to a number of pre-processing algorithms that, for example, detect edges, compute regions, reduce noise, and extract and enhance critical features. The data are then displayed in various ways. Multiple slices from different planes and at different depths are often displayed simultaneously. A three-dimensional surface description can be displayed as a high-quality two-dimensional projection onto a viewing screen. There is also the possibility of fully three-dimensional display techniques such as holograms or a display through stereoscopic viewers. Color can be used to differentiate regions, especially when there is no direct counterpart between acquired data and a visual feature (e.g., oxygen consumption or temperature).

New techniques are increasingly expanding the ways medical images are displayed as well as how they are intepreted with the aid of knowledge-based systems. The potential usefulness of automated interpretation is substantial, given the density of the imaged information and the complexity of the mapping between diagnostically relevant attributes of the body and visually apparent aspects of the image. Swets et al. (1991) have described a two-part interactive decision aid that first prompts the user to rate a given image according to a series of scales and then uses the obtained scale values to estimate the probability of malignancy, which is then communicated to the user. An analysis of breast-cancer diagnoses with and without the decision aid revealed a clear advantage for using the aid, an advantage that increased with the difficulty of the diagnosis.

The use of film-based radiological images is expected to decrease, owing to the advent of digital environments (Arenson et al., 1990). A rapidly developing tool is the picture archiving and communication system (PACS), a digital workstation that will allow physicians to call up and display stored images. The 1990 meeting on medical imaging of the International Society for Optical Engineering had 15 paper sessions and a poster session devoted to PACS (Dwyer and Jost, 1990). Among its many positive aspects, PACS offers the possibility for on-line image enhancement and interpretation aids; however, a negative aspect is the potential degradation in image quality due to digital display.

Techniques derived from perception and decision science have been widely used to evaluate the relative merits of imaging methods, as well as to assess the effectiveness of enhancement techniques and the degradation under systems like PACS. In particular, an analysis using the area under the relative operating characteristic (ROC; in the context of signal detection theory, this is also called the receiver operating characteristic) has proven highly effective. The ROC plots the proportion of true positive responses as a function of the proportion of false positives in a diagnostic test. The ROC measure is independent of the decision criterion and the frequency of diagnosed events (Metz, 1989; Swets, 1988). For example, Swets et al. (1979), using this measure, found computed tomography to be more accurate than radionuclide scanning in diagnosing brain lesions.

Much attention has already been given to human factors considerations in the design of medical imaging workstations (O'Malley and Ricca, 1990). A key to physician acceptance of these systems seems to be perceived savings in time and effort. In one survey study of factors influencing the acceptance of PACS systems by radiologists (Saarinen et al., 1989), speed of information delivery was found to be the primary concern. Thus, a particularly positive aspect was the time that might be saved in traveling to and from the site where radiological files were stored. Physicians also favored a system that would eliminate the need for multiple trips to retrieve films that had been misplaced or temporarily checked out by others. On the negative side, physicians were concerned with the possibility of significant downtime for the system. Somewhat surprisingly, interest in PACS was not affected by the age of the physician and was only slightly affected by prior experience with computers.

Biosensors

A biosensor is a device that places a transducer in close contact with molecules of a biological substance in order to produce an output that is correlated with concentration of the substance. The general sequence by which the sensor operates is that the measured substance interacts with a "biological mediator" (e.g., one or more enzymes, antibodies, or bacteria), leading to an output (e.g., electrochemical, optical, or calorimetric) that can be transformed into an electrical signal (Mascini et al., 1990). Reviews of biosensor technology can be found in Claremont (1987), Higgins and Lowe (1987), and Wilkins (1989). Schultz (1991) traces the modern biosensor to two antecedents: information technology, particularly the development of miniaturized components, and molecular biology, which identifies biomolecules that can serve as the mediators for recognizing the target substance.

The field of biosensor technology is now burgeoning at both basic and applied levels. Attractive features of this technique include its speed, the need for low-volume samples, and the possibility of in situ measurement and long-term implantation. The potential variety of biosensors is very large; sensors have been developed to measure such substances as urea, cholesterol, proteins, alcohol, and penicillin. These devices have widespread application beyond health care; for example, they are used in the food industry to measure contaminants.

One of the first areas of health care application of biosensors, and probably the most developed, is sensing of glucose concentrations (e.g., Ross et al., 1990; Schaffar and Wolfbeis, 1990). This has obvious important benefits to diabetics. For decades, diabetics have done home glucose monitoring by using visual assessment of color changes on reagent strips dipped in urine or blood. These methods have a high potential for error, and, particularly with urine testing, there is considerable temporal lag between a change in blood glucose and the ability to detect it. Subjective visual monitoring and urine tests for diabetes have given way to blood glucose monitors, pocket-sized electronic devices that assess the level of glucose from a blood sample by sensing results of its interaction with an enzyme. By performing periodic measurements and adjusting insulin injections or diet accordingly, diabetic patients can maintain acceptable glucose levels with minimal fluctuation.

Biosensors not only make possible more direct and accurate evaluation but also provide the possibility of long-term implantation. An implanted glucose sensor is critical to the development of an artificial pancreas that would monitor glucose levels and adjust insulin injections automatically; most of the components for such a system now exist. Expert-system technology can also be combined with glucose sensing to regulate glucose levels (Lougheed et al., 1987).

Medical Devices for Home Care

As was mentioned above, a clear trend in health care is to discharge patients after shorter hospitalizations, necessitating increased home care. Often, this is made possible by sending the patient home with a medical device. For example, ventilators may be placed in the home for patients with respiratory problems (Bach et al., 1992; Thompson and Richmond, 1990). Patients may receive medication at home with infusion pumps, electronically controlled devices used to regulate the flow of drugs along lines implanted in the body, for example subcutaneous catheters or venous access ports. These devices are used to supply antibiotics, drugs to reduce pain, and chemotherapy (New et al., 1991; Reville and Almadrones, 1989; Storey et al., 1990). Some patients receive "total parenteral nutrition," their entire daily calorie requirement, intravenously at home (Bisset et al., 1992; Viall, 1990). Home dialysis machines now enable kidney patients to undergo dialysis even while they sleep (Delano and Friedman, 1990; *Health Devices*, 1991). Although these devices have considerable benefits, not only economically but also in terms of patient welfare and satisfaction, they also put responsibility on home caregivers and patients alike.

Human Factors Challenges Related to Medical Instrumentation

The contributions that human factors researchers can make to the development and use of medical imaging seem enormous. New techniques for acquiring, displaying, and evaluating images call for increased interaction between the human user and the computer. User options for image enhancement and display adjustment increase the complexity of these systems. The same points apply as those that emerged in the evaluation of the PDQ system. The effectiveness of these systems will substantially depend on the extent to which the interface renders technology accessible, given the attitudes and training of physicians and technical staff.

Human factors researchers can also play a role in developing and evaluating systems for image enhancement and knowledge-based interpretation. Each new modality creates anew such questions as, which imaged features are most useful for diagnosis? Currently, the discovery of relevant features may be a piecemeal process taking several years; systematization of this process is badly needed (Swets, personal communication).

Advances in medical instruments to be used in hospitals and homes also create a substantial need for human factors research. Of course, any new instrumentation calls for training of operators and appropriate design of machine interfaces. The increased use of medical devices at home carries with it additional problems.

One is the adequacy of training. Written instructions and warnings should obviously be very clear, and hands-on training is likely to be necessary for all but the simplest devices. Lack of education and language skills by home users may hamper training efforts, as may advanced age. Another consideration is that stress from an incident such as machine malfunction could undermine the effects of training. Although the effects of long-term stress imposed on home caregivers and patients have received some attention (e.g., Smith et al., 1991; Wegener and Aday, 1989), the cognitive effects of acute stress merit more study. Clearly, patients and caregivers should have access to support personnel outside the home. Often, pharmacists are the ones responsible for advising patients about a device when they themselves have inadequate training (Kwan and Anderson, 1991).

Training is but one concern. Another is that the home environment may not be well suited for the device in, for example, the type and reliability of its power supply, the proximity of water and electrical connections, and the space available for the equipment and appendages like oxygen tanks. Ambient temperature may affect device function, and the home may not be adequately equipped to control it. Home use often demands that patients follow a regular schedule, indicating a need to provide records of past administration and to signal the next application. As was mentioned previously, the elderly may in particular need reminder systems to trigger self-care.

Glucose measurement devices have been closely evaluated from a human factors perspective (e.g., Kelly et al., 1990; McDonald, 1984; Moss and Delawter, 1986) and hence provide a test bed with which to evaluate home care more generally. Kelly et al. found a number of problems with even such a relatively simple device. In evaluating the instructional materials provided with blood glucose monitors, most diabetes educators judged them inadequate and believed that additional instruction was necessary. User error was identified in nearly three-fourths of reported problems. Patients often failed to clean the device or to calibrate it properly, possibly in order to reduce the time spent in operating it, and errors occurred at virtually every stage of operation. In efforts to cut costs, some patients split test strips so they could be used twice, with potentially adverse effects on sensing accuracy.

The case of the blood glucose monitor serves to emphasize the need for human factors research related to home use of medical devices. Whereas blood glucose monitors are simple devices and have been designed for patient use, other devices that are sent to the home may be far more complex and may have been designed for hospital use only (e.g., for ventilators, see *Health Devices*, 1988). These devices are even more likely to have inadequate written documentation for home caregivers and it is likely that any in-hospital training prior to discharge will be devised on an ad hoc basis.

ERGONOMICS ISSUES IN HEALTH CARE DELIVERY SYSTEMS

In this section we discuss some of the biomechanical problems encountered in health care, the magnitude of these problems, and the research topics they raise. Traditional concerns of human factors research, including design of the workplace, design and implementation of devices, and performance modeling, have substantial potential for application in medical contexts.

Physical Stress to Nursing Personnel

In any work system, including the health care delivery system, stresses are imposed on the body. These stresses can be (a) mechanical, involving the musculoskeletal system; (b) physiological, involving the cardiopulmonary system; or (c) psychological. Physically demanding tasks often produce mechanical stress for the musculoskeletal system, particularly the spine, resulting in low back injury. In health care delivery systems, handling of materials—particularly the handling of patients—results in high spinal stresses.

Back injuries account for approximately one of every five injuries and illnesses in the workplace (Bureau of Labor Statistics, 1982). Nursing aides and licensed practical nurses ranked fifth and ninth, respectively, in compensation claims for back injuries (Klein et al., 1984). Available statistics indicate that nursing personnel are as likely to suffer from a compensable back injury as are workers in occupations with heavy load-handling tasks (Jensen, 1987). Back injuries are the result of large stresses imposed on the spine. High levels of biomechanical stress imposed on the musculoskeletal system, particularly the spine, have been reported in tasks performed by nursing personnel (Gagnon et al., 1986; Stubbs et al., 1983; Torma-Krajewski, 1986). According to Lloyd et al. (1987), efficient and safe patient transfer practices should be based on sound biomechanical considerations.

Nursing personnel—and nursing aides in particular—often lift, move, and transfer patients whose weights range from 37 kg to over 100 kg. These weights are higher than the capacity of most females (National Institute for Occupational Safety and Health, 1981). Gagnon et al. (1986) reported that the compressive forces on the L5/S1 disc ranged from 5.74 to 7.95 kilonewtons (kN) for a single person handling a 72 kg mannequin. These values and the estimated compressive forces for handling a patient using two persons are approximately 4.44 kN; the values are significantly higher than the recommended National Institute for Occupational Safety and Health safe compression limits of 3.4 kN.

Physical Rehabilitation

The management of chronic pain and disability after an injury has always been considered the exclusive domain of medical professionals. Ergonomic involvement in the pre-injury (prevention) and post-injury (rehabilitation and return-to-work) stages has been shown to contribute significantly to the successful control and management of overall prevention and rehabilitation and the avoidance of disability recurrence (Khalil et al., 1985, 1988; Rosomoff, 1987; Rosomoff et al., 1981). Important issues in the rehabilitation of individuals with certain disabilities are measurement of their functional performance, accurate and objective examination and assessment of their capabilities and limitations, and quantitative descriptions of them in terms of human performance profiles (HPP) (Abdel-Moty, 1991).

In the evaluation of HPP, specific areas of interest are the following: (a) physical characteristics of the patient such as strength, flexibility, endurance, and posture; (b) functional capacity of the patient in performing certain activities such as lifting and walking; and (c) work-related capacities such as the ability to perform specific job tasks under prescribed conditions. In evaluating patients with chronic pain or certain physical disabilities, it is very important to use objective measurement of specific abilities (Khalil et al., 1990). The HPP can then be compared with that of healthy persons of equivalent age, sex, and work category in order to determine functional capacity.

Biomechanical Modeling

To enhance biomechanical research in health care delivery systems—particularly for back disorders—biomechanical modeling research can be invaluable. As representations of the real system, models can be useful in examining the behavior of the system under consideration (in this case, the human body) without exposing the body to a variety of hazardous conditions. Ergonomic models are discussed in another publication of the Committee on Human Factors (Kroemer et al., 1988).

Human Factors Challenges Related to Ergonomic Issues in Health Care Delivery Systems

In order to reduce and control back injuries in health care delivery systems, a major research effort is needed to evaluate handling of patients to reduce stress on the spine. The research must focus on several areas: (a) the training of personnel in methods of handling patients to reduce spinal stresses; (b) the evaluation of facility designs—which include beds, wheelchairs, and other devices to facilitate patient handling—for all systems, whether they use manual methods, assistive devices, or both; and (c) the evaluation of existing design and the ultimate redesign of assistive devices used in patient handling to reduce spinal stresses. The research should also consider several other variables related to spinal stress. These are (a) the time it takes to perform the handling activity, (b) patient comfort and safety, and (c) patient characteristics (Garg et al., 1991). Furthermore, it is important that the research pay careful attention to the differences between laboratory environments and the actual work site; laboratory results must be carefully evaluated (Garg et al., 1991) prior to field application.

The evaluation of the functional capacity of patients in need of physical rehabilitation is another area in which human factors research can play a role. Currently several techniques are being used to assess functional capacity. Some of these are quite subjective and rely heavily on data obtained through observation; others are more objective and gather data through quantitative measurement of such factors as mobility and strength. Even when the more objective measures are employed, there is a need to integrate and translate these measurements into functional capacity values. Therefore, research in the development and use of objective measures for the evaluation of functional capacity is sorely needed. This will make it possible to determine more accurately the level of individual abilities, which in turn can help determine the optimal program of rehabilitation (Abdel-Moty and Khalil, 1988; Abdel-Moty et al., 1989, 1990).

Research in biomechanical modeling is also needed. The current biomechanical models are quite basic at best and cannot adequately deal with the complicated structures of the musculoskeletal system. Three-dimensional biomechanical models that include, for instance, the effects of soft tissue and the individual muscle tension generated in performing typical patient handling could provide important insights into handling methods and design of work areas and equipment to minimize the stresses on the spine. In addition, good biomechanical models can be helpful in developing manual handling methods that minimize stresses on the body, particularly the spine.

CONCLUDING REMARKS

The primary aim of this chapter has been to call attention to the substantial contributions that could be made by human factors researchers in the area of medical care. This will be at least an initial step toward redressing the underutilization of human factors researchers in medical contexts. The topics that we have described include some traditional concerns but also point to a broad spectrum of relatively novel research needs that arise as a result of our changing societal and technological environment. Many of these problems will require that human factors personnel work with experts in other disciplines. We have identified a large number of research opportunities in the hope that their diversity will attract a larger number of researchers to the area. We suspect that many of these problems, unfortunately, will not be dealt with in the next decade.

Particularly important research goals noted in this chapter are the following:

- to identify and eliminate sources of error arising from the medical workplace and medical devices;
- to design health-promoting environments for the aged;
- to identify health risks in the workplace;
- to determine techniques for effective risk communication;
- to develop health risk reduction and illness prevention programs for groups;
- to design user-appropriate interfaces for new medical devices, such as decision aids, imaging systems, and biosensors;
- to identify ways to facilitate access to health information by medical personnel;
- to identify barriers to effective use of medical devices in the home and to redesign devices and home environments so as to promote effective home care;
- to identify and eliminate sources of biomechanical stress on the musculoskeletal system from health care practices;
- to develop techniques for measuring functional capacity of candidates for rehabilitation; and
- to develop biomechanical models to assist in determining potential health hazards.

As we indicated at the outset, the research needs that are reviewed here are far from exhaustive, and new technological developments are likely to expand the list. Additions will only reinforce the important role that human factors can play in fitting health care practices to those in need of care.

ACKNOWLEDGMENTS

We would like to acknowledge the help of John Swets and Sue Bogner with this chapter.

REFERENCES

Abdel-Moty, E.
1991 Ergonomics issues in low back pain: intervention strategies. *Proceedings of the Human Factors Society 35th Annual Meeting*. Santa Monica, Calif.: Human Factors Society.

Abdel-Moty, E., and T.M. Khalil
1988 Ergonomic considerations for the reduction of physical task demands of low back pain patients. Pp. 959-967 in F. Aghazadeh, ed., *Trends in Ergonomics/Human Factors*, Vol. IV. Amsterdam, Netherlands: North-Holland, Elsevier Science Publishing.

Abdel-Moty, E., T.M. Khalil, S.S. Asfour, M. Howard, R.S. Rosomoff, and H.L. Rosomoff
1989 Effects of pain on psychomotor abilities. Pp. 465-471 in A. Mital, ed., *Advances in Industrial Ergonomics and Safety*. New York: Taylor and Francis.

Abdel-Moty, E., T. Khalil, S. Asfour, M. Goldberg, R. Rosomoff, and H. Rosomoff
1990 On the relationship between age and responsiveness to rehabilitation. Pp. 49-56 in B. Das, ed., *Advances in Industrial Ergonomics and Safety*, Vol II. New York: Taylor and Francis.

Arenson, R.L., D.P. Chakraborty, S.B. Seshadri, and H.L. Kundel
1990 The digital imaging workstation. *Radiology* 176:303-315.

Bach, J.R., P. Intintola, A.S. Alba, and I.E. Holland
1992 The ventilator-assisted individual: cost analysis of institutionalization vs. rehabilitation and in-home management. *Chest* 101:26-30.

Ball, M.J., and J.V. Douglas
1990 Informatics programs in the United States and abroad. *MD Computing* 7:172-175.

Becker, M.H., and N.K. Janz
1987 Behavioral science perspectives on health hazard/health risk appraisal. *Health Services Research* 22(4):537-551.

Bisset, W.M., P. Stapleford, S. Long, A. Chamberlain, B. Sokel, and P.J. Milla
1992 Home parenteral nutrition in chronic intestinal failure. *Archives of Disease in Childhood* 67:109-114.

Bogner, S.
1991 Human factors and medicine. P. 682 in *Proceedings of the Human Factors Society 35th Annual Meeting*. Santa Monica, Calif.: Human Factors Society.

Breslow, L., J. Fielding, A.A. Herrman, and C.S. Wilbur
1990 Worksite health promotion: its evolution and the Johnson & Johnson experience. *Preventive Medicine* 19:13-21.

Bureau of Labor Statistics
1982 *Back Injuries Associated With Lifting*. Work Injury Report, Bulletin #2144:1. Washington, D.C.: U.S. Department of Labor.

Carleton, R.A., L. Sennett, K.M. Gans, S. Levin, C. Lefebvre, and T.M. Lasater
1991 The Pawtucket Heart Health Program: influencing adolescent eating patterns. *Annals of the New York Academy of Sciences* 623:322-326.

Claremont, D.J.
  1987   Biosensors: clinical requirements and scientific promise. *Journal of Medical Engineering and Technology* 11:51-56.
Cook, R.I.
  1991   How to do that voodoo that you do so well: medical human factors in the explicit context of use. P. 684 in *Proceedings of the Human Factors Society 35th Annual Meeting*. Santa Monica, Calif.: Human Factors Society.
Czaja, S.J., ed.
  1990   *Human Factors Research Needs for an Aging Population*. Panel on Human Factors Research Issues for an Aging Population, Committee on Human Factors, National Research Council. Washington, D.C.: National Academy Press.
Czaja, R., C. Manfredi, D. Shaw, and G. Nyden
  1989   Evaluation of the PDQ System: Overall Executive Summary. Report to the National Cancer Institute under Contract No. N01-CN-55459. Survey Research Laboratory, University of Illinois, May.
Delano, B.G., and E.A. Friedman
  1990   Correlates of decade-long technique survival on home hemodialysis. *ASAIO Transactions* 36:337-339.
Dowie, J., and A. Elstein
  1988   *Professional Judgment: A Reader in Clinical Decision Making*. Cambridge, England: Cambridge University Press.
Dwyer, S.J., III, and R.G. Jost, eds.
  1990   *Medical Imaging IV: PACS System Design and Evaluation*. Proceedings of SPIE, the International Society for Optical Engineering in cooperation with the American Association of Physicists in Medicine, Vol. 1234, parts 1 and 2. Bellingham, Wash.: SPIE.
Fink, D.J.
  1987   Preventive strategies for cancer in women. *Cancer* 60:1934-1941.
Gagnon, M., C. Sicard, and J.P. Sirois
  1986   Evaluation of forces on the lumbo-sacral joint and assessment of work and energy transfers in nursing aides lifting patients. *Ergonomics* 29:407-421.
Gail, M.H., L.A. Brinton, D.P. Byar, D.K. Corle, S.B. Green, C. Schairer, and J.J. Mulvihill
  1989   Projecting individualized probabilities of developing breast cancer for white females who are being examined annually. *Journal of the National Cancer Institute* 81:1879-1886.
Garg, A., B. Owen, D. Beller, and J. Banaag
  1991   A biomechanical and ergonomic evaluation of patient transferring tasks: bed to wheelchair and wheelchair to bed. *Ergonomics* 34:289-312.
Hadler, S.C.
  1990   Hepatitis B virus infection and health care workers. *Vaccine* 8(March Supplement):S24-S28; discussion:S41-S43.
Hayward, R.A., and M.F. Shapiro
  1991   A national study of AIDS and residency training: experiences, concerns, and consequences. *Annals of Internal Medicine* 114:23-32.
Health Devices
  1988   Portable volume ventilators. *Health Devices* 17(4):107-131.
  1991   Hemodialysis machines. *Health Devices* 20(6):187-232.
Health Services Research
  1987   October issue 22(4).

Helmreich, R.L., and H.G. Schaefer
1994   Team performance in the operating room. Pp. 225-254 in M.S. Bogner, ed., *Human Error in Medicine.* Hillsdale, N.J.: Erlbaum.

Higgins, I.J., and C.R. Lowe
1987   Introduction to the principles and applications of biosensors. *Philosophical Transactions of the Royal Society of London* 316:3-11.

Hubbard, S., J.E. Henney, and V.T. DeVita, Jr.
1987   A computer data base for information on cancer treatment. *New England Journal of Medicine* 316:315-318.

Jensen, R.
1987   Disabling back injuries among nursing personnel: research needs and justifications. *Research in Nursing and Health* 10:29-38.

Kelly, R.T., J.R. Callan, T.A. Kozlowski, and E. Menngola
1990   *Human Factors in Self-Monitoring of Blood Glucose.* Task 4 Final Report. FDA/CDRH-90/60. Springfield, Va.: NTIS.

Khalil, T.M., S.S. Asfour, E. Abdel-Moty, R.S. Rosomoff, and H.L. Rosomoff
1985   New horizons for ergonomics research in low back pain. Pp. 591-598 in R.E. Eberts and C.G. Eberts, eds., *Trends in Ergonomics/Human Factors.* Amsterdam, Netherlands: North-Holland, Elsevier Science Publishing.
1988   Quantitative assessment of outcome of a low back pain rehabilitation program. *Abstracts of the International Conference on the Study of the Lumbar Spine.* Miami, Fla. April 13-15.

Khalil, T.M., E. Abdel-Moty, and T.M. Asfour
1990   Ergonomics in the management of occupational injuries. Pp. 41-53 in B.M. Pulat and D.C. Alexander, eds., *Industrial Ergonomics: Case Studies.* Norcross, Ga.: Industrial Engineering and Management Press.

Kingsland, L.C., III, D.A.B. Lindberg, and G.C. Sharp
1986   Anatomy of a knowledge-based system. *MD Computing* 3:18-26.

Klatzky, R.L., J. Geiwitz, and S.C. Fischer
1994   Using statistics in clinical practice: a gap between training and application. Pp. 123-140 in S. Bogner, ed., *Human Error in Medicine.* Hillsdale, N.J.: Erlbaum.

Klein, B.P., R.C. Jensen, and L.M. Sanderson
1984   Assessment of workers' compensation claims for back strains/sprains. *Journal of Occupational Medicine* 26:443-448.

Kroemer, K.H.E., S.H. Snook, S.K. Meadows, and S. Deutsch, eds.
1988   *Ergonomic Models of Anthropometry, Human Biomechanics and Operator-Equipment Interfaces.* Committee on Human Factors, National Research Council. Washington, D.C.: National Academy Press.

Kwan, J.W., and R.W. Anderson
1991   Pharmacists' knowledge of infusion devices. *American Journal of Hospital Pharmacy* 48:10 Suppl 1, S52-S53.

Lasater, T.M., L.L. Sennett, R.C. Lefebvre, K.L. DeHart, G. Peterson, and R.A. Carleton
1991   Community-based approach to weight loss: the Pawtucket "weigh-in." *Addictive Behaviors* 16:175-181.

Ledley, R.S., and L.B. Lusted
1961   Medical diagnosis and modern decision making. Pp. 117-157 in *Proceedings of Symposia in Applied Mathematics*, Vol. 14. Providence, R.I.: American Mathematical Society.

Levi, L.
1990   Occupational stress: spice of life or kiss of death? *American Psychologist* 45:1142-1145.

Lloyd, P., C. Tarling, J.D.G. Troup, and B. Wright
    1987   *The Handling of Patients: A Guide for Nurses*, 2nd ed. London, England: The Royal College of Nursing.

Lougheed, W.D., A. Schiffrin, and A.M. Albisser
    1987   Stabilizing blood glucose with a novel medical expert system. *Biosensors* 3:381-389.

Mascini, M., D. Moscone, and G. Palleschi
    1990   Biosensor applications of continuous monitoring in clinical chemistry. Pp. 1429-1460 in D.L. Wise, ed., *Bioinstrumentation: Research, Developments and Applications*. Stoneham, Mass.: Butterworth.

McDonald, W.I.
    1984   Quality control of home monitoring of blood glucose concentrations. *British Medical Journal* 288:1915.

Metz, C.E.
    1989   Some practical issues of experimental design and data analysis in radiological ROC studies. *Investigative Radiology* 24:234-245.

Meyerowitz, B.E.
    1980   Psychosocial correlates of breast cancer and its treatments. *Psychological Bulletin* 87:108-131.

Morris, R.E., and E. Turgut
    1990   Human immunodeficiency virus: quantifying the risk of transmission of HIV to dental health care workers. *Community Dentistry and Oral Epidemiology* 18:294-298.

Moss, J.P., and D.E. Delawter
    1986   Self-monitoring of blood glucose. *American Family Physician* 33:225-228.

National Institute for Occupational Safety and Health
    1981   *Work Practices Guide for Manual Lifting*. NIOSH Technical Report No. 81-122. Washington, D.C.: National Institute for Occupational Safety and Health.

National Institutes of Health
    1991   *Clinical Electives Program for Medical and Dental Students 1992-1993*. NIH Publication No. 91-499. Bethesda, Md.: National Institutes of Health.

National Research Council
    1989   *Improving Risk Communication*. Committee on Risk Perception and Communication. Washington, D.C.: National Academy Press.

New, P.B., G.F. Swanson, R.G. Bulich, and G.C. Taplin
    1991   Ambulatory antibiotic infusion devices: extending the spectrum of outpatient therapies. *American Journal of Medicine* 91:455-461.

Newell, G.R., and V.G. Vogel
    1988   Personal risk factors: what do they mean? *Cancer* 62:1695-1701.

O'Malley, K.G., and K.G. Ricca
    1990   Optimization of a PACS display workstation for diagnostic reading. Pp. 940-946 in *Medical Imaging IV: PACS System Design and Evaluation*. Proceedings of SPIE, the International Society for Optical Engineering in cooperation with the American Association of Physicists in Medicine, Vol. 1234. Bellingham, Wash.: SPIE.

Orient, J.M.
    1990   Assessing the risk of occupational acquisition of the human immunodeficiency virus: implications for hospital policy. *Southern Medical Journal* 83(10):1121-1127.

Paryani, S.B., T.A. Marsland, P. Faucher, E. Fontanelli, M. Freeman, H. Johnston, W. Morrow, P. Prabhu, M. Stearman, and F. Vines
    1990   Breast cancer screening project in northeast Florida. *Journal of the Florida Medical Association* 77:29-31.

Pickett, R.M., and T.J. Triggs, eds.
    1974   *Human Factors in Health Care*. Lexington, Mass.: D.C. Heath.

Rappaport, M.
    1970   Human factors applications in medicine. *Human Factors* 12:25-35.

Regnier, S.J.
    1993   Symposium underscores value of OR teamwork. *American College of Surgeons Bulletin* 78:73-81.

Reville, B., and L. Almadrones
    1989   Continuous infusion chemotherapy in the ambulatory setting: the nurse's role in patient selection and education. *Oncology Nursing Forum* 16:529-535.

Rosomoff, H.L.
    1987   Comprehensive pain center approach to the treatment of low back pain. Pp. 78-85 in *Low Back Pain: Report of a Workshop*. Rehabilitation Research and Training Center, Department of Orthopaedics and Rehabilitation. Charlottesville, Va.: University of Virginia.

Rosomoff, H.L., C. Green, M. Silbert, and R. Steele
    1981   Pain and low back rehabilitation program at the University of Miami School of Medicine. In K.Y. Lorenzo, ed., *New Approaches to Treatment of Chronic Pain*. NIDA Research Monograph 36. Washington, D.C.: U.S. Department of Health and Human Services.

Ross, D., L. Heinemann, and E.A. Chantelau
    1990   Short-term evaluation of an electro-chemical system (ExacTech) for blood glucose monitoring. *Diabetes Research and Clinical Practice* 10:281-285.

Saarinen, A.O., G.L. Youngs, and J.W. Loop
    1989   The Attitude of Referring Physicians Towards PACS: A Pre-Installation Assessment. Report to the MITRE Corp. DIN/PACS Evaluation Project Contract N55-200. Department of Radiology, University of Washington DIN/PACS Evaluation Project, November 30.

Schaffar, B.P., and O.S. Wolfbeis
    1990   A fast responding fibre optic glucose biosensor based on an oxygen optrode. *Biosensors and Bioelectronics* 5:137-148.

Schucker, B., K. Bailey, J.T. Heimbach, M.E. Mattson, J.T. Wittes, C.M. Haines, D.J. Gordon, J.A. Cutler, V.S. Keating, and R.S. Goor
    1987   Change in public perspective on cholesterol and heart disease: results from two national surveys. *Journal of the American Medical Association* 258:3517-3531.

Schultz, J.S.
    1991   Biosensors. *Scientific American* 265(2)64-69.

Shortliffe, E.H.
    1976   *Computer-Based Medical Consultations: MYCIN*. New York: Elsevier.

Small, A.M.
    1987   Design for older people. Pp. 495-504 in G. Salvendy, ed., *Handbook of Human Factors*. New York: Wiley.

Smith, C.E., C.K. Giefer, and L. Bieker
    1991   Technological dependency: a preliminary model and pilot of home total parenteral nutrition. *Journal of Community Health Nursing* 8:245-254.

Smith, M.J.
　1987　Occupational stress. Pp. 844-860 in G. Salvendy, ed., *Handbook of Human Factors*. New York: Wiley.
Storey, P., H.J. Hill, Jr., R.H. St. Louis, and E.E. Tarver
　1990　Subcutaneous infusions for control of cancer symptoms. *Journal of Pain and Symptom Management* 5:33-41.
Stubbs, D.A., P.W. Buckle, M.P. Hudson, and P.M. Rivers
　1983　Backpain in the nursing profession, II: the effectiveness of training. *Ergonomics* 26:767-779.
Swanson, G.M.
　1988　Cancer prevention in the workplace and natural environment: a review of etiology, research design, and methods of risk reduction. *Cancer* 62:1725-1746.
Swets, J.A.
　1988　Measuring the accuracy of diagnostic systems. *Science* 240:1285-1293.
Swets, J.A., R.M. Pickett, S.F. Whitehead, D.J. Getty, J.B. Schnur, J.B. Swets, and B.A. Freeman
　1979　Assessment of diagnostic technologies. *Science* 205:753-759.
Swets, J.A., D.J. Getty, R.M. Pickett, C.J. D'Orsi, S.E. Seltzer, and B.J. McNeil
　1991　Enhancing and evaluating diagnostic accuracy. *Medical Decision Making* 11:9-18.
Thompson, C.L., and M. Richmond
　1990　Teaching home care for ventilator-dependent patients: the patients' perception. *Heart and Lung* 19:79-83.
Torma-Krajewski, J.
　1986　Analysis of Lifting Tasks in the Health Care Industry. Paper presented at the University of Washington Symposium on Occupational Hazards to Health Care Workers. Seattle.
Tversky, A., and D. Kahneman
　1981　The framing of decisions and the psychology of choice. *Science* 211:453-458.
Van Cott, H.P.
　1991　Human Error in Medical Devices. Paper presented at the Symposium on Human Factors and Medicine, 35th annual meeting of the Human Factors Society. San Francisco, September.
Viall, C.D.
　1990　Daily access of implanted venous ports: implications for patient education. *Journal of Intravenous Nursing* 13:294-296.
Walter, H.J.
　1989　Primary prevention of chronic disease among children: the school-based "know your body" intervention trials. *Health Education Quarterly* 16:201-214.
Wegener, D.H., and L.A. Aday
　1989　Home care for ventilator-assisted children: predicting family stress. *Pediatric Nursing* 15:371-376.
Wilkins, E.S.
　1989　Towards implantable glucose sensors: a review. *Journal of Biomedical Engineering* 11:354-361.
Williams, G.M.
　1991　Causes and prevention of cancer. *Statistical Bulletin of Metropolitan Insurance Companies* 72:6-10.
Wise, D.L., ed.
　1990　*Bioinstrumentation: Research, Developments and Applications*. Stoneham, Mass.: Butterworth.

5

Environmental Change

Raymond S. Nickerson and Neville P. Moray

THE PROBLEM OF ENVIRONMENTAL CHANGE

The subject of detrimental environmental change has received much attention in the news media for some time. Scientists, policy makers, and the public have become increasingly concerned about the threat that such change, if it continues unabated, poses for the future. Growing numbers of scientists from a variety of disciplines have been systematically studying specific aspects of this change and attempting to identify effective strategies for preventing or mitigating potentially catastrophic effects.

Human factors researchers have not focused much attention on this area in the past. Perhaps it has been assumed that the discipline has little to offer toward the solution of environmental problems. We believe it does have something to offer. This chapter represents an effort to stimulate and contribute to a dialogue that will help identify what some of the possibilities are.

Dimensions of the Problem

Some earth and atmospheric scientists have been documenting an increased concentration of carbon dioxide and other "greenhouse gases" in the atmosphere and have been attempting to better understand how a continuing accumulation will affect the future world climate (Houghton and Woodwell, 1989; National Research Council, 1983). Others have been studying such phenomena as "acid rain" and its effects on lakes and streams, forests, and materials (Baker et al., 1991; Mohnen, 1988; Schwartz, 1989), air pollution and urban smog (Gray and Alson, 1989; National Research Council, 1991; Office of Technology Assessment, 1988), and the thinning of ozone in the stratosphere (Stolarski et al., 1992; Stolarski, 1988). Studies have focused on the contamination and depletion of fresh-water supplies (la Riviere, 1989; National Research Council, 1977; Postel, 1985), on the depletion of the world's forests (Myers, 1989; Repetto, 1990) and wetlands (Steinhart, 1990; Wallace, 1985), and on the worldwide loss of arable land (Crossen and Rosenberg, 1989; National Research Council, 1990; Schlesinger et al., 1990). Biologists have been documenting the loss of wildlife habitat and the accompanying decrease in biodiversity (Soule, 1991; Wilson, 1989). More detailed discussions of the many facets of the problem are readily available (Gore, 1992; Nickerson, 1992; Stern et al., 1992).

Behavioral Causes of Environmental Change

Many of the most readily identified causes of these changes are human activities. Major contributors to the accumulation of greenhouse gases in the atmosphere include the burning of fossil fuels for heating and energy generation and the use of chlorofluorocarbons (CFCs) as coolants and aerosols. The burning of fossil fuels is also a major cause of acid rain, which is formed when airborne sulfur dioxide and nitrogen oxides combine with water vapor. Air pollutants include ozone, carbon monoxide, lead, sulfur dioxide, nitrogen dioxide, and particulates—all by-products of industrial and energy-generation processes. Stratospheric ozone thinning is believed to be a direct consequence of the accumulation of CFCs in the upper atmosphere.

Major threats to clean, fresh-water supplies include contamination not only from precipitation of chemical emissions that have accumulated in the atmosphere but also from agricultural runoff containing pesticides and fertilizers, from waste discharges into rivers, from salt used for highway de-icing, from hazardous wastes disposed of improperly, and from leachate from municipal dumps. Deforestation is the consequence both of converting forests to farmland and residential and business areas and of overharvesting timber. Wetland loss results from the "reclamation" of wetlands for commercial development. Desertification, the transformation of arable land into land on which crops will no longer grow, has a variety of causes, including overgrazing and the salinization of soil from excessive irrigation.

Since the human activities that are implicated in detrimental environmental change are aimed at satisfying human needs and desires, those activities can only be expected to increase as the population grows. And population growth, worldwide, is expected to continue for the near future at least, at something like its current rate, which would yield a doubling of the current number of about 6 billion before the middle of the twenty-first century. Moreover, if present trends continue, the pressures on the environment are likely to grow faster than the population. During the twentieth century, worldwide energy consumption has increased by a factor of about 15 and the total population has increased by a factor of about 3.5, which is to say that, compared with 1900, there are about 3.5 times as many of us now and each of us uses, on average, 4 times as much energy (Gibbons et al., 1989). There is now an enormous disparity between the per capita use of energy in the industrialized world and in developing countries; we can expect that the desire of the developing countries to close this gap will create a strong impetus to increase the average use worldwide.

In short, there is much evidence that human behavior can adversely affect the natural environment in a variety of ways and that the forces that motivate environmentally detrimental behavior are likely to become even stronger in the future. There is a need to better understand the coupling of behavior and environmental change and how to mitigate the undesirable effects.

POSSIBLE APPROACHES TO THE PROBLEM

The problem of detrimental environmental change is broad in scope and considerably beyond the ability of the human factors research community to solve. But human factors researchers can contribute greatly by working toward the goal of finding effective ways to modify, or mitigate the effects of, the human behavior that is a major cause of such change. It is useful to make a distinction between attempting to modify behavior directly and attempting to modify it indirectly by changing technology so that its use will be less detrimental to the environment.

Direct Behavioral Change

Possible ways to directly induce behavioral change include the use of coercion (legislation and regulation, backed up with the threat of civil or criminal sanctions), incentives (tax and other monetary incentives, public recognition, and awards), education (making people aware of problems and what can be done about them), and persuasion (appeals to moral responsibility or altruism—or the possibility of embarrassment or shame). All of these methods have been tried, many times in some cases, and in numerous variations.

Psychologists have done a considerable amount of research to assess the effectiveness of various strategies for behavior modification in the context of environmental concerns (Holahan, 1986; Russell and Ward, 1982;

Saegert and Winkel, 1990; Stern, 1992). Illustrative of this work are studies of the use of incentives, rewards, education and information campaigns, persuasion, and other techniques to motivate conservation in the use of energy or water, participation in recycling programs, decrease in waste generation and littering, and other behavior that would be desirable for environmental preservation (Baum and Singer, 1981; Coach et al., 1979; Cone and Hayes, 1980; Geller et al., 1982; Geller, 1986). This work has demonstrated that behavior can be changed with the use of incentives and other types of inducements, but the changes that have been effected have been modest in magnitude and have tended not to persist much beyond the duration of the experimental intervention.

Without questioning the need to continue this line of research, we note that behavior modification is not the only approach that can be taken to the problem of detrimental environmental change. Moreover, even assuming that much more effective means of changing behavior in desired ways will be discovered than have so far been found, it may be unrealistic to expect the problem to be solved by this approach alone. Effective and lasting behavior modification has proven very difficult to achieve. Efforts to effect behavioral change are unlikely to be very successful so long as the technologies and the products of technology that we use make it easy to behave in environmentally harmful ways (Crabb, 1992).

Changing Technology

One may work on the goal of water conservation by trying to persuade people to use less water when taking showers by, say, taking shorter showers, keeping the tap less than wide open, or keeping the tap closed while actually washing. Alternatively, or in conjunction with efforts to change behavior, one can attempt to design shower heads that automatically conserve water by limiting the flow when the tap is fully open but that still provide adequate water for showering. Ideally, one would like a water-miser shower head whose spray is preferred by users, who then will be motivated to use this head whether or not they are concerned about environmental change.

Finding ways to change the technology so that it is equally effective, if not more so, while doing less harm to the environment is a complementary alternative to attempting to modify behavior directly. This is the motivation for the concerted efforts to develop environmentally benign alternatives to the burning of fossil fuels for energy generation and for many other current research activities in the physical and biological sciences.

We wish to argue that human factors research has much to contribute to the goal of shaping technology so that the natural consequences of its use for human ends will be more environmentally benign.

HUMAN FACTORS AND ENVIRONMENTAL CHANGE

To date, human factors, as a profession, has not focused much on the problem of environmental change, at least as that problem is conceived here. What is sometimes referred to as "environmental ergonomics" has tended to focus on how one's immediate environment—temperature, humidity, noisiness—affects one's bodily and cognitive functions and performance. The interests of the Human Factors Society's Technical Group on Environmental Design (Human Factors Society, 1991:38), for example, "center on the human factors aspects of the constructed physical environment, including architectural and interior design aspects of home, office, and industrial settings." The Applied Experimental and Engineering Psychology issue of *PsycSCAN* has "environment" as one of six major topics under which the abstracts are organized. But each of the 12 subtopics in this section deals with the effects of some environmental factor (altitude, heat, noise) on human beings (performance, safety, or comfort). In general, the subject of the implications of human behavior for environmental change—as distinct from the effects of environmental variables on human behavior—has not been a focus of attention of the human factors community.

There is one major exception: the interest the field has shown in studying industrial accidents and near accidents, especially in the nuclear power industry, and in developing ways to decrease the probability and severity of such accidents (Reason, 1990; Senders and Moray, 1991). With this exception, most of what psychologists have done that relates directly to the problem of detrimental environmental change has not been done within the mainstream of human factors research, and the results of that work have not been published in the journals most strongly associated with human factors research. The problem of environmental change has not captured the imagination of the human factors research community as a whole.

As to why this is the case, we can only speculate. One possibility is that human factors researchers believe they have little to offer in this area. We think that human factors does have something to offer, and the main purpose of this chapter is to make that point.

Another possibility is that human factors researchers have assumed that the best way for the discipline to address the problem of environmental change is indirectly, through work on more generic problems, such as the design of displays, of person-machine interfaces, of work situations, and so on.

This view has considerable merit. When one designs a better interface for a computer system, or when one discovers and articulates principles that can help designers produce interfaces that are better suited to human use, one is facilitating the work of anyone who uses systems with these interfaces, including earth and atmospheric scientists working on the problem of global warming, modelers developing source-receptor models for predicting the dispersion of sulfur dioxide emissions, and agronomists attempting to balance variables in a plan for a sustainable-agriculture approach to the production of crops.

Similarly, when one designs an information-management system—or discovers characteristics of human beings as information processors that have implications for the design of such a system—one is contributing indirectly to the work of anyone who makes use of such a system, including a variety of people working on environmental problems. Just as it is not necessary for the materials scientist to have the building of better automobiles in mind in order to affect the automotive industry—by, for example, developing a new lightweight superstrong composite—one need not focus explicitly on the environment in order to have a beneficial impact on work on environmental problems.

This being said, we believe it is important to raise the question of whether there are opportunities for the human factors community to make more direct and explicit contributions to work on the problem of environmental change than it has done in the past. A major purpose of this chapter is to stimulate thought and discussion about this question.

RESEARCH NEEDS

We believe that the extremely important problem of detrimental environmental change represents a major challenge and opportunity for human factors research and that such research has something of value to contribute, especially when it is directed to the question of how technology might be developed so as to serve its human purposes without affording the means of environmental degradation.

In what follows, we suggest a few specific research questions that we believe deserve attention from the human factors community. We focus primarily on problems that fit reasonably well within human factors, as defined by the work that has traditionally been done in this field. The possibilities become more numerous when one considers problems that fall within the domain of applied psychology, broadly defined.

We offer our suggestions not as an agenda for research but as points of departure for further discussion. The most pressing need in this area is for human factors researchers who are concerned about environmental change to begin exchanging ideas about the problem and how the human factors profession might help address it. We believe that such discussion would identify many ways in which the community can productively involve itself in this important area.

Energy Production and Use

The production and use of energy effect environmental change through the methods used to extract energy sources from the earth, through the depletion of natural resources, and through the by-products of energy transduction and utilization. The per capita demands for energy vary greatly in different parts of the world. Not surprisingly, they are much greater in countries that are highly industrialized than in those that are not, but even within the industrialized world, there are large differences in energy use. There is a need to better understand why these differences exist. Some, but not all, of the difference can be attributed to differences in industrial productivity and in standards of living. There is a growing sense that attitudes—about efficiency and waste, about public versus private transportation, about personal conveniences and the public good—are important factors in the equation, but their role is not well understood.

There are numerous specific questions relating to energy use and environmental change that deserve the attention of human factors researchers. What determines when working from home or a satellite office can be an acceptable, if not a preferred, alternative to commuting to a centralized workplace? Under what conditions can teleconferencing be effectively substituted for face-to-face meetings? How can the effectiveness and acceptability of teleconferencing systems be increased?

As we noted earlier, human factors professionals have involved themselves in energy generation and use primarily by studying human error in power plant operations and developing methods to decrease the probability of mishaps and the potential severity of their effects. This research is of undoubted importance and should continue to be priority.

Improving Public Transportation Facilities

The personal automobile is, by far, the preferred mode of travel in the United States (Gray and Alson, 1989). This is due to a number of factors, but chiefly to the great convenience of the private auto, compared with other types of transportation in many parts of the country. Using private autos, however, puts a greater burden on the environment than does using public mass transit. An environmentally beneficial objective, therefore, is to enhance the attractiveness of public transportation so that it will more often be the preferred means of transport, especially in major urban areas.

This is, in part at least, a human factors problem. We need to better understand why people choose the modes of transportation they do when other modes are also available. Such knowledge could be applied to making public transportation more attractive, and, to the extent that the preference for private transportation is based on a lack of understanding of the pros and cons of the alternatives, it could be used to guide programs aimed at informing the public in this regard.

Substituting Resource-Light for Resource-Heavy Technologies

To the extent that needed or desired services can be provided by technologies that make relatively light demands on energy and other resources, the interests of environmental preservation will be well served by using such technologies. In particular, when the transmission of information can be substituted for the transportation of people and material, the environment benefits in a variety of ways.

Finding ways to substitute resource-light technologies for resource-heavy technologies is especially desirable in view of the rapidly increasing demands that underdeveloped countries are expected to put on resources by attempting to catch up, economically, with the more developed parts of the world (Stern et al., 1992). To the extent that these countries could be enabled to adopt energy- and resource-efficient methods of delivering desired goods and services, without first appropriating inefficient methods that have been utilized in much of the industrialized world, the benefits would be global as well as regional.

Telecommuting and Teleconferencing

Telecommunications technology has the potential to make it possible for more people to work, at least part of the time, from their homes rather than commute to offices. It also has the potential, via teleconferencing facilities, to reduce the need for travel to meetings. Although these possibilities were recognized by the earliest promoters of teleconferencing and telecommuting (Bavelas et al., 1963; Nilles et al., 1976), they have not yet been realized as much as might have been expected. Why that is so is not entirely clear, but there can be no doubt that many human factors issues are involved in the question of how to make these technologies attractive and effective from the user's point of view.

Electronic Substitutes for Paper

Substituting electronic means of storing and distributing information for methods that depend on the use of paper is one instance of the substitution of resource-light for resource-heavy technologies that deserves special attention. Paper and paper products account for about one-third of all solid waste in the United States. A considerable fraction of paper waste is from newspapers (whose daily circulation is about 63 million according to the Bureau of the Census, 1990) and magazines. Given that most buyers of newspapers are interested in only a fraction of the information that is in them, that buyers discard newspapers immediately after reading them, and that printing and distributing them are energy-intensive processes, this method of information distribution is extremely inefficient relative to technologically feasible alternatives. Similar observations apply to magazines.

Despite many predictions that less paper would be used as a result of the increasing use of electronic information exchange systems, there is little evidence of any such reduction. It may even be that computer technology has stimulated the use of more paper than ever before. Nevertheless, the potential remains for decreasing the use of paper by making more effective use of electronic information storage and distribution. Significantly realizing this potential would have the doubly beneficial effect of conserving the energy and natural resources used in the production of paper and of reducing the generation of solid waste.

The technology exists for distributing news and information electronically rather than in traditional newspapers and magazines; to date, however, that technology is not sufficiently widely installed in homes to be a feasible basis for replacing paper media. It seems highly likely, however, that in the not-distant future most homes will have the means of making electronic newspapers and magazines practical. There are likely to be nontechnical obstacles to their acceptance and use by the public, stemming from the fact that video displays, even if made to look something like a book, are very different from the types of print media with which people are familiar. It would be useful to know what would make electronic books, newspapers, and periodicals acceptable to people as replacements for their paper counterparts.

This is not to suggest that realization of the potential of information technology for reducing the need for paper awaits only a better understanding of the psychological deterrents to the use of electronic media. The practical usability and the actual use of communication facilities also depend on the existence of an adequate infrastructure, pricing policies that provide incentives for use, and general access to the critical facilities. User acceptability, however, is likely to be a key determinant of the extent to which this technology is appropriated when other impediments to its exploitation no longer apply.

Simulation and Virtual Reality Technology

Other special cases of substituting resource-light for resource-heavy technologies involve the use of simulated aspects of reality for a variety of purposes. Simulators have been used for flight training for many years. This was motivated in part by safety considerations, and in part by the fact that operating a simulator is much less expensive than operating real aircraft for training purposes. A major reason is that it requires less expenditure of energy and other resources, a fact that benefits the environment as well.

The development and use of SIMNET, the army's simulation system for training tank teams, illustrates what is possible by way of network-based systems that are capable of simulating situations involving many people and machines interacting in complex ways (Thorpe, 1993). A SIMNET-based training exercise is not only much less expensive than a comparable exercise involving real tanks but also much easier on the environment. How to ensure the effectivenss of this approach to team training remains a challenge.

Virtual reality technology carries simulation, in theory at least, to a new plateau. The goal of developing this technology is to simulate objects and situations in such a way that people can perceive and interact with the simulated realities very much as they would with whatever it is that is simulated, except without the inconvenience or, sometimes, the danger that would be involved in interacting with the real thing. There are many human factors questions relating to the development and use of virtual reality technology that represent challenges for research; the National Research Council has completed a study of some of these questions (Durlach and Mavor, 1995).

Recycling and Waste Handling

Improving the Technology of Recycling

Recycling of waste materials is a relatively new technology. It seems reasonable to assume that, as with any new technology, its effectiveness and efficiency could be improved. In particular, inasmuch as the energy required for some recycling operations limits their utility (Georgescu-Roegen, 1976), there is a need to find more efficient methods for processing recyclable materials, for example, new ways to separate trash into unrecyclable and the several recyclable categories and new ways to transform the recyclable types into reusable materials.

A major problem of waste recycling is getting sustained citizen participation in recycling programs (Geller et al., 1982). Education and advertising campaigns have not been very effective (Coach et al., 1979; Geller et al., 1975). Efforts to motivate people to recycle have sometimes met with modest short-term success but have not managed to effect lasting change (Geller, 1987; Humphrey et al., 1977). And in some cases, simple positive reinforcement schemes have even produced unwanted behavior (Geller, 1981). Planning and executing recycling programs that will effect the lasting changes in attitudes and behavior that are essential to make real progress on the problem of waste remains a significant unmet challenge.

Radioactive and Toxic Waste Handling

Radioactive and toxic wastes represent special problems and require some innovative approaches. Given that about a dozen U.S. nuclear reactors are ready for decommissioning now and about 50 will be ready for retirement in the Western world before the end of the century (Shulman, 1989), the problem of handling nuclear wastes will be getting much more attention than it has in the past. Because very few nuclear reactors have yet been dismantled, the technology for this undertaking is still being developed.

Much attention will be focused on the cleanup of the Hanford site in the state of Washington, where weapons-grade radioactive materials have been produced since the days of the Manhattan Project in the early 1940s. This cleanup project alone is planned to take 30 years to complete. It is also expected to require resolution of many human factors issues. Among these will be issues having to do with the design of control facilities for a vitrification plant in which liquid wastes will be solidified and issues relating to the design and use of tele-operator systems for remote handling of radioactive materials. Because of the industry's limited experience with dismantling and cleanup operations, there will need to be some innovative thinking about the allocation of functions to people and machines and the design of person-machine interfaces for this purpose (Wise and Savage, 1992).

Designing for Error Prevention

Human error is known to be a major cause of industrial accidents. The accident at Chernobyl occurred because of an interaction of poor plant design, poor management decisions, and violations of procedures (Reason, 1990). The Bhopal incident was caused by a combination of operator error, poor training, and bad policies, including the policy of storing large quantities of hazardous materials, thus increasing the chances that an accident, should it occur, would be on a very large scale (Hazarika, 1986). At Three Mile Island, inadequacies in training, operating procedures, control room interface design, and maintenance practice were all seen to be contributing factors. More generally, analyses of industrial accidents have revealed a great variety of human errors—in system design, regulation, operation, maintenance, communication and management (Rasmussen and Batstone, 1989; Reason, 1990).

Inasmuch as industrial accidents can have—and have had—serious environmental consequences, work on the problem of designing industrial control stations and operating procedures so as to minimize the possibility of human error is very much in the spirit of what this chapter is intended to promote. The most difficult challenge here is to identify vulnerable points in an industrial process before any disastrous errors occur. Although the occurrence of a disaster should always be a stimulus to research on how to prevent a recurrence, the greatly preferred objective is to prevent the initial disaster from happening. Unfortunately, the successful prevention of accidents of a type that has never occurred is likely to go unrecognized; until an accident has happened, people tend to be unaware of its possibility.

It has become increasingly evident that the traditional ergonomics of control room design is insufficient to prevent large-scale accidents. Accidents occur because of complex interactions among people at all levels of an organization and between people and plant hardware; they occur despite regulations, training, and operating procedures that are intended to minimize accident potential. Attitudinal variables may play a more important role than has been realized. Management's interest in developing a "safety culture" within a company plant is also a key factor. To understand the causes of accidents and how to prevent them, we need to understand the psychology of a system in its entirety, from the ergonomics of design to the social dynamics of "whistle blowing."

Automation is sometimes seen as a solution to the problem of human error because it removes human operators from the scene. But automation does not necessarily reduce the probability or severity of accidents. When highly skilled operators are removed from an industrial system, the system sometimes loses the protection against design errors that the workers' skill may provide, and the hazardous implications of those design errors may be very difficult to discover before an actual incident. In automated systems, the day-to-day role of humans tends to be the performance of maintenance, and we know that accidents can occur because of faulty maintenance. Zuboff (1988) has described how automation, if not introduced in an appropriate way, can reduce quality of performance.

Designing for Longevity, Recyclability, and Disposability

Safety and usability have long been major objectives of human factors engineers in equipment design. Other objectives that have implications for the management of environmental change include maintainability, repairability, recyclability, and disposability. Such design objectives should increase in importance if environmental issues become of greater concern. The special challenge to human factors is to find ways to satisfy the environmentally oriented objectives without compromising the traditional focus on user safety and convenience.

NEED FOR NEW PERSPECTIVES

Human factors concepts and methods can be applied to societal problems at many levels. One aid to thinking in these terms is the abstraction hierarchy proposed by Rasmussen (1986; see his Figure 4.1). At the lowest level of this hierarchy the focus is on "physical form." Examples of the application of ergonomic design at this level include switches that cannot be turned the wrong way, toilet handles that make it easy to use different amounts of water following urination or defecation (a design that is common in Australia where there is a chronic water shortage), and the interlock that requires one to cover the gas tank filler hole with a cap in some states so as to reduce vapor loss to the environment.

At Rasmussen's next level, "physical function," the emphasis is on localized systems. Designing for energy efficiency and resource conservation is a possibility at this level. An electrical system that automatically turns lights off when a room is unoccupied is one example of such a system. A central heating system that (except when overridden by manual control) adjusts temperature in different parts of a house according to patterns of use is another.

The next level, "general form," would include things like automated guideways for automobile traffic in specific locations and intelligent navigation systems that can reduce fuel consumption by optimizing travel routes. A fourth level, "generalized function," would include the design of complete living and communication systems, including "smart houses." At this level the application of information technology has the potential to change in fundamental ways how people work, travel, communicate, and live.

The top level concerns relatively global problems and goals—the control of global climate change would be a case in point. At this level, issues of politics, ethics, and perhaps philosophical or religious beliefs are likely to be encountered. (Several articles in *Science* in the 1960s cited instances in which people were given tractors and other equipment that would enable them to produce two harvests a year or to till more land; the advantages did not follow, because the dominant view in the culture was that fate, not technology, determines the provision of life's necessities). Difficulties occur because sometimes measures would benefit one country or region of the world at the expense of others. Ethical complications arise because people have different ideas about such questions as the moral responsibility of human beings toward other species and of this generation to future generations.

We have sketched Rasmussen's taxonomy here to make the point that different levels of problems require different kinds of approaches. This fact should be recognized in any consideration of how human factors might be applied to the problem of detrimental environmental change and the host of subsidiary problems that it subsumes. Traditionally, human factors has dealt primarily with problems at the level of the design of specific devices and person-machine systems. This will continue to be important, and such efforts can have significant environmental implications. There is, however,

PREDICTING THE BEHAVIORAL EFFECTS OF INTERVENTIONS

A generic problem is the difficulty of predicting the effects of efforts to change human behavior in the interest of environmental preservation, either directly or through the modification of technology. Actual effects often turn out to be different from what was desired. This point is illustrated by the consequences of some efforts to modify traffic patterns. Bypasses and beltways have been built to take traffic away from congested streets by providing alternative routes so that drivers do not have to go through the center of the city. Sometimes, however, these have increased traffic not just in the existing areas of congestion but in new areas as well. The provision of new roads has encouraged more people to drive into the city, and the appearance of large, empty roads has stimulated the development of housing to make use of them.

Highway safety provides another example of how an effort to reduce an undesirable effect of human behavior can itself have unanticipated consequences for human behavior. When antilock brakes were put into cars, the assumption that driving safety would be improved was sufficiently credible that insurance premiums were reduced for automobiles with this feature. Empirical evidence gathered by a taxi company in Munich indicated no significant decrease in accident frequency but a significant increase in driving speed (there is no speed limit on the autobahn) and in speed changes (Aschenbrenner et al., 1986). Apparently, people drove faster because they believed their braking systems were safer and that as drivers they would be able to cope more effectively with emergencies. As a result of this study, insurance implications were reconsidered.

The general problem was discussed in terms of cybernetic theory almost 40 years ago by Ashby (1956). The problem involves whether to treat self-organizing systems as deterministic or stochastic. It is very difficult to predict how a complex system will evolve if it has many self-organizing characteristics; this complicates the task of the planning of interventions involving such systems. "Self-organizing" here does not imply a conscious or purposive response by the system to the inputs of the planners; it simply connotes a system for which the list of states that can be entered changes suddenly and "spontaneously" from time to time or a system in which the transition probabilities from one state to another are not constant. In general, very large, complex, and tightly coupled systems with many subsystems show such self-organizing properties in response to disturbances, and such results may not be predictable, even stochastically (Moray, 1963).

From the above examples, one might draw the conclusion that the effects of countermeasures identified from a causal analysis may be offset by the effects of boundary-seeking human adaptation. The implication, for present purposes, is that anyone attempting to make design changes in a person-machine system to compensate for a pattern of behavior with undesirable environmental consequences should be aware that the design changes may evoke unanticipated changes in human behavior that could also have undesirable environmental consequences. This problem deserves more study than it has received; it is especially important for efforts to modify human behavior that harms the environment.

CONCLUDING COMMENTS

Many of the most troublesome aspects of environmental change are the direct consequence of human behavior, so it is appropriate that changing that behavior should be high on the list of goals for any program of environmental preservation. But behavior is seldom changed significantly or for very long simply in response to scolding, admonishment, pleading, or even rational explanation of why change is required. Describing what the environment could be in 100 years has little effect on people who feel little or no responsibility toward unborn generations. Systems must be designed that make it difficult to behave in ways that will make things worse.

We have focused here on how human factors research might contribute to shaping technology so that the natural consequences of its use for human ends will be more environmentally benign. We have pointed to a few examples of the kinds of specific research issues that should be addressed. Human factors research can be applied to the goal of reversing undesirable trends in environmental change in other ways as well. It can help extend our understanding of how human behavior causes environmental change; and it can contribute to the development of more effective tools for use in the study of environmental change; and it can help assess the effectiveness of efforts to modify undesirable current trends. Finally, it may be able to contribute in ways that will become clear only when a significant number of human factors researchers turn their attention to this area.

A major challenge for the design of environmentally benign systems is inducing people to see constraints less as constraints than as ways to afford something else. The constraint that we all drive on the same side of the road is recognized as an opportunity for safe travel. The constraint on the voltages available in domestic power supplies is an opportunity for safe and efficient use of appliances. Often, however, when constraints are introduced for safety or health purposes, people react negatively to what they see as encroachments on their freedom of choice. Mandatory seat-belt and motorcycle-helmet laws have sometimes been repealed despite conclusive evidence that seat-belt and helmet use prevents death and serious injury on the highways. Laws prohibiting smoking in public places encountered enormous initial resistance despite the evidence that smoking, including breathing secondhand smoke, is injurious to health. Dealing effectively with these kinds of issues is, at least in part, a human factors problem.

Detrimental environmental change is becoming perceived, by both the scientific community and the general public, as one of the most serious problems that is now faced—and that will continue to be faced for the foreseeable future—not only by individual nations but by the world as a whole. Because this problem has global implications, it should present unusual opportunities for international collaboration among researchers in many countries. If such collaboration is to be effective, the researchers must acquire some new skills and perspectives. It is not safe to assume that what works in one country or culture will work equally well in another. We must learn how to collaborate effectively if we are to have any hope of making real headway on problems that are global in extent. Otherwise, we run the risk of designing the behavioral analogs of very tall smokestacks and exporting various forms of cultural acid rain to other cultures in our efforts to help our own society.

A considerable amount of human factors research has addressed the question of how environmental variables affect human performance; however, the problem of environmental deterioration has not been a prominent focus of human factors research. Human factors, as a discipline, has much to offer to efforts to find solutions to various aspects of this problem. And the problem is sufficiently urgent that even a small probability of making a useful contribution justifies the attempt.

REFERENCES

Aschenbrenner, K.M., B. Biehl, and G.M. Wurm
   1986   Antiblockiersystem und Verkehrssicherheit: Ein Vergleich der Unfallbelaestung von Taxen mit und ohne Antiblockiersystem. (Teilbericht von die Bundesanstalt fur Strassenwesen zum Forschungsproject 8323.) Mannheim, F.R. Germany. Cited in Wilde, G.S. (1988). Risk homeostasis theory and traffic accidents: propositions, deductions, and discussion in recent reactions. *Ergonomics* 31:441-468.

Ashby, W.R.
   1956   *An Introduction to Cybernetics*. London, England: Chapman and Hall.

Baker, L.A., A.T. Herlihy, P.R. Kaufmann, and J.M. Eilers
   1991   Acidic lakes and streams in the United States: the role of acidic deposition. *Science* 252:1151-1154.

Baum, A., and J.E. Singer, eds.
   1981   *Advances in Environmental Psychology*, Vol 3. *Energy in Psychological Perspective* series. Hillsdale, N.J.: Erlbaum.

Bavelas, A., T. Belden, E. Glenn, J. Orlansky, J. Schwartz, and H.W. Sinaiko
1963 *Teleconferencing: Summary of a Preliminary Research Project.* Study S-138. Arlington, Va.: Institute for Defense Analysis.

Bureau of the Census
1990 *Statistical Abstract of the United States.* Washington, D.C.: U.S. Department of Commerce.

Coach, J.V., T. Garber, and L. Karpus
1979 Response maintenance and paper recycling. *Journal of Environmental Systems* 8:127-137.

Cone, J.D., and S.C. Hayes
1980 *Environmental Problems/Behavioral Solutions.* Monterey, Calif.: Brooks/Cole Publishing.

Crabb, P.B.
1992 Comment: effective control of energy-depleting behavior. *American Psychologist* 47:815-816.

Crossen, P.R., and N.J. Rosenberg
1989 Strategies for agriculture. *Scientific American* 261(3):128-135.

Durlach, N.I., and A.S. Mavor, eds.
1995 *Virtual Reality: Scientific and Technological Challenges.* Committee on Virtual Reality Research and Development, National Research Council. Washington, D.C.: National Academy Press.

Geller, E.S.
1981 Waste reduction and resource recovery: strategy for energy conservation. Pp. 115-154 in A. Baum and J.E. Singer, eds., *Advances in Environmental Psychology*, Vol 3. *Energy: Psychological Perspectives* series. Hillsdale, N.J.: Erlbaum.
1986 Prevention of environmental problems. Pp. 361-383 in *Handbook of Prevention.* New York: Plenum.
1987 Environmental psychology and applied behavior analysis: from strange bedfellows to a productive marriage. Pp. 361-388 in D. Stokols and I. Altman, eds., *Handbook of Environmental Psychology.* New York: Wiley.

Geller, E.S., J.L. Chafee, and R.E. Ingram
1975 Promoting paper-recycling on a university campus. *Journal of Environmental Systems* 5:39-57.

Geller, E.S., R.R. Winett, and P.B. Everett
1982 *Preserving the Environment: New Strategies for Behavior Change.* Elmsford, N.Y.: Pergamon Press.

Georgescu-Roegen, N.
1976 *Energy and Economic Myths: Institutional and Analytical Essays.* New York: Basic Books.

Gibbons, J.H., P.D. Blair, and H.L. Gwin
1989 Strategies for energy use. *Scientific American* 261(3):136-143.

Gore, A.
1992 *Earth in the Balance: Ecology and the Human Spirit.* New York: Penguin.

Gray, C.L., Jr., and J.A. Alson
1989 The case for methanol. *Scientific American* 261(5):108-114.

Hamilton, D.P.
1992 Envisioning research with virtual reality. *Science* 256:603.

Hazarika, S.
1986 *Bhopal: The Lessons of a Tragedy.* London, England: Penguin.

Holahan, C.
1986 Environmental psychology. *Annual Review of Psychology* 37:381-407.

Houghton, R.A., and G.M. Woodwell
  1989   Global climatic change. *Scientific American* 260(4):36-44.
Human Factors Society
  1991   *Human Factors Society 1991 Directory and Yearbook.* Santa Monica, Calif.: Human Factors Society.
Humphrey, C.R., R.J. Bord, M.M. Hammond, and S.H. Mann
  1977   Attitudes and conditions for cooperation in a paper recycling program. *Environment and Behavior* 9:107-124.
la Riviere, J.W.M.
  1989   Threats to the world's water. *Scientific American* 261(3):80-94.
Mohnen, V.A.
  1988   The challenge of acid rain. *Scientific American* 259(2):30-38.
Moray, N.
  1963   *Introduction to Cybernetics.* London, England: Burns and Oates.
Myers, N.
  1989   *Deforestation Rates in Tropical Forests and Their Climatic Implications.* London, England: Friends of the Earth.
National Research Council
  1977   *Drinking Water and Health.* Safe Drinking Water Committee, Commission on Life Sciences. Washington, D.C.: National Academy of Sciences.
  1983   *Changing Climate: Report of the Carbon Dioxide Assessment Committee.* Washington, D.C.: National Academy Press.
  1990   *The Improvement of Tropical and Subtropical Rangelands.* Board on Science and Technology for International Development, Office of International Affairs. Washington, D.C.: National Academy Press.
  1991   *Rethinking the Ozone Problem in Urban and Regional Air Pollution.* Committee on Tropospheric Ozone Formation and Measurement, Board on Environmental Sciences and Toxicology. Washington, D.C.: National Academy Press.
Nickerson, R.S.
  1992   *Looking Ahead: Human Factors Challenges in a Changing World.* Hillsdale, N.J.: Erlbaum.
Nilles, J.M., F.R. Carlson, P. Gray, and G.J. Hanneman
  1976   *The Telecommunications-Transportation Tradeoff.* New York: Wiley.
Office of Technology Assessment
  1988   *Urban Ozone and the Clean Air Act: Problems and Proposals for Change.* Washington, D.C.: U.S. Government Printing Office.
Postel, S.
  1985   Thirsty in a water-rich world. *International Wildlife* 15(6):32-37.
Rasmussen, J.
  1986   *Information Processing and Human-Machine Interaction: An Approach to Cognitive Engineering*, Vol. 12. New York: North Holland.
Rasmussen, J., and R. Batstone
  1989   *Why Do Complex Organizational Systems Fail?* Summary proceedings of a cross-disciplinary workshop in safety control and risk management. Washington, D.C.: World Bank.
Reason, J.
  1990   *Human Error.* New York: Cambridge University Press.
Repetto, R.
  1990   Deforestation in the tropics. *Scientific American* 262(4):36-42.
Russell, J.A., and L.M. Ward
  1982   Environmental psychology. *Annual Review of Psychology* 33:651-688.

Saegert, S., and G.H. Winkel
  1990   Environmental psychology. *Annual Review of Psychology* 41:441-477.
Schlesinger, W.H., J.F. Reynolds, G.L. Cunningham, L.F. Huenneke, W.M. Gerrell, R.A. Virginia, and W.G. Whitford
  1990   Biological feedbacks in global desertification. *Science* 247:1043-1048.
Schwartz, S.E.
  1989   Acid deposition: unraveling a regional phenomenon. *Science* 243:753-763.
Senders, J., and N. Moray, eds.
  1991   *Human Error: Cause, Prediction, and Reduction*. Hillsdale, N.J.: Erlbaum.
Shulman, S.
  1989   When a nuclear reactor dies, 98 million dollars is a cheap funeral. *Smithsonian* 20(7):56-69.
Soule, M.E.
  1991   Conservation: tactics for a constant crisis. *Science* 253:744-750.
Steinhart, P.
  1990   No net loss. *Audubon* July:18-21.
Stern, P.C.
  1992   Psychological dimensions of global environmental change. *Annual Review of Psychology* 43:269-302.
Stern, P.C., O.R. Young, and D. Druckman, eds.
  1992   *Global Environmental Change: Understanding the Human Dimensions*. Committee on the Human Dimensions of Global Change, National Research Council. Washington, D.C.: National Academy Press.
Stolarski, R.S.
  1988   The Antarctic ozone hole. *Scientific American* 258(1):30-36.
Stolarski, R.S., R. Bojkov, L. Bishop, C. Zerefos, J. Staehelin, and J. Zawodny
  1992   Measured trends in stratospheric ozone. *Science* 256:342-349.
Thorpe, J.
  1993   Synthetic Environments Strategic Plan. Draft 3B. Defense Advanced Research Projects Agency, Alexandria, Va.
Wallace, D.R.
  1985   Wetlands in America: labyrinth and temple. *Wilderness* 49:12-27.
Wilson, E.O.
  1989   Threats to biodiversity. *Scientific American* 261(3):108-116.
Wise, J.A., and S.F. Savage
  1992   Human factors in environmental management: new directions from the Hanford site. *Proceedings of the Human Factors Society 36th Annual Meeting*. Santa Monica, Calif.: Human Factors Society.
Zuboff, S.
  1988   *In the Age of the Smart Machine: The Future of Work and Power*. New York: Basic.

6

Communication Technology and Telenetworking

Raymond S. Nickerson

INTRODUCTION

Among the more significant events in the recent history of long-distance communication have been the building of computer-based communication networks and the development of technologies that have made possible the implementation and exploitation of these networks. In this chapter we focus on these technologies and on the challenges and opportunities for human factors research that they present.

We begin with a brief historical overview of computer-communications networking technology. We then focus on current trends, especially the phenomenon of "global connectivity" that networking is coming to represent and some of the implications this could have. In the remainder of the chapter we discuss some of the human factors issues and research needs that relate to networking and its future development and applications.

HISTORICAL OVERVIEW

Networks that link computers from different geographical locations are a relatively recent phenomenon. Since the first such networks were implemented, the technology has advanced rapidly. It appears that this rapid advance will continue and that applications of the technology will become increasingly widespread over the near term.

The first experimental networks connecting independent, nonhomogeneous, geographically distributed computers were established in the mid-1960s (Davies and Barber, 1973; Marrill and Roberts, 1966). The ARPANET, which was to become the largest operational network in the world and to remain so for many years, was started as a four-node system by the Advanced Research Projects Agency of the U.S. Department of Defense in 1969 (Heart, 1975; Heart et al., 1978). According to Pool (1993), its successor, the Internet, connected about 1.7 million host computers and between 5 million and 15 million users as of 1993, and the numbers have been doubling annually.

The establishment and proliferation of computer networks have been accompanied—indeed made possible—by an ever-increasing blurring of the distinction between computer and communication technologies. The Internet and the many smaller networks that connect to it depend on computing resources for all aspects of their operation and for the provision of the various services, such as electronic mail and bulletin boards, teleconferencing, information utilities, and the many others that they offer.

Today there are several types of networks: local networks, long-distance networks that use telephone lines, satellite networks that communicate by radio transmission, and network complexes that use a variety of means of transmission. Some networks are designed to connect only the terminals in a single building or office complex; at the other extreme are those that connect facilities in different countries and regions of the world. Networks have been established to serve the interests of government agencies, business corporations, educational institutions, and the general public.

Not only have networks been rapidly increasing in number and size; the bandwidth or "throughput" capacity of the individual links of which they are composed has been expanding greatly as well. Wide-area networks now typically operate at 1.5 megabits (million bits) per second, and many local-area networks have transmission rates of 10 megabits per second. Systems that use optical fibers as the transmission medium support rates of 100 megabits and, in a few cases, 1 gigabit (billion bit).

TECHNOLOGY TRENDS AND EXPECTATIONS

Network enhancements will come from the development of increasingly powerful computing devices, many of which are especially designed for network applications, as well as from improvements in the methods for transmitting information from point to point and from the development of new network configurations and new ways of linking networks together.

Rapidly Increasing Bandwidth

Probably the most significant predictable trend in networking is a continuing increase in network bandwidths at all levels of network operations.

Systems with transmission rates of 10 gigabits per second could be in place by the end of the century or soon thereafter (Kahn, 1987); and there is already speculation that terabit (trillion bit) capacity systems might be feasible in the not-distant future (Partridge, 1990).

Optical fiber will become increasingly used as the information conduit for future systems (Bell, 1989; Desurvire, 1992); it is expected that most homes will have access to broadband fiber networks within 10 to 20 years (Shumate, 1989). Wireless terminals, foreshadowed by cellular telephone technology, will also become more generally available, as will high-resolution color terminals with three-dimensional graphics capability.

If network bandwidth continues to increase at anything like its recent rate, it will soon be feasible to transmit enormous amounts of data (including digital voice and video) at very low cost. As more and more systems acquire the capacity to transmit high-quality speech with little or no compression, digital voice seems certain to be an increasingly practical mode of communication between one person and another through computer networks, and between people and computers (Makhoul et al., 1990). Speech recognition technology is also becoming sufficiently mature to be useful in a variety of contexts (Waldrop, 1988).

Although bandwidth limitations have been a problem for digital speech transmission, the consequences of these limitations have been more severe for picture transmission. Trying to transmit pictorial information through a channel that can handle, say, 50,000 bits per second is a little like trying to fill a swimming pool through a straw. But again, if network bandwidths increase at rates close to those experts have been predicting, digital video transmission will become a practical reality for many applications reasonably soon.

This is not to suggest that there will not be a desire for still greater bandwidth; to date the appetite for increased bandwidth has always managed to stay ahead of the technology's ability to deliver, and there is little evidence that this will change right away, if ever. For present purposes, the important point is that the technology is advancing rapidly, its applications and potential applications multiplying apace.

Global Connectivity

One way to characterize what is happening in telecommunications is to say that the degree to which people everywhere are connected, or could be connected, to other people and to information resources of all kinds, independently of location, is rapidly increasing. The National Research Council's (1988) Computer Science and Technology Board has recently called for the development of an integrated national computer network system that would permit communication between any two computers in the country. This call has been echoed in legislative proposals to make the establishment of high-speed data highways a matter of national priority (Gore, 1991). There is similar interest in the establishment of national networks in other countries around the world. Such networks could be linked by Internet gateways.

The evolving macro-system of interlinked networks can be thought of as one enormous global nexus that has the potential to increase by many orders of magnitude the degree to which individuals and information resources all over the world are interconnected and therefore accessible to each other. Some technologists envision, within the next decade or two, a single worldwide integrated services digital network that would be capable of handling digitally encoded information of any type (data, facsimile, voice, graphics, motion pictures) and that would link offices, schools, and homes to information resources of various sorts (libraries, museums, national and international data banks) throughout the world (Denning, 1989b; Forester, 1987).

What the continuing development of computer networking and the global connectivity it represents will mean is very difficult to say at this point, but it seems a safe bet that the implications—technological, political, social—will be profound. Denning (1989b) believes an emerging worldwide network of computers, which he refers to as "Worldnet," could be a pervasive reality by the year 2000, and he has argued that such a facility would quickly become indispensable to businesses that wish to remain competitive in a networked world. One suspects that the implications for education, for recreation, for politics, and for daily life will be equally great.

Access to Information and to People

A global wideband network has the potential to give individuals unprecedented access to information and information resources independently of their location. Such access will be used to provide the ability to browse through the world's libraries, dial up movies for home viewing, consult interactive encyclopedic information services (including process simulations and manipulable microworlds), make "virtual" visits to museums and other places of interest, study in classrooms without walls (including using international collaboratives for educational purposes), participate in instantaneous polls and referenda, enjoy interactive media ("tell [or show] me more" news and entertainment), and undoubtedly take advantage of possibilities that we cannot now imagine.

The kind of connectivity that computer networks are expected to provide in the future will increase not only access to information and information resources but also access to people, independently of their location. It is to be expected that new patterns of interpersonal communication will emerge from the widespread use of this technology. Already electronic mail, which accounts for a very substantial fraction of the total traffic over computer networks (Denning, 1989a), has significantly changed the communication patterns of many people who use it.

Unlike the telephone, radio, and television, computer networks are used for both point-to-point and broadcast communications. Electronic mail tends to be used for person-to-person communication; that is to say, messages are sent to specified individual recipients although they can also be readily sent to groups of recipients. Electronic bulletin boards are used to post messages that can be read by anyone who has access to the boards on which they are posted. Often this means a fairly large number of people.

New Methods of Information Distribution

The idea of electronic information repositories accessible to the general public is not new. For decades, forward-looking technologists have given visionary accounts of what interaction with such systems could be like (Bush, 1945; Licklider, 1965; Parker, 1973). Parker, for example, envisions the ability of the reader of an electronic newspaper to call up a bibliographic sketch of an individual who is the subject of a news story, to get tutorial information on a topic that is in the news, to do an automatic search of advertisements for items of interest, and, in general, to access information resources that could transform the reading of the news into a much richer experience than it now is.

Advantages of electronic communication of information include speed—information is communicated to everyone within a community of interest essentially instantaneously—and representational versatility—conventional text can be supplemented with animations and process simulations, including those with which the user can interact. In addition, the technology could provide users with a variety of tools and capabilities for working with very large information stores and getting what they want from them, without being burdened with a mass of data in which they have no interest.

Despite the considerable interest in the idea of electronic newspapers, magazines, and journals, not much has been done along these lines to date (although a great deal of "prepublication" information and data are exchanged among scientists via computer networks and much debate of topical scientific questions takes place on electronic bulletin boards that serve specific user communities). One wonders why the idea has not caught on more rapidly. One possibility is that not enough people have the terminal equipment and network access that is needed to make electronic distribution feasible on a large scale. Other impediments to the widespread use of electronic media for the distribution of news and technical information may be general resistance to change, distrust of (lack of confidence in) the medium, unacceptability of the quality of visual displays (as compared with conventional print media), and the intangibility of the medium (people—especially authors—may see printed pieces as more "real" than electronic ones). Research on the question of why people often object to reading print on visual display terminals as opposed to print on paper has provided some leads that need to be explored further (Gould et al., 1987).

Changing Roles and Functions

Given the type of connectivity global telenetworks are expected to provide and the information handling tools that are already beginning to appear, new methods of information distribution are likely to become increasingly widely used. These innovations will change our fundamental ideas about information packaging and dissemination and will have implications for the traditional roles of editors, publishers, librarians, and other information workers. Questions abound. What, in the age of electronics, will constitute a "document" or its electronic analogue? What will publication mean? Who will perform the functions of quality control historically performed by editors and publishers? What will be the nature of a library? What services will it provide?

It seems likely that there will be a continuing need for publishers, librarians, and other "information brokers" in a world in which information is increasingly gathered, stored, distributed, and used in electronic form. The daily tasks that information-handling professionals perform will change, however, as will the nature of the services they provide.

Implications for Energy and Resource Consumption

As we make greater use of electronic means of distributing information that has traditionally been distributed by newspapers, magazines, journals, and books, there could be a decrease in the need for and use of paper. This would be a desirable consequence from an environmental point of view, as it would reduce natural resource use, energy use, and waste production (Chapter 5). Whether increased use of electronic publishing will in fact decrease the demand for paper remains to be seen; to date, there is little evidence that computer technology generally, or the application of this technology to information distribution in particular, has done so (Herman et al., 1989; Nickerson, 1992).

It appears that people strongly prefer to read print on paper rather than words on an electronic visual display. It is not clear to what extent this is more than a matter of discomfort with change from the familiar. Given the potential that electronic distribution of information has for reducing our dependence on paper and the fact that, so far, increased use of electronic distribution has not been accompanied by a decrease in the use of paper, it seems important to attempt to understand better the causal relationships involved. In that way, effective steps can be taken to realize the potential for savings more fully.

Participatory Democracy

Because of the possibility of two-way communication, computer networks could have profound effects on the ways in which decisions are made at various levels of government and on the extent to which citizens participate directly in the making of those decisions (Lemelshtrich, 1990). Not only will instantaneous polls and referenda be possible, but also there will be new types of forums for debate on issues of public interest by anyone who wants to participate.

A hint of how network technology may facilitate spontaneous political communication can be seen from an inspection of the messages that are posted on general-interest electronic bulletin boards. One informal analysis of 1,000 messages posted over a five-week period on a company board revealed that about 14 percent of them were classifiable as political commentary, discussion, or debate. Other major message types were requests for information ("Does anybody know . . . ?"), 21 percent; advertisements (to sell or rent), 16 percent; and unsolicited information (meeting notices, news, service or product recommendations or warnings), 13 percent (Nickerson, 1994).

HUMAN FACTORS ISSUES

Human factors considerations pertaining to communications and telenetworking are important, in part, because eventually nearly everyone is likely to be a user of this technology. Many people will use it in their work, some in relatively traditional work settings and others in radically different contexts. Simply by making all types of information more generally accessible, computer networks greatly increase the possibility of doing "office work" in places other than traditional centralized offices. In some instances, information technology is changing the character of the work that gets done, and it will continue to do so. Telenetworking is itself a major industry that offers job opportunities to many people.

Not all the applications of this technology will be in the workplace. There will be many opportunities to apply it to education, to politics and governance, and to avocational and recreational activities as well. People will have access to computer networks in the future, whether they actively seek it or not, simply by virtue of the fact that televisions will be designed to serve not only as one-way delivery systems for broadcast TV but also as two-way terminals that can both bring information into the home and provide users with the means of transmitting to a network.

The opportunities for human factors research relating to this technology are many and varied. By way of illustrating the diverse nature of the problems, we will focus on three general problem areas: person-computer interaction, person-to-person communication, and what, for want of a better term, we will call "virtualization." We will mention a few research opportunities in each of these areas. The possibilities for human factors involvement in this field include these examples, but also extend way beyond them.

Person-Computer Interaction

Psychologists and human factors researchers have given a great deal of attention to the topic of person-computer interaction. Most of the research that has been done has focused on the question of how to facilitate it, how—through improvements in the designs of interfaces and interactive techniques—to make the interaction more natural and more effective. This problem is of undoubted importance and deserves continued attention.

Terminal and Interface Design

An obvious problem will be improving the design of terminals that give people access to computing resources and of the interfaces through which people communicate with information resources. This has been a major focus of human factors research since computer-based systems began to be used widely by people other than computer scientists and system developers (Nickerson, 1986).

Most terminals today depend primarily on two-dimensional visual displays as output devices and on typewriter-like keyboards, typically complemented with a pointing and drawing instrument, such as a mouse or trackball, for input. The effective exploitation of other input-output modalities and methods will become an increasingly important challenge as the community of users of terminals that provide access to computer networks continues to expand and as technology widens the range of practical input-output options. Speech will become an increasingly feasible option for both input and output, for example (Makhoul et al., 1990; Weischedel et al., 1990). So will "walk-around," three-dimensional, "virtual-reality" representations of objects and environments with which users can interact (Durlach and Mavor, 1995). How fully the technological possibilities in these and related areas are realized will depend, to a large extent, on how effectively the numerous human factors issues that pertain to them are addressed.

Information Finding and Utilization

Realization of the potential benefits that can come from the connectivity to information resources provided by computer networks will require the development of a variety of users' tools. Physical access to very large information repositories—the electronic equivalent of major libraries, for example—will be of little value unless users have effective methods for getting at the information they need or want without spending an undue amount of time and effort in unproductive search. As Bromley (1986:628) has put it:

> More information than we can ever conceivably want will be available to us within seconds. What do we do with it? How do we condense, correlate, and sort it so that humans can base decisions on it? This is one of the major challenges facing all science, all society.

There already exist many large electronic databases that serve a variety of purposes and special interests: electronic funds transfer, airline reservations, stock price quotations, credit card and check authorization, crime investigation, scientific research, and numerous others. Databases that hold the results of DNA sequencing research or the astrophysical data collected by space-probing satellites are growing very rapidly and are expected to continue to do so for the foreseeable future. Ensuring prospective users easy access to these databases is important, and achieving this goal becomes increasingly difficult as the amount of data in the databases grows.

There is a need for tools that will facilitate information access both in the narrow sense—access by specialists to the focused databases that serve their special interests—and in the broad sense—access by nonspecialists to information that is available to the public through news media, libraries, and general-purpose information services to which anyone can subscribe. There is also opportunity for innovative work on information representation and presentation—hypertext systems that include multimodal representations provide a hint of the possibilities—and on the design of navigation aids to help users move around effectively and efficiently in multidimensional data-rich environments. For more on the topic of information access and organization, see Chapter 7.

Personal Information Management

Most people, over the course of their lives, acquire a variety of types of information that they need or desire to retain for their personal use or reference: legal documents, medical records, financial papers, recipes, books, letters, pictures, and so on. All of this information is, in principle, storable in digital form, and in the future more and more of it will be delivered and retained electronically (or photonically). Also, because of the existence of computer networks and the connectivity to information sources they represent, people may acquire much more information that they wish to retain for personal use than they do now.

If people are to efficiently manage large amounts of information in personal electronic repositories, if they are to keep electronic files accurate, timely, and retrievable and not find this a burdensome chore, they will need some tools and methods designed for this purpose. Simply replacing paper files with electronic files does not ensure greater accessibility to information. Although the computer gives one the potential for handling much larger databases much more efficiently, it does not rule out the possibility of creating chaos. Moreover, because of the ease with which electronically stored information can be erased, either intentionally or inadvertently, special precautions must be taken to protect personally valued information.

The design of information systems to serve the personal needs of individuals represents a challenge to both software producers and human factors specialists. And this will become increasingly important as more and more people gain access to computer networks and to the ever-expanding collection of information sources to which they connect.

In addition to the need for tools to facilitate the organization of electronic information for retention, maintenance, and ease of access and use, there will be a need for new methods for coping with the information overload that connectivity to electronic mail, electronic bulletin boards, and other information resources can produce. Heavy users of such resources often develop their own techniques for keeping the amount of information they have to attend to within acceptable bounds. The need for approaches that are demonstrably effective and usable by nontechnical users will grow as the potential connectivity of the ordinary citizen to information sources of various sorts increases.

Attitudes and Beliefs About Computers

One question that has not received the attention it deserves is, what attitudes and beliefs do people have about the computer systems with which they interact and how are those attitudes and beliefs are affected by their interactions? Weizenbaum (1976) has expressed amazement at, and considerable discomfort with, the seriousness and intensity of the interactions that many people had with his Rogerian "Eliza" program. He and others have questioned the advisability of giving people the impression that the systems with which they communicate are smarter than they really are.

There is some evidence that people are more willing to disclose information, including information about socially frowned-upon behavior or attitudes, on a computer-administered questionnaire than in a face-to-face interview (Kiesler and Sproull, 1986; Sproull and Kiesler, 1991). There are several possible explanations for this, but so far they must be considered speculative.

Several observers have raised the question of how the ubiquity of computers and computer-based systems, and their increasing ability to do things that were once considered uniquely human, will affect the way we perceive ourselves (Turkle, 1984; Roszak, 1986). The more that people with little understanding of how computers work have occasion to interact with these systems, the more relevant and important this question becomes. What can be done to increase the likelihood that people's conceptualizations of what computer-based systems can do and how they do it are reasonably accurate, or at least not inaccurate in destructive ways?

Person-to-Person Communication

In the past, human factors researchers gave more attention to person-computer communication than to computer-mediated communication between and among people. This is not surprising, as the facilitation of interpersonal communication was not seen as a primary application of computer technology until fairly recently. With the development and rapid growth of computer networks, however, it has become apparent that facilitating communication between and among people is one of the most powerful applications that this technology can have. Among the obvious examples of such applications are electronic mail, electronic bulletin boards, and computer-based teleconferencing.

Electronic Mail

Electronic mail (E-mail) is a new form of interpersonal communication that has been made possible by the existence of computer networks. It should be studied from a human factors point of view. To date, E-mail facilities have been used primarily by people who are "computer literate," in particular those who use computers regularly either in their work or for avocational purposes. It seems highly likely, however, that the use of E-mail will increase in the future and that the user community will include many people who do not now use computers for either work or play.

There are many research questions relating to E-mail and its use. Who uses it? And for what purposes? How accessible is it? How much training do people who do not use computers for other purposes need in order to be able to use E-mail effectively? How do the usability and usefulness of specific E-mail systems depend on the details of the systems' designs and operating characteristics? Does computer-mediated communication have any unique characteristics, and if so, what are they? To what extent does it replace other means of communication for its users? What effects, if any, does E-mail have on the operation of organizations whose members use it daily? How does interpersonal communication through E-mail resemble or differ from communication via other media?

Clearly, electronic messages do not contain many of the nonlinguistic cues (e.g., "body language") that often convey nuances of feelings and attitudes of participants in face-to-face or voice interactions, and they typically lack most of the clues to senders' status or position that are often found in letterheads or signature blocks (Sproull and Kiesler, 1991). Computer-mediated communication systems also tend to be opaque to personal characteristics—unattractive physical appearance, speech impediments, behavioral anomalies—that sometimes, unfortunately, inhibit effective communication on a person-to-person basis (Zuboff, 1988). Similarly, they filter out the advantages that aspects of appearance—physical attractiveness, an authoritative voice, an imposing demeanor—can provide. In general, they have been seen by many observers as user "equalizers" in the sense that they mask many of the factors that differentiate people in face-to-face contact (Hiltz et al., 1980; Vallee et al., 1974; Zuboff, 1988).

Sproull and Kiesler point out that the weakening of social differences is not unique to computer-mediated communication, but is seen, albeit to lesser degrees, in other technologies (1991a:43):

> The telephone eliminates visual cues and therefore reduces one's ability to deduce the other person's social position and to grasp the importance of social differences in the interaction. Over the telephone, though, one retrieves some social information in nonvisual form. The secretary who answers or places calls, variations in standard ways of greeting, and pauses and tone of voice all convey social information.

Several years ago, Uhlig (1977) suggested the need for an etiquette of computer-based message technology that would probably differ from the etiquette that applies to more traditional forms of communication. Has such an etiquette emerged? Is one emerging? Is E-mail less constrained by traditional social conventions than is face-to-face or voice communication? Does it evoke more extreme positions and the venting of anger more openly (Kiesler, 1984; Kiesler et al., 1984; Short et al., 1976)?

Sproull and Kiesler (1991) suggest that because E-mail is relatively impoverished in social cues and shared experience, it lends itself to communication in which the participants produce messages that display less social awareness than face-to-face or voice communication. Reduced social awareness is seen in "messages characterized by ignoring social boundaries, self revelation, and blunt remarks" (p. 39). People tend to be more open and less inhibited when using electronic mail than in face-to-face or voice communication, Sproull and Kiesler argue, because the possibility of a critical audience is less apparent in the former case: "Because a person composing an electronic message lacks tangible reminders of his or her audience, the writer can easily forget the forms appropriate for communicating with that audience" (p. 49).

We know that interpersonal communication—face to face, by telephone, or in writing—is by far the most common form of office activity (Bair, 1987; Helander, 1985; Panko, 1982). We know too that a large part of the typical manager's job is communication, much of which occurs in meetings and group settings (Mintzberg, 1973). Whether or not widespread use of E-mail will affect the amount of time devoted to communication, patterns of communication that did not exist before will probably emerge. One can speculate about this, but it would be useful to have some observational studies of the patterns of use of actual E-mail communities, of which many already exist. Because this form of communication is new, new observational and experimental techniques may be needed to study it. The general question of how computer and communication technologies will affect the ways in which people communicate with each other is extremely important and deserves considerable research effort (National Research Council, 1984).

One question of special interest is how the use of E-mail and related technologies can be expected to affect the productivity of their users. At this early stage in the development and use of these facilities, it is not possible to say anything definitive on this question. One might guess that use of E-mail would improve productivity, since E-mail can greatly facilitate communication among all the members of a functional group. Of course, it would be naive to assume that all the E-mail communication that is done directly serves the goals of the organization; however, there may be indirect benefits even from the facilitation of communication for purely personal purposes. Some preliminary data on the question have been reported. Sproull and Kiesler (1991) found a high correlation between the degree to which a group used E-mail to coordinate its activities and the group's productivity. They note that correlation does not demonstrate a cause-effect relationship, but they appear to believe there to be one.

Electronic Bulletin Boards

An electronic bulletin board can be thought of as a special form of electronic mail, and much of what has been said in the preceding section about E-mail applies to electronic bulletin boards as well. The electronic bulletin board is a sufficiently important innovation, however, to warrant special mention.

Like a conventional bulletin board, an electronic bulletin board is a public medium; messages posted on it can be read by anyone who has access to it. Also like conventional bulletin boards, electronic bulletin boards serve a variety of communities of users. The users of a conventional bulletin board might be the occupants of a building, the patrons of a store, the residents of a neighborhood; those of an electronic bulletin board might be the employees of a corporation, the students and faculty of a university, the users of a computer network service.

Perhaps the most obvious advantage of electronic bulletin boards over conventional bulletin boards is that users do not have to go to a particular place to read or post messages. Because boards can be accessed from essentially anywhere, the user community for a particular board can be defined by common interests rather than by a common location. Other advantages include the ease with which the postings on an electronic bulletin board can be scanned, excerpted, corrected, elaborated, and processed in many other ways.

Individuals can use electronic bulletin boards in a variety of ways and to varying extents. One may read notices only, or one may post some as well. Some users read all the notices that are posted, checking the postings at fairly regular intervals; others scan them only occasionally. Replies to "Does anybody know?" questions are sometimes posted; often they are sent privately to the questioners via E-mail. Patterns of communication that spontaneously emerge include multiperson dialogues, perhaps initiated by a question or comment that evoked responses from several people.

The existence of electronic bulletin boards could have a great effect on the way people communicate with each other in the future. As more and more people acquire access to computer networks, they will also be acquiring access to widely distributed communities with common interests. What this will mean to people's daily lives remains to be seen. One opportunity for human factors research is the observational study of the evolution of this new mode of communication. Another is developing an understanding of this technology and of the needs and preferences of its users as a step toward promoting design decisions that will enhance the usefulness of electronic bulletin boards to the general population.

Teleconferencing

The idea of using network technology to hold conferences among "attendees" located in different places has been of interest since the early 1960s (Bavelas et al., 1963). Despite this long-standing interest and despite the implementation of several experimental systems, teleconferencing has not yet become widely used, even when face-to-face meetings involve the expense and inconvenience of considerable travel. This may be due to a combination of technical and nontechnical factors (Nickerson, 1986).

Some of the technical limitations of teleconferencing systems are being eased as networks acquire sufficient bandwidth to support the real-time transmission of sufficiently high-fidelity video representations of participants' images to create a realistic impression of an actual gathering. Whether this will ensure the maturing of teleconferencing into an effective and widely used technology remains to be seen. It seems likely that the acceptability of this technology will depend, to a large degree, on the kinds of users' tools that teleconferencing systems offer to facilitate group problem solving and decision making.

The dynamics of electronic "meetings" differ from those of face-to-face meetings in a variety of ways. For example, many of the factors that determine which participants play dominant roles in face-to-face meetings appear to be less influential in electronic meetings. These include gender (McGuire et al., 1987) and status (Dubrovsky et al., 1991). One might expect similar diminution of the influence of age, race, and physical appearance.

Sproull and Kiesler (1991) suggest that consensus is more difficult to achieve with electronic than with face-to-face groups. Is this because people are more susceptible to peer pressure in a face-to-face group than when communicating electronically? If so, the possibility arises that a face-to-face meeting is more likely than an electronic one to result in what *appears* to be a consensus, but in fact is not.

It seems reasonable to expect that electronic meetings will differ substantially from face-to-face meetings for some time to come, perhaps indefinitely. We would also expect, however, that the dynamics of electronic meetings will change as the technology continues to be developed and enhanced. Ideally, we would like the technology to develop in such a way that electronic meetings come to have all, or most, of the advantages of face-to-face meetings and some others that come from the availability of facilitative tools as well. Realization of this desire will require a better understanding than we now have of the factors that can determine the relative effectiveness of working groups.

Virtualization

As used here, the term *virtualization* connotes the development of capabilities that make it possible for people to function at a distance from needed resources or to interact with simulated representations of objects and environments much as they would with the real things. Capabilities of these types are basic to such ideas and phenomena as telecommuting, computer-supported cooperative work, and virtual or artificial realities.

Telecommuting

Whether information technology, generally, has yet had a significant impact on productivity in the workplace is a matter of some debate (see Chapters 1 and 8). The picture is clouded in part by (a) the difficulty of measuring productivity—especially white-collar productivity—in completely unambiguous ways and (b) the fact that information technology, by virtue of its pervasiveness, has many indirect effects that are hard to isolate and track.

There seems little doubt, however, that information technology, particularly computer-based communication networks, has the potential to enhance productivity greatly. The need to maintain large inventories of materials, parts, and finished products has a negative effect on productivity because it represents a major component of the cost of getting products into the hands of users. "Just-in-time" manufacturing reduces the need for such inventories, but it depends on fast and effective communication techniques. The substitution of the transmission of information for the movement of people and material is another way in which computer networks can contribute to increased productivity, because this means the delivery of the same services at the expenditure of fewer natural or economic resources.

Using computer networks to enable people to work at home or in "virtual offices" outside traditional centralized office buildings has been of interest for some time. The idea of teleworking, or "telecommuting," was promoted by Nilles et al. (1976) shortly after the first oil crisis of the early 1970s, partly on the grounds that bringing jobs to people electronically, rather than transporting people to jobs, would have the doubly beneficial effect of conserving energy and saving transportation costs. It would also help the environment by reducing air pollution from vehicle emissions, traffic congestion, and office space requirements.

According to one estimate, perhaps as many as 15 or 16 million people in the United States could be considered teleworkers as of 1989 (Martin, 1989). However, the percentage of the workforce, especially the white-collar workforce, that could work from home is believed to be many times larger than is currently doing so (Harkness, 1977; Kraut, 1987). How fully the potential for telework will be realized will depend, in part, on how effectively a variety of human factors issues are addressed. Some of these issues have to do with the design of devices and software—the tools that make telework technically feasible. Others, however, relate to less tangible, but no less important, aspects of the work situation that help determine not only productivity but worker satisfaction.

Telework has become a symbol of liberation to some people and of isolation to others (Huws et al., 1990). How it is perceived depends on the individual worker's circumstances and the details of the arrangement between worker and employer. Among the issues that appear to contribute to worker satisfaction with telework are whether the worker is considered an independent contractor or a company employee, whether work is monitored electronically, and how compensation is determined. This is not to suggest that all workers prefer the same arrangement (Chamot and Zalusky, 1985; Gregory, 1985). Telework situations differ from more traditional work arrangements in many respects, some obvious, some subtle; studies are needed to provide a better understanding of the variables that determine productivity and job quality from the worker's point of view.

Computer-Supported Cooperative Work

With the help of computer networks, colleagues can cooperate at a distance in ways that were impossible until fairly recently. People in widely separated locations can collaborate, for example, in real time on writing a paper; all the authors can critique the same draft and have the benefit of all the critiques as soon as they are made. They can, in effect, share the same "writing surface" despite their geographical separation.

Network technology has the potential to bring expertise to bear on problems that are located someplace other than where the expert or experts happen to be. A team of experts, all located at different places, might collaboratively address a problem requiring their expertise. In theory, at least, high-resolution displays, coupled with tele-operator control technology could make it possible for a remotely located surgeon not only to give advice to on-site personnel but even to perform operating procedures.

How far such innovation can be taken remains to be seen. There is little doubt that the ubiquity of computer networks will make expertise, like many other things, less constrained by space and time than it now is. The effectiveness with which geographically separated individuals will be able to collaborate, via computer networks, on complex tasks that draw on their combined skills will depend, to no small degree, on how well the many human factors issues relating to the design of the underlying systems are resolved.

Telepresence

The telephone created a telepresence of sorts. When two people talk on the phone, they are in each other's presence in a real, if rudimentary, sense, even if they are located half a world away from each other. The higher the quality of the voice transmission and the fewer the interruptions, the greater the sense of presence is. Film and television can also create a sense of presence; watching a film or televised event, one can sometimes get the sense of "being there" to the extent of forgetting that what one is looking at is a picture on a screen.

One goal of virtual reality technology is to increase the sense of presence considerably beyond what the current state of the art of communication technology permits. It would add, for example, the sense of touch and would also give the recipient of the sensory information the ability to move the sensing devices around in the experienced context. By moving one's head and eyes, for example, one would cause the optical sensors to move in a corresponding way. This would give the recipient control over where to look, much as one would have if at the remotely located scene.

As the bandwidth of computer networks continues to increase, it will be possible to transmit an increasingly detailed and veridical representation of a physical situation to a remotely located individual. It seems unlikely, however, that it will be possible, at least anytime soon, to represent most nontrivial situations in sufficient detail that one could not tell the virtual reality from the real thing. Fortunately, this degree of realism is not necessary for most applications, but the question of how real (in appearance) is real enough is open and probably must be answered on a case-by-case basis. Representations of reality that would be more than adequate for some applications might be inadequate for others. There is a need for work on the question of how to determine the degree of fidelity required in specific instances.

CONCLUDING COMMENTS

Our main purpose in this chapter has been to point out that computer networking and associated developments have profound implications for human communication and represent some significant challenges and opportunities for human factors research. We have considered only a few of the research questions that arise and these only in a cursory way. The need to focus more attention on computer-based networking will increase as the technology for it continues to develop and its applications continue to expand.

In keeping with the overall theme of this book, we have focused primarily on the future of networking and the prospects of extensive connectivity of people to information resources of many types and of people to people independently of geographical location. It should also be noted, however, that there are many opportunities for human factors work on the networking systems that currently exist. The evidence suggests that the currently operating electronic information resources are not utilized as effectively as they could be even by people who have easy access to them and could presumably benefit from making greater use of them (General Accounting Office, 1989). In this regard, see also Chapters 4 and 7.

There can be little doubt that the increasingly widespread use of communications technology in the workplace is changing the nature of many jobs and the knowledge and skill requirements for performing them. This technology is affecting jobs throughout industry and at all levels of responsibility—from clerical workers who have to master a variety of computer-based document preparation and information management tools to high-level managers who have to learn to deal with new patterns of intra- and intercompany communication, more fluid organizational structures, and challenges stemming from the innovative exploitation of communication technology by competitors. Studies are needed both to help us understand the changes that are taking place and to anticipate the likely job-skill requirements of the future.

But the effects of computer networking will be felt far beyond the workplace. We need to better understand its implications for education; for civic, political, and personal decision making; for leisure and daily life outside the workplace. What might it mean for people who are confined to their homes, to hospitals, to nursing homes, to prisons? How can it be exploited to give shut-ins greater access to other people, to information resources, and to the world in general?

What must be done to ensure the usefulness and usability of this technology? Special and general-purpose databanks will be increasingly accessible through computer networks; information services will proliferate, as will network-accessible expert systems, electronic consultants, and electronic advice givers. How can such resources be made approachable for people who are intimidated by computer technology, for people who cannot type, for people who are disadvantaged in one way or another? How can we minimize the chances that these resources will widen the gap between haves and have nots? Between better-educated and less well-educated people? Between people who are well connected to computer resources and people who are not?

What are some of the major risks inherent in this technology? What are the computer network analogues to nuisance telephone calls? To mail and telemarketing fraud? What safeguards can be built against them? How should people be encouraged to think about networks, data structures, and information spaces? What sorts of metaphors or mental models will be useful and not misleading and detrimental?

Communication is a fundamental human activity. Technology has affected it in numerous ways in the past. The proliferation of computer networks has the potential to affect it greatly in the future. Already it is possible to identify many human factors questions that are raised by telenetworking and associated developments; many more are likely to arise as the impact of the continuing development of these technologies on our daily lives becomes increasingly evident.

REFERENCES

Bair, J.H.
    1987   User needs for office systems solutions. Pp. 177-194 in R.E. Kraut, ed., *Technology and the Transformation of White Collar Work.* Hillsdale, N.J.: Erlbaum.

Bavelas, A., T. Belden, E. Glenn, J. Orlansky, J. Schwartz, and H.W. Sinaiko
    1963   *Teleconferencing: Summary of a Preliminary Research Project.* Study S-138. Arlington, Va.: Institute for Defense Analysis.

Bell, T.E.
    1989   Telecommunications. *IEEE Spectrum* 26(1):41-43.

Bromley, B.A.
    1986   Physics: natural philosophy and invention. *American Scientist* 74:622-639.

Bush, V.
    1945   As we may think. *The Atlantic Monthly.* 176(July):101-108.

Chamot, D., and J.L. Zalusky
    1985   Use and misuse of workstations in the home. Pp. 76-84 in M.H. Olson, ed., *Office Workstations in the Home.* Board on Telecommunications and Computer Applications, National Research Council. Washington, D.C.: National Academy Press.

Davies, D.W., and D.L.A. Barber
    1973   *Communications Network for Computers.* New York: Wiley.

Denning, P.J.
    1989a   The science of computing: the ARPANET after 20 years. *American Scientist* 77:530-534.
    1989b   The science of computing: Worldnet. *American Scientist* 77:432-434.

Desurvire, E.
    1992   Lightwave communications: the fifth generation. *Scientific American* 266(1):114-121.

Dubrovsky, V., S. Kiesler, and B. Sethna
    1991   The equalization phenomenon: status effects in computer-mediated and face-to-face decision making groups. *Human Computer Interaction* 6:119-146.

Durlach, N.I., and A.S. Mavor, eds.
    1995   *Virtual Reality: Scientific and Technological Challenges.* Committee on Virtual Reality Research and Development, National Research Council. Washington, D.C.: National Academy Press.

Forester, T.
    1987   *High-Tech Society: The Story of the Information Technology Revolution.* Cambridge, Mass.: MIT Press.

General Accounting Office
    1989   *Cancer Treatment: National Cancer Institute's Role in Encouraging the Use of Breakthroughs.* Washington, D.C.: General Accounting Office.

Gore, A.
    1991   Infrastructure for the global village. *Scientific American* 265(3):150-153.

Gould, J.D., L. Alfaro, R. Finn, B. Haupt, and A. Minuto
    1987   Reading from CRT displays can be as fast as reading from paper. *Human Factors* 29:497-507.

Gregory, J.
    1985   Clerical workers and new office technologies. Pp. 112-124 in M.H. Olson ed., *Office Workstations in the Home.* Board on Telecommunications and Computer Applications, National Research Council. Washington, D.C.: National Academy Press.

Harkness, R.C.
1977 *Technology Assessment of Telecommunications-Transportation Interactions.* Menlo Park, Calif.: Stanford Research Institute.

Heart, F.
1975 The ARPANET network. In R.L. Grimsdale and F.F. Kuo, eds., *Computer Communication Networks: 1973 Proceedings of the NATO Advanced Study Institute.* Leyden, Netherlands: Noordhoff International Publishing.

Heart, F., A. McKenzie, J. McQuillan, and D. Walden
1978 *ARPANET Completion Report.* Cambridge, Mass.: Bolt Beranek and Newman.

Helander, M.G.
1985 Emerging office automation systems. *Human Factors* 27(1):3-20.

Herman, R., S.A. Ardekani, and J.H. Ausubel
1989 Dematerialization. Pp. 50-69 in J.H. Ausubel and H.E. Sladovich, eds., *Technology and Environment.* National Academy of Engineering. Washington, D.C.: National Academy Press.

Hiltz, S.R., K. Johnson, C. Aronovitsh, and M. Turoff
1980 *Face-to-Face Versus Computerized Conferences: A Controlled Experiment.* Report 12. Newark: New Jersey Institute of Technology.

Huws, U., W.B. Korte, and S. Robinson
1990 *Telework: Towards the Elusive Office.* New York: Wiley.

Kahn, R.E.
1987 Networks for advanced computing. *Scientific American* 257(4):136-143.

Kiesler, S.
1984 Computer mediation of communication. *American Psychologist* 39:1123-1134.

Kiesler, S., and L.S. Sproull
1986 Response effects in the electronic survey. *Public Opinion Quarterly* 50:402-413.

Kiesler, S., J. Siegel, and T.W. McGuire
1984 Social psychological aspects of computer-mediated communication. *American Psychologist* 39:1123-1134.

Kraut, R.E.
1987 Social issues and white-collar technology: an overview. Pp. 1-21 in R.E. Kraut, ed., *Technology and the Transformation of White-Collar Work.* Hillsdale, N.J.: Erlbaum.

Lemelshtrich, N.
1990 The expression of opinions through the new electronic mass media: an experimental and cybernetic view. In N. Moray, W.R. Ferrell, and W.B. Rouse, eds., *Robotics, Control and Society.* New York: Taylor and Francis.

Licklider, J.C.R.
1965 *Libraries of the Future.* Cambridge, Mass.: MIT Press.

Makhoul, J., F. Jelinek, L. Rabiner, C. Weinstein, and V. Zue
1990 Spoken language systems. *Annual Review of Computer Science* 4:481-501.

Marrill, T., and L.A. Roberts
1966 Cooperative network of timesharing computers. Pp. 425-431 in *Proceedings of the AFIPS 1966 Sprint Joint Computer Conference.* Arlington, Va.: American Federation of Information Processing Societies Press.

Martin, A.
1989 There's no place like home . . . to work. *Human Resource Executive* July:50-51.

McGuire, T., S. Kiesler, and J. Siegel
1987 Group and computer-mediated discussion effects in risk decision making. *Journal of Personality and Social Psychology* 52:917-930.

Mintzberg, H.
 1973 *The Nature of Managerial Work.* New York: Harper and Row.
National Research Council
 1984 *Research Needs on the Interaction Between Information Systems and Their Users: Report of a Workshop.* Committee on Human Factors. Washington, D.C.: National Academy Press.
 1988 *Toward a National Research Network.* National Research Network Review Committee, Computer Science and Technology Board. Washington, D.C.: National Academy Press.
Nickerson, R.S.
 1986 *Using Computers: Human Factors in Information Technology.* Cambridge, Mass.: MIT Press.
 1992 *Looking Ahead: Human Factor Challenges in a Changing World.* Hillsdale, N.J.: Erlbaum.
 1994 Electronic bulletin boards: a case study in computer-mediated communication. *Interacting With Computers* 6:117-134.
Nilles, J.M., F.R. Carlson, P. Gray, and G. Hanneman
 1976 Telecommuting—an alternative to urban transportation congestion. *IEEE Transactions on Systems, Man, and Cybernetics* 6:77-84.
Panko, R.
 1982 Serving managers and professionals. Pp. 97-103 in *AFIPS Office Automation Conference Proceedings.* Arlington, Va.: American Federation of Information Processing Societies Press.
Parker, E.
 1973 Technological change and the mass media. Pp. 619-645 in I. Pool, W. Schramm, F. Frey, N. Maccoby, and E. Parker, eds., *Handbook of Communication.* Chicago, Ill.: Rand McNally.
Partridge, C.
 1990 A faster data delivery. *Unix Review* 8(3):43-48.
Pool, R.
 1993 Beyond databases and e-mail. *Science* 261(August):841-843.
Roszak, T.
 1986 *The Cult of Information: The Folklore of Computers and the True Art of Thinking.* New York: Pantheon Books.
Short, J., E. Williams, and B. Christie
 1976 *A Social Psychology of Telecommunications.* New York: Wiley.
Shumate, P.W., Jr.
 1989 Optical fibers reach into homes. *IEEE Spectrum* 26(2):43-47.
Sproull, L., and S. Kiesler
 1991 *Connections: New Ways of Working in the Networked Organization.* Cambridge, Mass.: MIT Press.
Turkle, S.
 1984 *The Second Self: Computers and the Human Spirit.* New York: Simon and Schuster.
Uhlig, R.P.
 1977 Human factors in computer message systems. *Datamation* 23(4):120-126.
Vallee, J., R. Johansen, R.H. Randolph, and A.C. Hastings
 1974 *Group Communication Through Computers.* Vol. 2: *A Study of Social Effects.* Report R-33. Menlo Park, Calif.: Institute for the Future.

Waldrop, M.M.
  1988   A landmark in speech recognition. *Science* 240:1615.
Weischedel, R., J. Carbonell, B. Grosz, W. Lehnert, M. Marcus, R. Perrault, and R. Wilensky
  1990   Natural language processing. *Annual Review of Computer Science* 4:435-452.
Weizenbaum, J.
  1976   *Computer Power and Human Reason: From Judgment to Calculation.* San Francisco: Freeman.
Zuboff, S.
  1988   *In the Age of the Smart Machine: The Future of Work and Power.* New York: Basic Books.

7

Information Access and Usability

Christopher D. Wickens and Karen S. Seidler

INTRODUCTION

We are a society increasingly driven by information. Accurate, up-to-date information underlies the effective and progressive functioning of governmental, economic, and social institutions, impacting everything from the focus and quality of public policy decisions to the competitiveness of industrial research and development. In this modern reality—sometimes labeled the information society (e.g., Salvaggio, 1989) or the postindustrial society (Bell, 1973)—knowledge and information become the primary resources, displacing manufactured goods and agricultural products as key commodities (Stonier, 1983).

A number of trends associated with the evolution of an information society beg attention from the human factors community. One trend is the greatly expanded quantity of information available to people in all professions. For the scholar, this translates into a 3 percent growth per year in the United States alone in the number of journals, professional conferences, and proceedings (King et al., 1981). Worldwide, the growth is even more rapid. This makes keeping abreast of current research on even a restricted topic daunting, and scholars are increasingly unsure of the extent to which they have located relevant literature, let alone digested it. Lawyers have witnessed a comparable overwhelming growth in the records they can consult for legal precedents. In the medical profession, the limited ability of physicians and other health professionals to extract from the literature relevant information on patient care, teaching, and research has engendered so much concern (e.g., Huth, 1989) that a new subfield—medical informatics—has been created to address issues related to the development and operation of medical information systems (Hewins, 1990). Somewhat ironically, this subdiscipline is generating its own growing body of literature (see Chapter 4).

In industry, the exponential increases in system complexity have also resulted in voluminous documentation and procedures to support system operations and maintenance. The documentation required by the F-18 aircraft alone contains 300,000 pages. In nuclear facilities, it is not unheard of for manuals of procedures for dealing with accidents to exceed hundreds of pages. For the consumer, a growing wealth of information on products and costs is at least potentially available. In the face of this data-rich environment, users are experiencing frustration in locating, accessing, tracking, and interpreting the information that is germane to their concerns.

New information technologies, though, offer viable and powerful ways to manage information collections. Among these innovations are: (1) greatly enhanced storage and access capabilities (e.g., CD-ROM), (2) expanded communication systems that allow faster and greater connectivity, (3) hypermedia developments that allow greater flexibility in information representation and database "navigation" (Gluschko, 1990), (4) the advanced graphics capabilities that make direct manipulation interfaces and data visualization possible (Newby, 1992), and (5) developments in artificial intelligence to support "intelligent" systems and interfaces that can, for example, infer user information needs and support dialogue in more natural language (Allen, 1991).

Yet advances in collection, storage, and dissemination technology do not ensure successful use of electronic information repositories. Muckler (1987) has documented the low "hit rate" obtained in typical library database searches, in terms of both the relevant documents that fail to be located (misses) and the unnecessary retrieval of many documents of limited usefulness or relevance (false alarms). In one case a computer search through 10 databases retrieved 16,816 abstracts that were identified as relevant for a particular topic. Only 166 were. More disturbing was the fact that in an independent, nonelectronic search through the literature, a user located 177 useful references that had not been found by any of the database searches. In a study of user success with a legal database, Blair and Maron (1985) found a similar pattern of results, with only a 20 percent recall of relevant documents. Even more alarming, though, was the finding that searchers believed they were recovering 75 to 80 percent of relevant documents. This finding is consistent with a general conclusion reached by MacGregor et al. (1987) that users overstated the success of their information retrieval from electronic databases.

Of course, there have been positive interactions as well. Wilson et al.

(1989) report a dramatic example: in eight incidents the lives of patients were saved because a search of the MEDLINE database uncovered information that led the physicians to change the course and direction of their treatment procedures. As we increasingly attend to the difficulties users have with information technologies, such a positive synergy between user and machine will become more common.

Yet this attention will need to extend beyond the technically sophisticated user. A growing proportion of the labor force will perform information-handling activities (i.e., production, maintenance, and interpretation) in the future. We have already seen a substantial growth in the number of white-collar workers (information handlers), from less than 18 percent of the workforce in 1900 to over half today (Koenig, 1990). And since communication and computing technologies have allowed data to be made directly available in the office, workers are increasingly being asked to confront the information stores without the aid of expert human search intermediaries (e.g., librarians) familiar with the database systems. Further, the growth in numbers of end users is not limited to the workplace. Electronic information databases are becoming more common in everyday private life. With the purchase of a modem and an on-line service, the average person can use a personal computer to tap into large databases at home and carry on many personal activities directly (e.g., make airline reservations, shop at home, conduct bibliographic searches). This means that a wide range of skills and tasks are being brought to the computer interface. These more casual or infrequent users are less likely to possess the intellectual resources or motivation to invest in a time-consuming learning process. The implications are that information system designers cannot rely on user perseverance or high technical skills to overcome poor interface design. Instead, user friendliness will become critical in attracting and keeping end users.

In summary, the issues may be represented by three major concerns. First, users must often confront enormous quantities of information in order to identify and utilize the small portion that is ultimately relevant and meaningful to them. Second, the interface and the database representation designed to manage the vast information collections may often not be compatible with the user's mental model of how the system should work, leading to user frustration and abandonment of the system. Third, even if a user is satisfied with the on-line operation of a system, he or she may be less satisfied with its output if many unwanted items are retrieved (i.e., high recall but low precision). Alternatively, although the user may be satisfied with the information retrieved, that satisfaction could be unwarranted because of an unsuspected high miss rate.

From a human factors perspective, we may think about the problems of information access and retrieval outlined above as resulting from the user's requirement to map an information need (sometimes ill defined) onto a very large database whose contents and structure may or may not be immediately apparent. That is, information is somewhat "hidden" in the database. Lucky (1989) describes this as a problem of acquiring meta-information (information about information). That is, users must determine whether desired information is contained within the database, how it is represented, where it is located, and how to access it. Exploiting the new information technologies to address these problems defines a host of important human factors research issues for the next decade, such as the following: What is the optimum display interface through which the user can understand what information is available? What is the appropriate control/display interface through which the user can explore the information base? How strongly and how well can intelligent systems draw correct inferences about human needs and wishes and how well can those systems help the user formulate and refine those wishes? How can one adequately assess performance with information systems, since user satisfaction may not correlate with recall and precision indices?

In the following pages, we discuss these issues as they pertain to three selected classes of databases: (1) the restricted database, the well-structured database for a particular system in a restricted domain; (2) the fluid database, the less-structured database of scientific, academic, or general knowledge; and (3) the base constituted by scientific data. We recognize that the three classes are not discrete; that is, they share many overlapping features relating to information accessibility and interpretability. In fact, we view the restricted and fluid databases along a continuum that reflects the degree to which the database organization can be determined by a user prior to actual use. For convenience and parsimony of expression, we also take the liberty of using the term *electronic database* in its broadest sense, that is, to represent any type of electronic information repository (e.g., an on-line bibliographic system, a videotext system, a computerized maintenance manual). We conclude this chapter by focusing in detail on the set of human factors research issues that are common to all of these areas.

DATABASE REPRESENTATION ISSUES

Restricted Domains

The most structured information-representation domain, and the one most readily studied from a human factors viewpoint, is one in which the knowledge pertains to the functioning of a single, well-structured system—a restricted domain. Examples include the on-line help system for a word-processing system, the airline flight reservation systems (Boehm-Davis et al., 1992), the onboard in-flight library of information about an aircraft and its surrounding airspace conditions (Curran, 1991), all of the entries in a maintenance document for a complex piece of equipment, and the contents of an industrial inventory. These systems represent the electronic counterparts to such conventional paper-based items as manuals, descriptions of procedures, schedules, encyclopedias, and handbooks.

In the restricted domain, the contents of the database often are less of a mystery to the user than is the location of an information item within the system and the means of accessing it. In the hard-copy counterparts, users may consult a table of contents or an index to locate information; if the document is alphabetically arranged (e.g., a dictionary), its structure is immediately apparent and users can infer the location of information. Alternatively, users might choose to thumb through the document or browse to find information of interest. To keep track of their place when cross-referencing, they might bend down corners of pages or use highlighters or bookmarks. Advanced electronic systems sometimes have analogous capabilties.

In the electronic version of an information system, the issue is how to design the interface so that the user can (a) readily locate needed items of information and (b) access the items efficiently (this might include needing to shift rapidly between sets of "related" items of information). The first need has to do with the *organizational structure* of the database (e.g., hierarchical, matrix, network; see Durding et al., 1977), and the second, with the *navigational tools* for moving from entry to entry (Seidler and Wickens, 1992). The organizational structure may or may not be independent of the navigational tools. For example, a database may be organized in a strict hierarchical fashion, but navigational tools may allow the user to directly access any node in the database from any other node with a single command, thus bypassing the hierarchical relationships. This contrast is closely related to that between menu-driven search and key word search.

Much of the empirical investigation relevant to these areas has focused on such issues as menu organization and key word systems (Norman, 1991; Shneiderman, 1987). For instance, the issue of the trade-off between menu depth (the number of levels in the hierarchy) and breadth (the number of items per level) has been a popular research topic (e.g., Kiger, 1984; Miller, 1981; Snowberry et al., 1983). The general consensus of these studies has been that search time lessens and accuracy improves as breadth is increased and depth decreased. Increasing depth seems to increase the likelihood of becoming "lost" in the information space. Other menu studies (e.g., Card, 1982; Giroux and Belleau, 1986) have looked at the ordering and organization of menu items within a page (e.g., alphabetical, semantic, or random) and have found systematic arrangements better than random ordering of items. There have also been studies looking at the advantages of adding menu descriptors (e.g., Latremouille and Lee, 1981; Snowberry et al., 1985); these have been found to improve performance when applied to upper levels of the menu hierarchy only. Most of these studies of menu organization, though, have used relatively small, homogeneous, and often abstract databases (Fisher et al., 1990) unlike those typically found in operational settings. Also, search questions were generally goal-directed, which means that results may be inappropriate for understanding browsing behavior.

Research is still needed on how to organize both the structure and the navigational tools in a way that is compatible with both the users' mental model of the database and the task-defined needs to access the information in a particular sequence. A classic study by Roske-Hofstrand and Paap (1986) revealed that F-16 pilots used an in-flight information menu more readily when its organizational structure corresponded to their mental model of which systems were related to each other; the pilots were less likely to use the menu when the menu designers' own plan guided the structure. Prior to obtaining user input on the question of structure, the designer should be able to define the set of options. This can be difficult because there are so many different ways in which groups of "information nodes" could be organized. For an aircraft, they might be related in terms of system components: thus, all entries pertaining to a common system (e.g., fuel-use instruments) would be close together or commonly grouped. Alternatively, they might be grouped according to phase of flight, so all instruments that are frequently consulted in the same flight phase (and hence frequently used in sequence) would be located together, even if they relate to entirely different systems. There is little or no guidance as to whether one organizational structure is preferable, how either might relate to the users' mental model, how homogeneous these mental models may be across different users, or whether the organizational structure for one task (e.g., normal operation) might differ substantially from the structure needed for a different task (e.g., fault diagnosis and trouble-shooting). It is possible that these different structures can exist in parallel through implementation of networks or redundant nodes.

Allowing the user to tailor or structure the organization adaptively as he or she sees fit is an attractive approach. Yet this solution has three potential dangers not always considered when making design recommendations:

- The user might not know the ideal organization because all possible scenarios in which information is needed cannot be anticipated a priori.
- There may be undesirable set-up costs in establishing configurable displays.
- Individual structuring will lead to a lack of consistency across different databases.

This last concern is crucial, as has been understood by those considering the lack of consistency across such systems as word-processing applications, keyboards, and aircraft (Andre and Wickens, 1992; Wiener, 1989). What should be the trade-off, then, between flexibility and adaptability on the one hand, and consistency on the other? It may be that having too much freedom to structure information differently in a system is as bad as having too little. Human cognition is somewhat adaptable, and it may well be that the user can adapt to a non-ideal but fixed information structure more rapidly and effectively than the user can mold or adapt the structure to his or her ideal.

Identifying the elements that foster this adaptation is the challenge. It requires making the organizational structure and navigational procedures readily apparent to the user, so that users always know where they are in the database relative to where they want to be. This issue of visibility of database structure to the user is addressed by the implementation of electronic database "maps," whose utility has been well validated (e.g., Vincente and Williges, 1988). Major design challenges still confront the human factors community in addressing how such maps can spatially represent the vast number of nodes and entries in large databases in a way that allows the user to visualize the full structure readily (Mackinlay et al., 1991; Shneiderman, 1987).

Fluid Domains

When dealing with information in a restricted domain, the user generally must access or "unpack" specific pieces of information that exist at nodes of the database and whose identity is known before the search begins. For example, a technician might need to look up a particular maintenance procedure. The contents of the restricted-domain databases are fairly well defined in advance by the system designer. In contrast, many domains have boundaries that are far less concretely defined, and the organizational structure within these domains is much less well described by either a systems hierarchy or a simple n-dimensional matrix. Instead, the relationships between information sources in the fluid database, such as a library's holdings, derive meaning from the information needs and use context. The task in using such databases may extend beyond "unpacking" information within a node to understanding the relationships between items in the database. As examples, we may consider the on-line bibliographic retrieval of articles about a particular topic (Newby, 1992); books of a user's preference (Pejtersen, 1989); the set of legal decisions, statutes, and regulations bearing on a particular type of case; the set of aircraft incident reports with common features listed in the Aviation Safety Reporting System (Williams et al., 1992); and a set of medical cases with similar typologies (Hubbard et al., 1987; see also Chapter 4).

A database of this sort is characterized by the fact that a single organizational framework is *not* typically imposed at the outset of data generation, or, if one is, then it must evolve to accommodate the unpredictable growth of information. This contrasts with the computer help system or the maintenance information system mentioned in the previous section; there the components and functions are specified in advance by design and, therefore, possess a fairly concise hierarchical or network structure. Instead, the fluid databases are contributed to by various people (or events) over time, have no tightly constrained structure or boundary, and are employed by users who may have a very divergent set of task needs. Any efforts by one individual to impose a particular parsimonious taxonomy will probably defeat other users with different needs and information uses.

Consider, for example, a database of human factors research. Different people will have very different criteria regarding the boundaries of such a database. Should it include the social psychology of multiperson interaction? Personality types as they affect this interaction? Physiology? Management and organizational practices? Correspondingly, how should the information sources in the data be organized? Their multidimensionality defies classification into a simple hierarchical framework. Four logical dimensions of organization of the human factors data immediately come to mind. That is, such data can be organized according to:

- the domain of applicability (e.g., computer, transportation, consumer product),
- the functions addressed (e.g., control, communications, monitoring),
- the relevant psychological components (e.g., perception, memory, attention), and
- the chronological order of the study.

Other dimensions are also possible.

The difficulty in organizing and condensing the entire contents of the fluid database makes a menu structure generally unwieldy as a search mechansim (although menus have sometimes been employed in front-end database aids and some very specialized domain databases, such as CANSEARCH, an expert system to search cancer therapy literature, described by Pollitt, 1987). Furthermore, menus do not offer search power for complex queries. Instead, the command mode, usually via key word entry, places the user in an active search role (Shneiderman, 1987) and is commonly used to enact the search request. In order to generate sets and subsets of information items, a user constructs a query statement that is then compared against a document representation via some information retrieval technique (Belkin and Croft, 1987). The information system may employ any one of a number of information retrieval techniques. The most common method used today in large operational systems is the exact match, whereby the users' query statement must exactly match the document representation, often described by key words, in order for a document to be recovered. This method may involve Boolean, string, or full-text search. Exact match techniques generally result in high-precision retrieval but low recall. Other retrieval methods—partial match techniques—compare the query with a document or group of documents represented as sets of features or index terms. These terms are assigned a priori by either human indexers or machine-automated processes. Documents are then ranked or recovered based upon some measure of the similarity (e.g., probabilistic, vector-space, cluster, fuzzy set, spreading activation) between the query and document representation.

Some key word searches are performed relatively easily. These generally encompass what has been termed known-item (Blair, 1984) or data retrieval (Vigil, 1985) searches, such as when a user seeks a specific known title, author, date, or ISBN number of a record. Other types of searches (topical or subject searches), requiring more complex strategies and problem-solving behaviors, are generally less effective (Muckler, 1987). This is largely because users have trouble translating concepts into the formal logic of the computer. In indexed databases, the key words may not correspond to the user's representation of the terms. In full-text databases, there may be problems with synonymy and imprecise language. In both, key word systems users often experience difficulty in constructing and manipulating Boolean statements.

To assist users in their search tasks, artificial intelligence techniques are increasingly being applied to information system design. Successful prototype systems to date have already automated many of the mechanical aspects of the search process that were problematic for users. Examples include guidance in connecting to an on-line service or database, downloading results, and translating the user query into a Boolean statement (Hawkins, 1988). The more intellectual components of the search process, such as the formulation of the search query and selection of a search strategy, however, are much less well understood, although they have received attention. For these latter research attempts, a major focus has been describing the knowledge people use when searching for information, with the goal of incorporating this knowledge into the expert system (e.g., Croft and Thompson, 1987).

Other research has examined various aspects of the cognitive processes involved in information search, including learning (e.g., Polson and Friedman, 1988), problem solving (e.g., Chen and Dhar, 1991), memory (e.g., Prasse et al., 1988), and comprehension, again with the aim of supporting information systems design. For instance, in their research on problem-solving behavior in information search, Chen and Dhar (1991) identified five strategies used during on-line catalog search and reported improved performance with a system redesigned to respond to these strategies. Other problem-solving research has included identification of search tactics (Bates, 1986), development and selection of search strategy (Brooks et al., 1986), identification of heuristics used in search formulation (Harter and Peters, 1985), and identification of query formulation subtasks (Harter, 1986). Some research has also focused on developing user models that can be incorporated into an intelligent system to enhance its capability to respond to the varied needs and the inherent flexibility of individual users (e.g., Brooks et al., 1986). Modeling knowledge acquisition has also been of interest as a way to facilitate machine learning that will allow intelligent systems to build their own capabilities over time.

Drenth et al. (1991) noted, however, that these models have rarely focused on anything beyond task identification and documentation of procedures. Our understanding of these cognitive processes remains limited. We have no comprehensive picture of the knowledge and expertise underlying the intellectual component of the search process, in part because of the failure to attend to methodological issues. Much of the behavior involved in human-computer interaction is covert. If we are to explore information retrieval behavior within a research paradigm reflecting realistic operational settings and constraints, we must identify and use nontraditional techniques (e.g., verbal protocol analysis, critical incident technique; Flanagan, 1954). Little attention to date has been devoted to exploring these alternative approaches or to developing innovative techniques to elicit and analyze data.

Other considerations must also be addressed before flexible, powerful, and intelligent intermediaries can truly be implemented and made available for use by the general population. A critical but rarely addressed issue concerns the value of information. What attributes define the subjective value or usefulness of information to the user (Morehead and Rouse, 1985)? How much responsibility should the computer be given for deciding (and, therefore, providing) what is of value, and what is not? How much information (title? abstract?) should be provided to users to help define the relevant set? Can the user specify the necessary attributes of value in advance of a search in such a way that an expert system could meaningfully use these to guide the search and retrieval process?

A somewhat related issue concerns the evaluation of retrieval system performance. We have noted hints and some documentation of the less than adequate performance of many database searches (Mann, 1987) but no formal procedure for evaluating their performance level is available. Even a simple signal detection theory analysis of hits and misses begs the question of what is considered signal and what is noise (Granda and Halstead-Nussloch, 1988; Morehead and Rouse, 1985). It may be that the evaluation of retrieval performance needs to encompass the *use* of the information, that is, how the information accessed affects task decisions (see Hewins, 1990).

Evaluating retrieval performance becomes particularly complex when the users' goals are not restricted to locating a number of cases with certain attributes and examining each in detail but include understanding more global aspects of the database, such as the correlations or constraints within the database—for example, whether human factors studies involving group processes are related primarily to training or to performance. This kind of understanding is often achieved through browsing, and good performance metrics have not been established for browsing.

Our discussion of fluid domains has focused on the information search process. As in restricted domains, it is reasonable to suppose that search performance will be supported if effective graphic displays to support visualization of the database are available. However, three features of fluid databases present challenges for the design of such displays. First, the size of fluid databases may be overwhelming, so that very little of the relevant database can be viewed at any one time, and still less of it can be reviewed in any detail (e.g., the text that might describe a particular case). Consider, for example, how extensive is the human factors database of all articles published in the *Human Factors Journal*, *Human Factors Proceedings*, *Ergonomics*, and *Applied Ergonomics*. Second, database browsing is best done flexibly, in real time, so that sometimes a user can "unpack" and examine a particular item or set of items while at other times he or she might merely want to note the location of the items in relationship to other items (i.e., the user should be able to "zoom in" and "zoom out" to various levels of detail). This capability might, for example, help define "clusters" or "use groups," or it might allow the user to determine the correspondence (or lack of correspondence) between the independent data sets of the same underlying space. For example, the density of researchers actively working in certain areas within a particular field could be overlaid on a representation of the frequency of problems calling for research in that field, in order to determine if there is a mismatch and how that mismatch might best be addressed. Third, the structure of such data may both be ill defined (i.e., different from user to user) and have a complex relational structure that does not easily lend itself to a two-dimensional visual rendering. For example, it is clear that many of the terms relating items within a database have "analog" characteristics so that the spatial proximity between items is a meaningful semantic concept. The degree of similarity between the domain represented by two cases (e.g., studies or incidents) is one example. Two legal cases may deal with identical circumstances or with circumstances that have varying levels of similarity. Analog relations might also be defined in terms of the date of court proceedings, the level of the court, and so on. These analog characteristics make "distance" a relevant concept for the database study. But how distance in a multidimensional space is best displayed in a way that is meaningful to and interpretable by the user is not well understood. A three-dimensional structure calls into play issues of three-dimensional graphics (to be discussed below), while spatial dimensionality greater than three will not easily be envisioned. Recent advances in computer graphics displays, such as the Information Visualizer System, the "data wall," and the "cam tree," give innovative solutions to these problems (Card et al., 1990; Clarkson, 1991). The data wall, for example, is a three-dimensional rendering of a wall (actually a perspective building) upon whose surfaces reside visible information nodes. These can be organized according to any two-dimensional structure and examined at any level of distance (or detail) by zooming in and out. The cam tree allows visual representation and examination of a three-dimensional hierarchical database. Newby (1992) has explored issues in navigating through the three-dimensional spatial representation of bibliographical databases.

The issue of the visual representation of large-scale databases brings us to the third form of database: one in which the data are those revealed by scientific inquiry (e.g., scientific experiments, meta-analysis of experimental results, survey data); this defines a set of research issues related to *scientific visualization*.

Scientific Visualization

The availability of massive amounts of scientific data in domains such as meteorology, geology, demographics, and microbiology, coupled with the power of computer graphics, has enabled the scientist to visualize data in a variety of sophisticated forms (Zorpette, 1989). Scientific visualization both allows and exploits such capabilities as flexibility of representation (e.g., two-dimensional vs. rotating three-dimensional views, various color-coding options), high-dimensionality renderings, dynamic updating, and multi-format representation such as windowing text, graphics, and sound. However, the rapid proliferation of such graphics capabilities has somewhat outstripped the development and application of human factors principles for their appropriate employment.

In fact, a review of the literature reveals very few empirical studies that have evaluated the effectiveness of different forms of data representation for the understanding of complex data sets (Jensen and Anderson, 1987; Liu and Wickens, 1992; Merwin and Wickens, 1991). This paucity of data is, in part, understandable. The advanced display technology is often complex, expensive, and hard to tailor in a way that is consistent with experimental protocol. Furthermore, it is extremely challenging to develop experimental questions that would correspond to the kinds of loosely structured, often exploratory questions that scientists typically ask of their data. The "task analysis" of scientific inquiry is not well understood.

HUMAN FACTORS ISSUES

In the previous section, we described three general domains of information representation, each of which can potentially be served by advanced interactive display technology. We now directly consider the human factors and human performance research issues associated with this technology.

Cognitive Task Analysis

Some research effort must be devoted to carrying out comprehensive task analyses of users interacting with complex databases. This includes users of complex, ill-structured information databases performing tasks of a browsing nature as well as scientists attempting to gain insight from complex data (Langley et al., 1987). The need for cognitive task analysis has direct relevance for the research issues on mental models, automation, and expert systems that are suggested below.

Mental Models

Inquiry must continue on the mental models that users have of both restricted and fluid databases (Carroll and Olson, 1987). This remains a key issue because of our current awareness that the effectiveness of database retrieval depends on how congruent the organization of the database and the navigational mechanism for traversing it are to the user's mental models (Allen, 1991; Seidler and Wickens, 1992). Yet how these models should be assessed, how flexible they are within and between users, and how different mental models should be "averaged" across users all remain issues of considerable uncertainty.

User Models, Automation, and Expert Systems

Distinct from the user's model of the system is the system's model of the user. This knowledge is essential for the effective development and implementation of automation and expert systems that will support information accessibility. The issue cuts across all three domains discussed above. From a computer's point of view, the issue is: "What is the user trying to achieve at this moment and how can I best satisfy those needs?" It is clear that such answers must be based on insight gained from research in the two previous areas.

Flexibility Versus Consistency

Automated systems, as we described above, will attempt to adapt themselves flexibly to suit the momentary needs of the human user. Yet too much adaptability may well be counterproductive. If a well-intentioned, adaptive change in carrying out a command occurs when the user does not expect (or need) it, considerable harm could result. Yet the research community to date has provided little guidance on the level of sophistication at which intelligent adaptability may stop being beneficial and may actually be harmful. When, in short, is consistency better than flexibility?

Methodology

In order to grasp the intricacies underlying information retrieval in all three domains, we must continue to develop and assess ways to explore user behavior in contexts reflecting realistic operational constraints. Also acutely needed are meaningful, practical metrics that can be used to consistently evaluate performance within and across systems. These metrics will have to be sensitive to system usability (see Shneiderman, 1987), quality of information, problem context, and resolution.

Greater attention to these methodological issues will allow studies to complement one another, leading, over time, to the development of comprehensive theories of human-computer interaction that can be used to guide effective system design.

The Spatial Metaphor

In each domain, we have argued that there are analog dimensions of relatedness, similarity, and ordering that may often be best served within a spatial framework. This, of course, is definitely true with many aspects of scientific visualization, such as the geosciences, in which Euclidian space is an underlying dimension of the data.

Given also that humans have extensive familiarity navigating and manipulating in space, a strong argument can be made for exploiting the spatial designs of databases and information networks. Yet these design decisions bring with them a number of unresolved issues relating to navigation and lostness (Billingsley, 1982; Gluschko, 1990; Mackinlay et al., 1991). Some examples of research issues are the following:

• How should information for very large databases be portrayed, when users may need the flexibility to zoom in to higher levels of detail? What levels of detail and/or abstraction are necessary?
• What sort of options should a user have to navigate through an information base? Should these be defined by spatial coordinates, nodes, or key words? Should there be constraints on motion (e.g., only along x, y, and z coordinate axes)? How many user options for navigation should be

available? What are the best control devices for an operator to navigate through this "virtual information space" (Card et al., 1990)?

• What are the best ways of conveying information in a dimensionality greater than two? If some representation of three-dimensional space is used, what depth cues are necessary to provide a salient sense of depth for different kinds of databases (Wickens et al., 1989)? Should depth be absolute or ordinal? When should perspective be used to convey information regarding a third dimension and when should color or intensity be employed instead? When should the user be presented with two two-dimensional plans or views? When the dimensionality expands beyond three, the display-formatting issues grow exponentially. Other important research issues relate to the assignment of display dimensions to their referents (i.e., semantic dimensions). Are some assignments better than others? Are some display dimensions best for representing categorical, rather than continuous, semantic dimensions?

• The issue of spatial representation is closely tied to the related concept of virtual reality (Eglowstein, 1990; Pausch, 1991), the attempt to render all aspects of database interaction, not just the visual display characteristics, in a three-dimensional spatial mode. These use such features as direct hand position sensing with simulated tactile feedback, spatially localized sound sources, and an "inside-out" perspective. Scores of potential applications have been proposed and, in some instances, demonstrated: from exploratory surgery to architectural design to scientific inquiry to education. Certainly human factors concerns relate to the "level of reality" that best supports performance. It is, for example, well recognized in both the training and the system design community that there are times when more reality does *not* support better performance; at those times performance is better supported by more abstract displays or discrete controls (Hutchins et al., 1985; Jones et al., 1985; Wickens, 1992). Other human factors issues pertain to fidelity trade-offs that may be imposed by hardware limitations. Can some image resolution be sacrificed to obtain faster speed? Or should the opposite trade-off be preferred (Pausch, 1991)? There is also an intriguing possible trade-off between performance and learning. To what extent do the features of virtual reality that support better performance within a virtual world actually inhibit the learning and long-term retention about the more abstract properties of that world (Wickens, 1992)?

Training Issues

The issue of training overlaps with issues of information access and utilization in at least three respects. First, it is apparent that adequate use of information technology will often depend upon users being trained early in their careers about the value and importance of computer-based information systems. This lesson has been well learned in the medical profession, in which large numbers of health care professionals, untrained in medical information services, now fail to make use of the advantages those services provide (see Chapter 4).

The second issue relates back to user models. Different levels of user experience may dictate very different structuring needs of information systems (Allen, 1991). Hence, intelligent interfaces should be quite attentive both to the user's level of knowledge about the domain represented in the database and to the user's knowledge of the interface itself. Finally, there is the issue of the use of the interface tool itself to train or educate learner-users about the information it contains. As we have reported elsewhere (Wickens, 1992), there really is very little hard evidence that the extensive automated flexibility of an information database improves users' ability to master the information in it.

REFERENCES

Allen, B.L.
    1991   Cognitive research in information science: implications for design. *Annual Review of Information Science and Technology* 26:3-37.

Andre, A.D., and C.D. Wickens
    1992   Compatibility and consistency in display-control systems: implications for aircraft decision aid design. *Human Factors* 34(6): 639-653.

Bates, M.J.
    1986   Subject access in online catalogs: a design model. *Journal of the American Society for Information Science* 37:357-376.

Belkin, N.J., and W.B. Croft
    1987   Retrieval techniques. *Annual Review of Information Science and Technology* 22:109-145.

Bell, D.
    1973   *The Coming of the Post-Industrial Society*. New York: Basic Books.

Billingsley, P.A.
    1982   Navigation through hierarchical menu structures: does it help to have a map? Pp. 103-107 in *Proceedings of the 26th Annual Meeting of the Human Factors Society*. Santa Monica, Calif.: Human Factors Society.

Blair, D.C.
    1984   The management of information: basic distinctions. *Sloan Management Review* 26(1):13-23.

Blair, D.C., and M.E. Maron
    1985   An evaluation of retrieval effectiveness for a full-text document retrieval system. *Communications of the ACM* 28(3):289-299.

Boehm-Davis, D.A., R.W. Holt, and A.C. Schultz
    1992   The role of program structure in software maintenance. *International Journal of Man-Machine Studies* 36:21-63.

Brooks, H.M., P.J. Daniels, and N.J. Belkin
    1986   Research on information interaction and intelligent information provision mechanisms. *Journal of Information Science: Principles and Practice* 12(1):37-44.

Card, S.K.
1982 User perceptual mechanisms in search of computer command menus. Pp. 190-196 in *CHI '82 Conference on Human Factors in Computer Science*. New York: Association for Computing Machinery.

Card, S.K., J.D. Mackinlay, and G.G. Robertson
1990 The design space of input devices. Pp. 117-124 in *CHI Proceedings*. New York: Association for Computing Machinery.

Carroll, J.M., and J.R. Olson, eds.
1987 *Mental Models in Human-Computer Interaction*. Committee on Human Factors, National Research Council. Washington, D.C.: National Academy Press.

Chen, H., and V. Dhar
1991 Cognitive process as a basic for intelligent retrieval systems design. *Information Processing and Management* 27(5):405-432.

Clarkson, M.A.
1991 An easier interface. *BYTE* February:277-282.

Croft, W.B., and R.H. Thompson
1987 I^{3}R: a new approach to the design of document retrieval systems. *Journal of the American Society for Information Science* 36(6):389-404.

Curran, J.
1991 Multidimensional ELS User Interface. SAE Technical Paper Series 912108. Presented at the Aerospace Technology Conference and Exposition, September 23-26, Long Beach, Calif.

Drenth, H., A. Morris, and G. Tseng
1991 Expert systems as information intermediaries. *Annual Review of Information Science and Technology* 26:113-153.

Durding, B.M., C.A. Becker, and J.D. Gould
1977 Data organization. *Human Factors* 19:1-14.

Eglowstein, H.
1990 Reach out and touch your data. *BYTE* July:283-290.

Fisher, D.L., E.J. Yungkurth, and S.M. Moss
1990 Optimal menu hierarchy design: syntax and semantics. *Human Factors* 32:665-683.

Flanagan, J.C.
1954 The critical incident technique. *Psychological Bulletin* 51(4):327-358.

Giroux, L., and R. Belleau
1986 What's on the menu? The influence of menu content on the selection process. *Behavior and Information Technology* 5:169-172.

Gluschko, R.J.
1990 *Hypertext: Prospects and Problems for Crew System Design*. CSERIAC SOAR-90-22. Wright-Patterson Air Force Base, Ohio: CSERIAC.

Granda, R.E., and R. Halstead-Nussloch
1988 *The Perceived Usefulness of Computer Information Sources: A Field Study*. IBM Technical Report TR00.3495. Poughkeepsie, N.Y.: IBM.

Harter, S.P.
1986 *Online Information Retrieval: Concepts, Principles, and Techniques*. Orlando, Fla.: Academic Press.

Harter, S.P., and A.R. Peters
1985 Heuristics for online information retrieval: a typology and preliminary listing. *Online Review* 9(5):407-424.

Hawkins, F.
1988 *Human Factors in Flight*. Brookfield, Vt.: Gower.

Hewins, E.T.
 1990 Information need and use studies. *Annual Review of Information Science and Technology* 25:145-172.
Hubbard, S., J.E. Henney, and V.T. DeVita, Jr.
 1987 A computer data base for information on cancer treatment. *New England Journal of Medicine* 316:315-318.
Hutchins, E.L., J.D. Hollan, and D.A. Norman
 1985 Direct manipulation interfaces. *Human-Computer Interaction* 1(4):311-338.
Huth, E.J.
 1989 The underused medical literature. *Annals of Internal Medicine* 110(2):99-100.
Jensen, C.R., and L.A. Anderson
 1987 Comparing three dimensional representation of data to scatterplots. Pp. 1174-1178 in *Proceedings of the 31st Annual Meeting of the Human Factors Society*. Santa Monica, Calif.: Human Factors Society.
Jones, E.R., R.T. Hennessy, and S. Deutch, eds.
 1985 *Human Factors Aspects of Simulation*. Committee on Human Factors, National Research Council. Washington, D.C.: National Academy Press.
Kiger, J.L.
 1984 The depth/breadth trade-off in the design of menu-driven user interfaces. *International Journal of Man-Machine Studies* 20:201-213.
King, D.W., D.D. McDonald, and N.K. Roderer
 1981 *Scientific Journals in the United States: Their Production, Use and Economics*. Stroudsburg, Pa.: Hutchinson Ross.
Koenig, M.E.D.
 1990 Information services and downstream productivity. *Annual Review of Information Science and Technology* 25:55-86.
Langley, P., H.A. Simon, G.L. Bradshaw, and J.M. Zytkow
 1987 *Scientific Discovery: Computational Explorations of the Creative Processes*. Cambridge, Mass.: MIT Press.
Latremouille, S., and E.S. Lee
 1981 The design of videotex tree indexes: the use of descriptors and the enhancement of single index pages. Pp. 65-112 in D. Phillips, ed., *Telidon Behavioral Research II: The Design of Videotex Tree Indexes*. Ottawa, Canada: Department of Communications.
Liu, Y., and C.D. Wickens
 1992 Visual scanning with or without spatial uncertainty and divided and selective attention. *Acta Psychologica* 79:131-153.
Lucky, R.W.
 1989 *Silicon Dreams: Information, Man and Machine*. New York: St. Martin's Press.
MacGregor, D., B. Fischhoff, and L. Blackshaw
 1987 Search success and expectations with a computer interface. *Information Processing and Management* 23(5):419-432.
Mackinlay, J.D., G.G. Robertson, and S.K. Card
 1991 The perspective wall: detail and context smoothly integrated. Pp. 173-179 in S.P. Robertson, G.M. Olson, and J.S. Olson, eds., *Human Factors in Computing Systems. Reaching Through Technology. CHI '91*. New York: Association for Computing Machinery.
Mann, T.
 1987 Computer searches. Pp. 80-102 (Chapter 9) in *A Guide to Library Research Methods*. New York: Oxford University Press.

Merwin, D.H., and C.D. Wickens
1991 Comparison of 2D planar and 3D perspective display formats in multidimensional data visualization. In *Proceedings of the International Society for Optical Engineering.* Bellingham, Wash.: SPIE.

Miller, D.P.
1981 The depth/breadth tradeoff in hierarchical computer menus. Pp. 296-300 in *Proceedings of the 25th Annual Meeting of the Human Factors Society.* Santa Monica, Calif.: Human Factors Society.

Morehead, D.R., and W.B. Rouse
1985 Computer-aided searching of bibliographic databases: on line estimation of the value of information. *Information Processing and Management* 21:387-399.

Muckler, F.A.
1987 The human-computer interface: the past 35 years and the next 35 years. In G. Salvendy, ed., *Cognitive Engineering in the Design of Human-Computer Interaction and Expert Systems.* Proceedings of the Second International Conference on Human-Computer Interaction, Honolulu, Hawaii. Amsterdam, Netherlands: Elsevier.

Newby, G.B.
1992 An Investigation of the Role of Navigation for Information Retrieval. ASIS Meeting.

Norman, K.
1991 *The Psychology of Menu Selection.* Hillsdale, N.J.: Erlbaum.

Pausch, R.
1991 Virtual reality on five dollars a day. Pp. 265-269 in *Proceedings CHI '91.* New York: Association for Computing Machinery.

Pejtersen, A.M.
1989 *The Book House: Modeling Users' Needs and Search Strategies as a Basis for System Design.* RISO-M-2794. Roskilde, Denmark: Riso National Library, Department of Information Technology.

Pollitt, A.S.
1987 CANSEARCH: an expert systems approach to document retrieval. *Information Processing and Management* 23(2):119-138.

Polson, M.C., and A. Friedman
1988 Task-sharing within and between hemispheres: a multiple-resources approach. *Human Factors* 30(5):633-643.

Prasse, M.J., M. Dilon, M.J. Gordon, B. Mortland, and A. Repka
1988 F-TAS: a full-text access system. Pp. 327-332 in M.E. Williams and T.H. Hogan, comps., *Proceedings of the 9th National Online Meeting.*

Roske-Hofstrand, R.J., and K.R. Paap
1986 Cognitive networks as a guide to menu organization: an application in the automated cockpit. *Ergonomics* 29(11):1301-1311.

Salvaggio, J.L.
1989 *The Information Society: Economic, Social, and Structural Issues.* Hillsdale, N.J.: Lawrence Erlbaum.

Seidler, K.S., and C.D. Wickens
1992 Distance and organization in multifunction displays. *Human Factors* 34:555-569.

Shneiderman, B.
1987 *Designing the User Interface: Strategies for Effective Human-Computer Interaction.* Reading, Mass.: Addison-Wesley.

Snowberry, K., S.R. Parkinson, and N. Sisson
1983 Computer display menus. *Ergonomics* 26(7):699-712.

    1985   Effects of help fields on navigating through hierarchical menu structures. *International Journal of Man-Machine Studies* 22:479-491.

Stonier, T.
    1983   *The Wealth of Information: A Profile of the Post-Industrial Economy.* London, England: Methuen London in association with Thames Television International.

Vigil, P.
    1985   Computer literacy and the two cultures revisited. Pp. 240-242 in C.A. Parkhurst, ed., *Proceedings of the American Society for Information Science (ASIS) 48th Annual Meeting.* White Plains, N.Y.: Knowledge Industry Publications, Inc.

Vincente, K.J., and R.C. Williges
    1988   Accommodating individual differences in searching a hierarchical file system. *International Journal of Man-Machine Studies* 29:647-668.

Wickens, C.D.
    1992   *Engineering Psychology and Human Performance*, 2nd ed. New York: Harper Collins.

Wickens, C.D., S. Todd, and K. Seidler
    1989   *Three-Dimensional Displays: Perception, Implementation, and Applications.* University of Illinois Institute of Aviation Technical Report ARL-89-11/CSERIAC-89-1. Also CSERIAC SOAR 89-001, AAMRL, December. Wright-Patterson Air Force Base, Ohio. Savoy, Ill.: Aviation Research Laboratory.

Wiener, E.
    1989   Errors and error management in high technology aircraft. In *Proceedings of the 7th Meeting on Aeronautic and Space Medicine*, June, Paris.

Williams, H.P., M.P. Tham, and C.D. Wickens
    1992   *Resource Management and Geographic Disorientation in Aviation Incidents: A Review of the ASRS Database.* Technical Report ARL-92-3/NASA-92-2. Savoy: University of Illinois, Aviation Research Laboratory.

Wilson, S.R., N. Starr-Schneidkraut, and M.D. Cooper
    1989   *Use of the Critical Incident Technique to Evaluate the Impact of MEDLINE.* Final Report AIR-64600-9/89-FR; NLM/OPE-90/01. Palo Alto, Calif.: American Institutes for Research in the Behavioral Sciences.

Zorpette, G.
    1989   The main event. *IEEE Spectrum* 26(1):28.

8

Emerging Technologies in Work Design

Paul A. Attewell, Beverly M. Huey, Neville P. Moray, and Penelope M. Sanderson

INTRODUCTION

The last 15 years have seen dramatic changes in the competitive situation of many sectors of the U.S. economy and also in the levels of investment in and deployment of new technologies. These developments present new challenges for human factors specialists. They raise issues about how to design jobs to make sure that the latest technologies fulfill their promise of raising industrial productivity and competitiveness. In addition, they stimulate conceptual questioning about the proper relationship of humans, machines, and systems. Experiences, both good and bad, with emerging technologies have stimulated new philosophies of design, have highlighted gaps in our research knowledge, and have suggested new avenues for human factors research. In this chapter we examine the implications of recent technological and economic changes for job design and for research in the human factors community.

THE ECONOMIC AND TECHNOLOGICAL CONTEXT

Over the last two decades, the world economy has been rapidly shifting from a structure in which a few geographic centers dominated world trade to a more complex, multicentered structure. The rapid expansion of world trade has been accompanied by the emergence of new industrial powers and the erosion of dominance of older centers (National Research Council, 1990).

In the United States, this was experienced, during the 1980s, as a crisis of competitiveness (President's Commission on Industrial Competitiveness, 1985). In many industries, U.S. firms lost substantial domestic market shares to foreign imports, and a similar loss occurred in export markets.

Industrial technology is viewed by many as critical for success in this environment of intensified competition, and a backwardness in utilizing advanced technologies in manufacturing has been identified as one cause of the earlier U.S. decline (Jaikumar, 1986; National Academy of Engineering, 1988; Adler, 1991). Out of these concerns came several policy recommendations for U.S. firms: among these were that U.S. firms make changes in traditional product development processes, accelerate the implementation of new technologies, adopt strategies of continuous improvement, and embrace practices that encourage employee involvement. More recently, the federal government has embraced total quality management (TQM), not only as a strategy for itself but also as a requirement for its industrial suppliers.

The recommendations and diagnoses of several commissions emerged just as many U.S. firms were implementing a host of interlinked organizational changes—from outsourcing to de-layering to using high-performance work teams—while making substantial technological investment, most notably in computers, communications, and robotic technologies. The implementation of this mix of technological and organizational change was far from smooth. Although there are examples of increased productivity, there are also troubling indications that some technology investments have failed to boost productivity to the extent expected (Adler, 1991; Attewell, 1994; Computer Science and Telecommunications Board, 1994).

The problems of design, implementation, and operation of new technologies have therefore moved to center stage. Human factors specialists are faced with technologies that are complex in purely engineering terms but whose effectiveness also appears to depend on teamwork and a host of other organizational innovations (National Research Council, 1986; Adler, 1991; National Academy of Engineering/National Research Council, 1991). Job designers therefore face a dual task: to design jobs to take advantage of new technologies while fitting these jobs into new organizational structures and strategies.

THE CHANGING GOALS OF JOB DESIGN

Early in this century Frederick Taylor (1911) and the Gilbreths ([1917] 1973) were the first to argue that job design should be undertaken in a systematic manner, drawing upon scientific knowledge of human motor and cognitive capacities, of fatigue, and of attention span. Later, the sociotechnical systems perspective added the idea that effective work design must take into account the interactions among individuals in the workplace as well as the interplay of human and machine.

Drawing upon these insights, various methodologies have developed for analyzing jobs to discover their constituent tasks, and rules of best practice have emerged about how to combine tasks most effectively, where to draw boundaries between jobs, and how to join jobs together.

However, the principles and practices of job design have not remained constant over the last half century, in part because the goals of design have shifted. *Different design philosophies or theories focus on optimizing different aspects of work. Over time, design philosophies have shifted as different goals or dimensions of work rise and fall in importance.* This point can best be illustrated by a brief historical review of the major approaches to job design and a consideration of some new goals of work design that have emerged or have at least become more salient in recent years.

Taylorism or scientific management dominated job design prior to World War II. As a theory of design, it was centrally concerned with the optimization of physical effort in order to increase speed of production. Its various offshoots, such as time and motion study, sought to eliminate superfluous movement through design of both the job and its attendant machinery (jigs, machine tools, feeding mechanisms, etc.). But Taylorist attempts to optimize on physical effort and speed had important consequences for other aspects of the job:

• There was a separation of "indirect labor" (planning, preparation, maintenance, quality control) from direct labor.
• Jobs were subdivided as far as possible, so that each job encompassed a narrow range of repetitve tasks. This was intended to enhance work speed and reduce the time needed to learn the job.
• Worker discretion was as far as possible eliminated in favor of "one best way" (the designer's way) to accomplish the tasks (Knights et al., 1985).

The introduction of the assembly line, intended to optimize the efficiency and flow of production, led to further changes in job design. Jobs were designed around the need for continuous (uninterrupted) production. Machine tools became highly specialized, and feed and conveying operations were automated whenever possible. Assembly line jobs encompassed only a few repetitive tasks requiring minimal discretion or knowledge. The emphasis was on speed, simplicity, and stamina. The resulting work design philosophy, an extension of Taylorism, is often called "Fordism."

Taylorism and Fordism were fairly successful as theories of work design, *in terms of their professed goals*. But they eventually came under great criticism because they unintentionally degraded important aspects of work and, therefore, undermined other goals of production. From the 1950s onward, sociologists demonstrated that Taylorized assembly line jobs were boring yet physically demanding, leading to stress, low worker motivation, frequent absenteeism, and labor conflict (Walker and Guest, 1952; Kornhauser, 1965). Ultimately, this led to the emergence of a new work design movement, the human relations approach (McGregor, 1960), which stressed a very different goal for job design—improving job satisfaction or psychological fulfillment as a means of increasing workers' motivation.

This goal implied very different design principles:

• To encourage meaningful work and a sense of achievement, multiple tasks should be grouped into a single job in such a way that an individual could gain a sense of completion.
• Task variety and job rotation are desirable.
• Some choice over the sequencing of work, methods, and speed should be left to the employee.

The human relations approach to work design expanded from the 1950s on and became linked to a movement within job design known as the "quality of work life" movement. The approach remains especially salient where high levels of worker motivation (high initiative, good judgment, and carefulness) are important requirements of production processes.

Quality

The very success of Fordism in maintaining the high speed of the assembly line led, in many cases, to chronic problems with the quality of products. Alienated automobile assembly line workers who were required to work at a fast pace had a tendency to allow poorly constructed cars to go forward even when there were obvious flaws. Workers would take shortcuts that enabled them to keep a fast pace but that could undermine the quality of the product.

With intensified international competition from better-quality imported goods in the 1980s, the inferior quality of some U.S.-made goods became a major worry. Attaining and maintaining high standards of quality consequently became a central preoccupation in many U.S. manufacturing firms (Deming, 1986). This in turn has influenced job design. Quality testing has been reintegrated into many machine operators' and assembly line workers' jobs: instead of a separate quality-testing staff at the end of the production process, assembly line and other workers have been encouraged to "build quality in" and prevent bad work from going forward. There has been a renewed emphasis on job rotation (even during a single day) and on paying workers for developing maintenance and repair skills. Some production jobs have been broadened to include cleaning one's machine and surrounding areas, under the logic that this increases a sense of ownership and pride in one's work.

A commitment to high-quality products places new or increased performance demands on production employees. There is a greater need for relatively intangible skills such as taking responsibility and showing good judgment: for example, many of today's assembly line employees are empowered to pull the "andon cord" that brings the assembly line to a halt if they see a quality problem. This was not the case in Taylorist or Fordist job designs. Other new skills are intellectual, such as the substantial statistical and conceptual skills expected of workers who participate in quality circles. Cole (1979) delineates the mathematical knowledge required for statistical control procedures, including familiarity with elementary probability theory, statistical variance, and quasi-experimental approaches to diagnosing sources of error.

The growth of the total quality management movement has resulted in an increased respect for the intellectual contributions of shop-floor and office workers. In many workplaces they are encouraged to join quality circles to look for ways to improve production processes and to save resources. They may also be trained in group problem-solving techniques and in teamwork more generally.

In addition to changing the work lives of many employees, TQM's emphasis on quality has highlighted certain gaps in the human factors' scientific knowledge base. One important kind of quality flaw stems from operator error. Yet we know surprisingly little about the causes and sources of human error and even less about designing jobs and processes to reduce the frequency and seriousness of errors. Research is also needed to integrate the knowledge base on human error into design principles for machinery, displays, work procedures, and communications processes. This is one of several challenges that emerging technologies and the current economic climate pose for human factors research.

Workplace Health and Safety

Another equally important goal of work design, made more prominent by recent technological developments, is the maintenance and improvement of workplace health and safety. Recent concerns over workplace safety and occupational health reflect Occupational Safety and Health Administration legislation in the United States and, in Scandinavia and elsewhere in Europe, the influence of trade unions, which have targeted workplace health as a major issue (Butera and Thurman, 1990). Employees are demanding safer workplaces; legislation has declared that to be their right and has placed an obligation upon employers to provide safe jobs. In the United States, burgeoning health insurance and disability costs related to workplace illnesses provide an additional powerful impetus for employers to attend to these problems; this will ultimately result in demands for safer, healthier workplaces via better-engineered and designed work (see Chapter 4).

Designing jobs and work processes with health and safety in mind builds on traditional ergonomic concerns. Machine guards, good lighting and air quality, and concerns about protection from noise, lifting and stretching, fatigue and attention span, and legibility of signs and displays are all long-standing human factors issues. They will remain basic to design for tomorrow's workplaces. But the range of factors to be considered in work design will probably need to be broadened to include other, more social and psychological stressors in the workplace.

Consider, for example, the recent upswing in musculoskeletal disorders found among employees who work at computer terminals. In certain U.S. industries, such as telephone companies and newspaper offices, epidemiologists have documented recent mini-epidemics of both wrist pain and neck and upper-back problems, so-called "repetitive strain disorders." Such complaints may affect from 20 to 40 percent of workers in certain workplaces. The problems span a spectrum from severe conditions—such as carpal tunnel syndrome, which can be so painful that sufferers are unable to work and therefore often resort to surgery—to less dramatic, but still consequential, intermittent pain, which can affect workplace productivity and employee morale.

The traditional ergonomic response to such repetitive strain disorders is to study and reengineer workplace machinery, utilizing our knowledge about vision, posture, hand motion, and other areas. Thus, we have seen the design of better chairs, video display terminals, lighting, and keyboards, drawing upon the human factors knowledge base.

However, while giving traditional ergonomic factors their due, the epidemiological research suggests that additional sociopsychological factors may be at work to create job stress that is in turn related to the physical symptoms of repetitive strain disorders. These disorders can be found even in offices that have invested in ergonomically sound equipment. Epidemiologists at the National Institute of Occupational Safety and Health and elsewhere have documented that a variety of job stressors—from fear of job loss to the variability of work, from the existence of deadlines to the nature of supervision and surveillance—are related to these disorders in certain workplaces. It will take considerable design ingenuity to minimize such stressors.

Current design responses to repetitive strain disorders include enhancing work-group dynamics and scheduling rest breaks and job rotation, as well as designing better keyboards, monitors, and other equipment. Unfortunately, there is little research that examines whether these responses are effective in countering the spread of repetitive strain disorders. Beyond studying the efficacy of these particular interventions, there is a need for human factors specialists to orient research toward the intersection of ergonomic knowledge, sociotechnical factors, and epidemiology with the goal of understanding the links between sociotechnical design, workplace stress, and worker health.

Intellectual Work and Teamwork

Intellectual Work

For several decades the proportion of the labor force in white-collar jobs has been growing, with managerial, professional, and technical occupations leading the way. As "knowledge work" has come to dominate the information economy, job design has had to embrace new concepts and approaches especially suited to intellectual and informational jobs. Because knowledge workers work at machines, most notably computers, traditional ergonomic concerns with seating, keyboards, and displays remain important. Much attention has also been focused on such human-computer interfaces as displays of information and cognitive maps. But beyond this, computing technologies have changed—and continue to change—the division of labor and the content of jobs, often in dramatic ways.

Before computerization, most information jobs in large firms were subject to Taylorist design. In insurance offices, for example, paper flowed from department to department in a sequence: first to the claims examination department, then to the accounts payable department, then to the check-writing department, then to bookkeeping. Within each department, tasks were narrowly subdivided, routinized, and linked in a chain. Each department typically kept its own databases and paper files. Thus, a single business process, for example accounts payable or assessing an insurance claim, involved sending the originating paper through a long multistep multi-department chain, each link of which consisted of a specialized clerical worker perfoming a narrow repetitive task. There tended to be backlogs of work in each department, and each department expected a substantial period of time (e.g., one week) to clear a claim through its part of the process.

Taylorist design could efficiently process very high volumes of business items, but was slow because of the many handoffs from person to person. A given claim or payment often took weeks to process from start to finish, and there was no easy way to locate a particular piece once it had entered the stream.

Interactive or real-time computing was widely adopted from the early 1970s on, and it made this fine division of labor obsolete. Real-time computing allowed previously distinct databases to be linked. As soon as a piece of information was entered, all the relevant databases were instantly updated to reflect the new datum. A further consequence was that previously separate clerical or informational tasks could now be integrated. So, when an order arrived, a clerk could call up on the computer a database that would first determine the customer's credit, that is, whether any bills were outstanding. If the buyer was found creditworthy, the software would present inventory data to see whether the requested item was in stock. If sufficient stock was available the clerk could have a "pick slip" printed at the warehouse giving the order to ship the goods. At the same time the accounts receivable database would note that the customer should be charged, and an invoice would be prepared. And the company's books would be updated to note the additional amount receivable. If the items were not in stock, the clerk could initiate an order for the shop floor to produce the items and could even (via materials requirements planning, MRP software) set in motion orders for more raw material needed before production could commence.

In sum, the interlinked databases allowed for a reintegration of previously fragmented tasks, each of which was previously a separate job in a long clerical sequence. Instead of simplifying steps, the new design logic was to give each order-taking clerk access to all of the steps and activities needed to complete an order, to make a job coincide with the complete range of informational tasks associated with an order. New complementarities also arose between data retrieval, customer service, and data entry. Because databases were now easily accessible, it made sense to give computer clerks the role of intermediary between the information system and the customer. So a customer could phone and ask the clerk whether certain goods were available for shipment, and if not, when they could be produced and delivered. At the same time, the clerk could enter data on the specfics of a new order, payment method, and shipping details. Thus, one-time highly specialized and routinized clerks became more multifaceted "customer service representatives" (Attewell, 1992:71-74).

At first, job designs like these evolved spontaneously. But experiences with interactive computer systems have become codified and today inform a new practice of job design known as "business process reengineering" or (more grandly) "reengineering the corporation" (Hammer and Champy, 1993). This approach looks for previously fragmented tasks to integrate; it "delinearizes" and resequences tasks, creates new, broader jobs, and also restructures (typically reduces) managerial controls to fit the new, broader division of labor. Proponents of the approach claim that it increases productivity, improves service, and enhances job satisfaction.

While new information technologies have created many of these multifaceted clerical positions at the lower end of information work, they have also stimulated change in higher-paid information work. Today's managers use management information systems—streams of production and financial data—to survey and understand their areas of jurisdiction. This in turn can lead to rather different styles of managerial control and decison making than those found a generation ago (Attewell, 1992), styles that are heavily dependent on quantitative data and computerized decision tools. This can have both positive and negative consequences. Studies of the use of decision tools such as spreadsheet models (e.g., Kotteman and Remus 1987; Kotteman et al., in press) indicate that managers can become dependent on the use of these techniques even when they fail to produce better (or even adequate) decisions. In particular, the researchers cited above have found that managers systematically overestimate the effectiveness of these methods. They have coined the phrases *cognitive conceit* and *the illusion of control* to describe these effects.

Such studies indicate an emerging trend in human factors research to examine *trust* in machines (and in computers and data) as well as the emotional and cognitive consequences of dependence on cognitive tools such as computers. (This is clearly related to the issue of human error discussed earlier.) We shall return below to ways these concerns have become reflected in job design and philosophies of appropriate levels of automation.

Teamwork

A rather different aspect of the impact of computer technologies on jobs involves the rising importance of teamwork in modern workplaces. Teamwork among small groups of skilled individuals predates the diffusion of information technology: one thinks of physican-nurse teams in a hospital operating room, the flight team in an aircraft cockpit, and the team of sailors taking bearings and steering a ship into port. However, several commentators have argued that small-group teamwork of this kind has become increasingly widespread in the modern workplace, going beyond the traditional professions and into new work domains. Frequently, the tasks involved in knowledge work in modern workplaces are highly complex and therefore benefit from coordinated teams of highly skilled individuals, often with different specialties, working together (Peters and Waterman, 1983; Savage, 1990; Drucker 1993:83-89; Mills, 1991; see also Hill, 1992).

This increasing emphasis on teamwork presents a challenge to human factors researchers and the behavioral science community generally. First, there are lacunae in our basic knowledge of small-group dynamics. As a recent review (Simpson and Wood, 1993) put it: "Despite the widely recognized importance of groups, basic social processes underlying group dynamics have received scant and intermittent attention. This has been particularly true within social psychology." (However, see McGrath, 1984, and Hackman, 1987, for a more optimistic view.)

Second, the research literature on work groups in situ—in real work settings as distinct from laboratory experiments or simulated workgroups—is sparse. There are, however, notable exceptions, for example, R. Helmreich's research on flight cockpit crew dynamics and on operating room surgical teams.

The potential importance (and practical difficulties) of applying human factors approaches to the job design of group work or teamwork can be illustrated by the example of software development, a steadily growing area of employment in our postindustrial economy. Most software development projects are carried out by small teams of systems analysts, programmers, and coders. The work is creative, and for that reason uncertain: up to 25 percent of projects fail, and cost overruns and missing deadlines are endemic (DeMarco and Lister, 1987; see also Computer Science and Telecommunications Board, 1990).

Given the high cost of software development projects, great efforts have been made to increase the productivity of programming efforts. Metrics of productivity and output have been developed (Boehm, 1987). New programming languages and software tools abound; it is claimed that many of these increase productivity. Unfortunately, differences in programming technology do not explain observed differences in software engineering productivity (DeMarco and Lister, 1987), and the introduction of new programming tools has not solved problems with software productivity: several studies have found that the most modern of Computer Aided Software Engineering (CASE) tools are associated with *lower* productivity (Banker et al., 1991; Orlikowsky, 1988.) Thus, the attention of researchers and job designers has been drawn to sociotechnical factors, most notably the processes of coordination and division of tasks among team members.

Perhaps because of the incompleteness of scientific research, one finds dramatically different approaches to job design in these settings. Some authorities, viewing coordination as the Achilles' heel of software teams, pursue design strategies aimed at removing ambiguity and complexity from teamwork. Tasks are decomposed, simplified whenever possible, and organized as modules to reduce the necessity of coordination or interaction among team members. Lines of authority are made as specific as possible, and standard operating procedures are spelled out. Meetings are formalized and scheduled regularly (Boehm, 1987; Brooks, 1987), and standardized methods of documentation and testing are prescribed. This approach has come to be known as "structured development" or simply as "programming methodologies."

One difficulty with this approach is that by creating formal structures to minimize the burden of coordination and communication among team members, designers may reduce *necessary* or fruitful commmunication, resulting in inferior performance. Kiesler et al. (1994:226) argue that the most productive combination is moderate levels of formalization and structure combined with moderate levels of team communications. At either extreme—high structure or high communication—team productivity suffers.

In contrast, those who take an ethnographically informed approach to job design for software teams are quite hostile to using formalized methodologies to structure team dynamics and programming. DeMarco and Lister (1987:114) give the following critique:

> A Methodology is a general systems theory of how a whole class of thought-intensive work ought to be conducted. It comes in the form of a fat book that specifies exactly what steps to take at any time, regardless of who is doing the work, regardless of where or when. . . . Methodology is an attempt to centralize thinking. All meaningful decisions are made by the Methodology builders, not by the staff assigned to do the work. . . . They do this by trying to force the work into a fixed mold that guarantees:
> - a morass of paperwork,
> - a paucity of methods,
> - an absence of responsibility, and
> - a general loss of motivation.

DeMarco and Lister (1987) turn elsewhere for factors that influence effective teamwork. They argue that software productivity requires deep thinking as well as communication. They find that modern workplaces rarely provide good environments for thinking, especially for "flow," a kind of thinking requiring extended periods of concentration and focus. In many workplaces there are frequent interruptions from phone calls and colleagues, and noise levels are high—phenomena inimical to flow. DeMarco and Lister find, for example, that being able to silence one's phone makes a substantial difference in productivity. Teams do better if they have receptionists who shield them from interruption. DeMarco and Lister make a series of suggestions regarding the design of work spaces, noise reduction, recruitment of team members, and style of leadership, which include traditional human factors concerns as well as sociotechnical factors.

Software development is but one of the numerous contexts in which teamwork is the preferred strategy for organizing work. But the contrasting prescriptions and beliefs about how to best design teamwork prove that the knowledge base and science behind job design of teamwork are at very rudimentary stages. This is an area of great promise and great practical importance, but one in which relatively little has yet been accomplished.

ADVANCED AUTOMATION

As information technologies become increasingly capable of performing a wide range of functions in both manufacturing and white-collar work, attention has turned to the proper role of humans in the system. There are two extreme design philosophies (Kantowitz and Sorkin, 1987): (1) the human should be eliminated entirely from systems, or if that is not possible, the human role should be minimized (i.e., automate as much as is possible) and (2) the human operator should be involved as much as possible in the system, even if artificial tasks must be created to accomplish this. The first philosophy, which is technology-driven, is defined by decisions about when and how to automate (Air Force Studies Board, 1982), which, in turn, define the role of the operator in the system.

Researchers have reported that system designers often believe that the greater the degree of automation the better (Bainbridge, 1982; Boehm-Davis et al., 1983; Wiener, 1985; Wiener and Curry, 1980). However, this moves the human further from direct contact with the system; this has been found to sometimes result in negative consequences when the operator is required to intervene (Air Force Studies Board, 1982). New types of errors or accidents have often been created because automation has changed the nature of the human-machine relationship in unforeseen ways (e.g., Hirschhorn, 1984; Moray and Huey, 1988; Nobel, 1984). A workshop on automation (Boehm-Davis et al., 1983) identified five problems associated with highly automated systems: (1) newly automated systems do not usually provide all the anticipated benefits; (2) when automated equipment fails, the system loses credibility; (3) automation usually increases training requirements; (4) designers often fail to anticipate new problems that will be created by the automated systems; and (5) automation changes the role of the human from active controller to system monitor. Numerous other researchers have also identified both pros and cons of automation (Bainbridge, 1982; Wiener, 1985; Wiener and Curry, 1980; Parsons, 1985).

Thus, many researchers have come to the view that in order to achieve high levels of productivity, quality, safety, and worker satisfaction and motivation, human and machine skills and intelligence must be integrated. There is a need to complement, not replace, human skills and abilities (Havn, 1990.) New technologies should be skill-enhancing rather than skill-replacing (Kidd, 1990).

Although there has been great investment in advanced automation and robotics in U.S. industry, it has not resulted in a dramatic U.S. lead in manufacturing technology. One reason seems to be that in the United States automation has largely been seen as a chance to reduce the labor force; another reason is the belief that automated equipment is sufficiently intelligent in itself to be run by relatively unqualified, and hence cheap, labor.

By contrast, in both Europe and Japan, there has been a tendency to integrate highly qualified labor (e.g., even graduate engineers) into the control of automated manufacturing equipment.

Recently, both industrialists and academics (Brodner, 1986; Kellso, 1989; Kuo and Hsu, 1990; De Greene, 1991) have claimed that the attempt to substitute machine for human intelligence in industry has failed, and they have called for a better understanding and implementation of a human-machine symbiosis in manufacturing. In a classical paper, Bainbridge (1983) pointed out the "ironies of automation": only those processes that are well understood can be automated; those processes that are poorly understood are then left to humans, so that progressively the humans face more and more difficult tasks when systems falter or fail.

It is now being realized that advanced automation places great demands on the human workforce and that automation should not amount to "designing the human out of the system" (Shaiken, 1984). The question is, rather, how the special characteristics of humans can be integrated into technological systems so that productivity is optimized from the point of view of production level, quality, and safety (Wall et al., 1987).

Sanderson (1989) pointed out a number of aspects of discrete manufacturing in which humans play a central role that is only imperfectly understood, even at the level of the individual operator. To support the integration of humans into manufacturing systems requires that designers, trainers, procedure writers, and all levels of management understand the nature of the skills of the human operator at many levels. This means understanding traditional problems of interface design at the lowest level, through complex activities such as planning, scheduling, expediting, and maintenance, up to the most global level of policy setting by management. Indeed, the concept of macro-ergonomics has recently been introduced. This refers to the extremely important role played by managerial and organizational factors in the efficiency of automated and hybrid systems, and to the dynamics of groups, teams, and crews. It provides a top-down ergonomics to complement the traditional bottom-up ergonomics of design.

The European Economic Community (EEC) has undertaken a massive investment in the application of human factors to industrial and manufacturing systems of the future. This can be expected to give the Europeans a large lead in integrating human intelligence into manufacturing production, something that will have a great impact on their global competitiveness. In the United States there are few signs of a change in thinking; many still see automation as nothing more than a way to employ inexpensive and relatively unskilled labor to run sophisticated machines. But all the indicators are that this will lead to a disastrous failure in competitive manufacturing.

There are two main themes that arise from the considerations discussed above. The first is a drive for research that will improve productivity and efficiency in industry. The second is a need for research toward a work climate, culture, and environment that will be humane and fulfilling to the workforce. Thus, all levels of ergonomics will have to be studied. These include the interface design for flexible manufacturing systems and computer-integrated manufacturing, as well as the role of humans in planning and scheduling, in design, in maintenance, and in supervisory control as applied to discrete manufacturing systems. In addition, we need models of humans in discrete manufacturing situations, and a consideration of macroergonomic factors, both theoretical and applied, to support design and systems development.

Skill Requirements in Advanced Manufacturing

As automation advances, the workforce required to operate and maintain it splits into two job classes. One class requires a few highly skilled workers. Tasks for these workers are often intellectually challenging. These workers perform the high-technology jobs called for in, for example, advanced flexible manufacturing technology. The other job class requires more workers but places few demands on them for specialized training and education. Their jobs are generally unrewarding, either intellectually or financially. Such jobs may remain in the overall job inventory even in the wake of complete automation.

In the present state of the art, one problem is that, when systems falter or malfunction, humans are called on to perform tasks that may be beyond their capabilities. In such cases, a boring job may suddenly become extremely demanding. It is not clear what level or type of education will best serve manufacturing industry so as to match the task to the human. If the sophistication of advanced machinery calls for highly qualified operators, how are the jobs to be designed so that such people will not be bored and dissatisfied during normal operation? What steps are needed to enable people with a suitable level of education to be satisfied in operating a highly automated plant?

Although relatively few manufacturing jobs can give rise to severe hazards (in contrast to chemical industries, the nuclear industry, etc.), there are certainly problems of both safety and economics when systems fail. It has been said that plants may be designed on the assumption that they will be down for maintenance and reprogramming for as much as 30 percent of their operating time, a situation that is clearly undesirable. We need to know how to maintain operator skills so that time lost due to accidents and system faults is minimized. Supervisory control can cause skills to be lost and leave operators unable to cope with abnormal conditions. Our understanding of what makes operators intervene to take charge of a faltering automated system is rudimentary (Lee and Moray, 1992), and the dynamics of trust between humans and their machines is far from adequate to define allocation of function or to design systems that optimize their interaction (Zuboff, 1988).

Hirschhorn (1984), Adler (1986), and others have argued that even low-level jobs involving computers require greater responsibility on the part of employees. Errors ramify rapidly through interconnected databases and information systems, and such errors are often hard to reverse. Consequently, care and high-quality work, with attention to detail, are especially important in these jobs. This again raises the issue of sources of error and error prevention in computer-related jobs. We need more empirical studies of performance errors in a range of computer-based jobs in order to design job procedures and whole job systems that prevent or minimize error.

In transaction processing, continuous-flow manufacturing, and several other jobs using information technologies, computers are programmed to take care of routine cases, leaving human operators with a greater mix of trouble-shooting and handling exceptions—employees become monitors, maintenance people, and troubleshooters, rather than "doers." This kind of monitoring work can require different skills: more recognition of patterns, more logical or abstract decision making, more learning from rare events rather than via frequent repetition (Clark et al., 1988:Chapter 4).

Thus, in addition to ergonomic concerns, human factors specialists who seek to understand the demands of these jobs as an input into work design must find new methods of measuring the cognitive workload and cognitive skills of the jobs. Some aspects of cognitive workload and skills are relatively well researched: information overload and competencies in absorbing numerical data and graphical representations. Others are less well understood. For example, many people working with information systems need to visualize the logical structure of the system in order to understand the implications of their own actions, their effects on others, and how to diagnose and correct mistakes. Researchers are at an early stage in understanding system visualization as a skill and as an aspect of cognitive workload (see Carroll and Olson, 1987).

As blue-collar work has become automated, fewer jobs are "hands on"; more and more involve the employee in monitoring production via dials, gauges, and so forth. As Zuboff (1988) has explained, this often leads to a loss of data received directly through the physical senses, resulting in shifts in skills and additional cognitive workload. Machinists, for example, used to depend heavily on tactile skills and sense of vibration when feeding and cutting on traditional machine tools. They cannot use the same senses now that programmable machine tools ("machining stations") are automated and are surrounded by heavy metal hoods. However, some machinists have developed hearing skills to sense when a cutting or milling procedure is going wrong. Even though their machines are muffled by hoods, they seem able to identify slight changes in tone in what seems to outsiders to be a very noisy machine-shop environment. These clues enable them to stop jobs that are about to go wrong (Attewell, 1992; Zicklin, 1987).

This example demonstrates that skill changes resulting from new technologies can be quite subtle. A simple task analysis would undoubtedly note that the automated machinist needs new programming skills, but might well miss the tactile-to-aural shift. Researchers have hardly begun to describe and make an inventory of these higher-level cognitive skills, let alone begun to design job systems with them in mind.

Summary

The widespread introduction of information technologies and automation introduces a number of significant problems that affect employment, the nature of work, the configuration of the workplace, the performance of workers, and the well-being of the workforce.

Although it was once believed that automation would increase unemployment in the United States, this has generally not been the case. Some traditional jobs—particularly those with low skill and knowledge demands—have declined as functions have been automated; however, new jobs have been created. This has raised a number of areas that human factors researchers have the knowledge to contribute to. Examples include:

- making technology more human-centered;
- designing systems to make proper use of unique human capabilities and designing aids to support task performance;
- making work meaningful (e.g., acceptable workload levels; reducing boredom, fatigue, and job stress; giving a sense of control over the work process); and
- attempting to satisfy the need for autonomy, education, and training.

Research is needed to address such questions as the following: How can we monitor and enhance worker trust in automation? How can operators be kept in the loop so that they can respond when their skills are needed? How can we monitor and measure human performance to detect symptoms of stress and performance impairment? How do we avoid operator skill obsolescence from the use of rapidly advancing technologies? What are the effects of individual differences in cognitive style on worker performance with automated systems?

CONCLUSION

We have argued that computer technologies and the shift to an information economy present a considerable challenge for those involved in work design and for those researchers in human factors and related disciplines who try to provide a scientific basis for work design. Although a number of topics and areas of research have been raised in this chapter, four central themes stand out:

First, more systematic and detailed descriptive research about the skills—especially the higher-level cognitive skills—that are used by today's high-technology workers is an indispensable base for future design of jobs and technologies. The kinds of task and skill inventories used in the past have not adequately captured these kinds of cognitive and organizational skills—leaving designers to "fly blind."

Second, further research is needed on performance and error characteristics of emerging technology jobs. We must find out what causal mechanisms or features underlie observed differences in performance, especially in errors and error rates, so that work systems can be designed to avoid error where possible and so that errors that cannot be avoided can at least be more easily identified and remedied.

Third, research must investigate informal learning processes and skill acquisition among those working with new technology; this is a relatively neglected but important research topic. Insights gathered from such research should be used both to plan formal training procedures and to improve ease of on-the-job learning.

Finally, additional research is required to identify the full range of stressors within emerging-technology workplaces and their consequences for employee performance, morale, and health. Some of these stressors will be ones that ergonomics is well equipped to study. Others will necessitate adding sociopsychological studies to traditional human factors methodologies.

In the long run one might think that the increase in highly intelligent computers, robotics, and microelectronics will continue to displace human operators. Yet in Europe and Japan this does not seem to be the case. In recent years, factories built in the United States by foreign companies have reached high levels of productivity with surprisingly little automation and robotics. And one manufacturer who opened a state-of-the-art car assembly plant stated that only about 30 percent of operations are worth automating. Beyond that, automation is relatively cost-ineffective, and it is better to use human skills appropriately. If that is the case, human factors and management psychology may be as important as automation engineering in ensuring high productivity. But there is little understanding of these factors, and what understanding there is comes largely from foreign experience, which for cultural reasons may not be transplantable to the United States.

The challenge of automation to the human factors community in future years is to learn how to adapt automation to people in ways that are both human-centered and productive. Human-centered technology is being actively pursued in Japan and Europe. Until about five years ago, little of this approach was practiced in the United States, but a still timely research question is how to produce human-centered designs in the context of American culture.

REFERENCES

Adler, P.
- 1986 New technology, new skills. *California Management Review* 29(Fall):9-28.
- 1991 Capitalizing on new manufacturing technologies: current problems and emergent trends in U.S. industry. Pp. 59-88 in National Academy of Engineering/National Research Council, *People and Technology in the Workplace*. Washington, D.C.: National Academy Press.

Air Force Studies Board
- 1982 *The Effectiveness of the Air Force Nondestructive Inspection Management*. Panel on Nondestructive Inspection, Committee on Mechanical Reliability, Assembly of Engineering, National Research Council. Washington, D.C.: National Academy Press.

Attewell, P.A.
- 1992 Skill and occupational change in U.S. manufacturing. Chapter 3 in P. Adler, ed., *Technology and the Future of Work*. New York: Oxford University Press.
- 1994 Information technology and the productivity paradox. Pp. 13-53 in D.H. Harris, ed., *Organizational Linkages: Understanding the Productivity Paradox*. Panel on Organizational Linkages, Committee on Human Factors, National Research Council. Washington, D.C.: National Academy Press.

Bainbridge, L.
- 1982 Displays as a source of task load. *Ergonomics* 25(4):335.
- 1983 Ironies of automation. Pp. 129-135 in G. Johannsen and J.E. Rijnsdorp, eds., *Analysis, Design and Evaluation of Man-Machine Systems: Proceedings of the IFAC/IFIP/IFORS/IEA Conference*, Baden-Baden, Federal Republic of Germany, 27-29 September 1982. New York: Pergamon.

Banker, R.D., S.M. Datar, and C.F. Kemmerer
- 1991 A model to evaluate variables impacting the productivity of software maintenance projects. *Management Science* 17:1-18.

Boehm, B.W.
- 1987 Improving software productivity. *IEEE Software Society* 20:43-57.

Boehm-Davis, D.A., R.E. Curry, E.L. Wiener, and R.L. Harrison
- 1983 Human factors of flight-deck automation—NASA/industry workshop. *Ergonomics* 26:953-961.

Brodner, P., ed.
- 1986 *Proceedings of the IFAC Workshop on Skill-Based Automated Manufactoring*. Karlsruhe, West Germany: IFAC.

Brodner, P.
- 1986 Skill-based manufacturing vs "Unmanned Factory": which is superior? *International Journal of Industrial Ergonomics* 1:145-153.

Brooks, F.P.
- 1987 No silver bullet: essence and accidents of software engineering. *IEEE Computer Society* 20:10-18.

Butera, F., and J. Thurman
- 1990 *Automation and Work Design*. A study prepared for the International Labour Office. New York: North-Holland.

Carroll, J.M., and J.R. Olson, eds.
1987   *Mental Models in Human-Computer Interaction.* Committee on Human Factors, National Research Council. Washington, D.C.: National Academy Press.

Clark, J., I. McLoughlin, H. Rose, and R. King
1988   *The Process of Technological Change.* New York: Cambridge University Press.

Cole, R.E.
1979   *Work, Mobility, and Participation: A Comparative Study of American and Japanese Industry.* Berkeley: University of California Press

Computer Science and Technology Board
1990   Scaling up: a research agenda for software engineering. National Science Foundation. *Communications of the ACM* 33:281-293.

Computer Science and Telecommunications Board
1994   *Information Technology in the Service Society.* National Research Council. Washington, D.C.: National Academy Press.

De Greene, K.B.
1991   Emergent complexity and person-machine systems. *International Journal of Man-Machine Studies* 35(2):219-234.

DeMarco, T., and T. Lister
1987   *Peopleware: Productive Projects and Teams.* New York: Dorset House.

Deming, W.E.
1986   *Out of the Crisis.* Cambridge, Mass.: MIT Press.

Drucker, P.
1993   *Post-Capitalist Society.* New York: Harper.

Gilbreth, F.B., and L.M. Gilbreth
[1917] 1973 *Applied Motion Study.* Easton, Pa.: Hive Publishing.

Hackman, J.R.
1987   The design of work teams. Pp. 315-342 in J.W. Lorsch, ed., *Handbook of Organizational Behavior.* Englewood Cliffs, N.J.: Prentice-Hall.

Hammer, M., and J. Champy
1993   *Reengineering the Corporation: A Manifesto for Business Revolution.* New York: Harper and Row.

Havn, E.
1990   Designing for cooperative work. Pp. 35-42 in W. Karwowski and M. Rahimi, eds., *Ergonomics of Hybrid Automated Systems II.* New York: Elsevier.

Hill, G.W.
1992   Group versus individual performance: are N + 1 heads better than one? *Psychological Bulletin* 91:517-539.

Hirschhorn, L.
1984   *Beyond Mechanization.* Cambridge, Mass.: MIT Press.

Jaikumar, R.
1986   Post-industrial manufacturing. *Harvard Business Review* 63(November/December):69-76.

Kantowitz, S.H., and R.D. Sorkin
1987   Allocation of functions. Pp. 355-369 in G. Salvendy, ed., *Handbook of Human Factors.* New York: John Wiley and Sons.

Kellso, J.R.
1989   CIM in action: microelectronics, manufacturer charts course towards true systems integration. *Industrial Engineering* 21:18-22.

Kidd, P.T.
1990   Information technology: design for human involvement or human intervention? Pp. 417-424 in W. Karwowski and M. Rahimi, eds., *Ergonomics of Hybrid Automated Systems II.* New York: Elsevier.

Kiesler, S., D. Wholey, and K.M. Carley
- 1994 Coordination as linkage: the case of software development teams. Pp. 214-239 in D. Harris, ed., *Organizational Linkages: Understanding the Productivity Paradox*. Panel on Organizational Linkages, Committee on Human Factors, National Research Council. Washington, D.C.: National Academy Press.

Knights, D., H. Wilmott, and D. Collinson
- 1985 *Job Redesign: Critical Perspectives on the Labor Process*. Brookfield, Vt.: Gower Publishers.

Kornhauser, A.
- 1965 *The Mental Health of the Industrial Worker*. New York: Wiley.

Kotteman, J., and W. Remus
- 1987 Evidence and principles of functional and dysfunctional decision-support systems. *International Journal of Management Science* 15(2):135-144.

Kotteman, J., F.D. Davis, and W. Remus
- In Press Computer-assisted decison-making: performance, beliefs and illlusion of control. In *Organizational Behavior and Human Decision Processes*. New York: Academic Press.

Kuo W., and J.P. Hsu
- 1990 Update: simultaneous engineering design in Japan. *Industrial Engineering* 22:23-28.

Lee, J.D., and N. Moray
- 1992 Operators' monitoring patterns and fault recovery in the supervisory control of a semi-automatic process. Pp. 1143-1147 in *Proceedings of the Human Factors Society 36th Annual Meeting*. Santa Monica, Calif.: Human Factors Society.

McGrath, J.E.
- 1984 *Groups: Interaction and Performance*. Englewood Cliffs, N.J.: Prentice-Hall.

McGregor, D.
- 1960 *The Human Side of the Enterprise*. New York: McGraw Hill.

Mills, D.Q.
- 1991 *Rebirth of the Corporation*. New York: John Wiley.

Moray, N.P., and B.M. Huey, eds.
- 1988 *Human Factors Research and Nuclear Safety*. Panel on Human Factors Research Needs in Nuclear Regulatory Research, Committee on Human Factors, National Research Council. Washington, D.C.: National Academy Press.

National Academy of Engineering
- 1988 *The Technological Dimensions of International Competitiveness*. Committee on Technology Issues That Impact International Competitiveness. Washington, D.C.: National Academy Press.

National Academy of Engineering/National Research Council
- 1991 *People and Technology in the Workplace*. Washington, D.C.: National Academy Press.

National Research Council
- 1986 *Human Resources Practices for Implementing Advanced Manufacturing Technology*. Manufacturing Studies Board. Washington, D.C.: National Academy Press.
- 1990 *The Internationalization of U.S. Manufacturing: Causes and Consequences*. Manufacturing Studies Board. Washington, D.C.: National Academy Press.

Nobel, D.F.
- 1984 *Forces of Production: A Social History of Industrial Automation*. New York: Alfred A. Knopf.

Orlikowsky, W.J.
- 1988 Information Technology in Post-Industrial Organizations. Unpublished PhD dissertation. Stern School of Management, New York University.

Parsons, H.M.
   1985   Automation and the individual: comprehensive and comparative views. *Human Factors* 27(February):99-111.

Peters, T., and R. Waterman
   1983   *In Search of Excellence*. New York: Warner Books.

President's Commission on Industrial Competitiveness
   1985   *Global Competition: The New Reality*. Washington, D.C.: U.S. Government Printing Office.

Sanderson, P.M.
   1989   The human planning and scheduling role in advanced manufacturing systems: an emerging human factors domain. *Human Factors* 31(6):635-666.

Savage, C.
   1990   *Fifth Generation Management*. Burlington, Mass.: Digital Press.

Shaiken, H.
   1984   *Work Transformed: Automation and Labor in the Computer Age*. New York: Holt Reinhart Winston.

Simpson, J., and W. Wood
   1993   Where is the group in social psychology? An historical overview. In S. Worchell, W. Wood, and J. Simpson, eds., *Group Process and Productivity*. Newbury Park, Calif.: Sage Press.

Taylor, F.
   1911   *The Principles of Scientific Management*. New York: Harper and Row.

Walker, C., and R. Guest
   1952   *Man on the Assembly Line*. Cambridge, Mass.: Harvard University Press.

Wall, T.T., C.W. Clegg, and N.J. Kemp
   1987   *The Human Side of Advanced Manufacturing*. New York: Wiley.

Wiener, E.L.
   1985   Beyond the sterile cockpit. *Human Factors* 27:75-89.

Wiener, E., and R.E. Curry
   1980   Automation in the cockpit: some generalizations. In *Proceedings of the Annual Meeting of the Human Factors Society*, October. Santa Monica, Calif.: Human Factors Society.

Zicklin, G.
   1987   Numerical control machining and the issue of deskilling: an empirical view. *Work and Occupations* 14:452-466.

Zuboff, S.
   1988   *In the Age of the Smart Machine: The Future of Work and Power*. New York: Basic Books.

9

Transportation

Herschel W. Leibowitz, D. Alfred Owens, and Robert L. Helmreich

INTRODUCTION

The transportation field comprises vehicles ranging from submarines to spacecraft and technologies that span the iron age and the computer age. Human factors issues are important because individuals must interact with the large, complex systems of modern forms of transportation. Safety is critical, as most citizens utilize several modes of transportation, and transportation disasters are highly publicized. Investigations of transportation mishaps typically invoke contributory human factors issues, thus enhancing research opportunities.

The range of areas within transportation illustrates the expanded definition and role of human factors research. Traditional ergonomic issues clearly inform the design of vehicles and transportation systems. Operationally relevant human factors research encompasses cognitive, perceptual, and engineering psychology, as well as organizational, social, personality, educational, and cross-cultural disciplines. Levels of analysis range from the individual to the group to the organization to the culture; many research questions cut across disciplinary boundaries and demand a multidisciplinary approach.

Rather than document the full array of transportation research questions, we will focus on opportunities for behavioral research in two areas representative of the broad scope of human factors: vehicular traffic safety and aviation safety. In the first area, we discuss traditional concerns with the individual and the machine, while in the second we describe the role of team performance and training in a high-technology environment. The human factors research relevant to aviation includes organizational, social and personality, educational, and cross-cultural psychology—as well as the more traditional cognitive, experimental, and engineering concerns. Many of the research issues raised here are not limited to vehicular and air domains, but apply in whole or part to other domains of transportation, including maritime and rail operations and space flight.

VEHICULAR TRAFFIC SAFETY

In dealing with traffic safety, the human factors community has the opportunity to contribute to the solution of a major public health problem. In the United States, motor vehicle accidents are the leading cause of death for people between the ages of 1 and 38 and are responsible for more deaths than all other causes combined between the ages of 15 and 24 (National Safety Council, 1993). Currently 1.7 million disabling injuries and 43,000 fatalities occur annually. Because traffic accidents take a disproportionate toll on the younger population, the average number of years of life lost per death is 2 and 3 times higher than for cancer and heart disease, respectively, and the mortality cost per death is 3.7 and 6.2 times higher.

On the basis of mortality costs, Sivak (1993) has estimated that the United States invests 15 times more heavily in research on cardiovascular disease and nearly 24 times more for cancer. Perhaps this discrepancy in expenditures occurs because traditionally traffic safety has been viewed—along with other issues of transportation—as a matter of commerce and technology. Certainly, our modern transportation systems exemplify the great benefits of technology for commerce. However, they also have major consequences for public health and welfare; these consequences arise not from the machinery per se but from the interaction of human users and technological systems. It is at this interface that human factors research comes into play.

A central requirement for progress in this area is the development of a coherent theoretical framework for addressing the task of driving. We must develop a theory of driving behavior that is both empirically grounded and ecologically valid. Present theoretical frameworks often consist of commonsense judgments (e.g., "speed kills") or abstract flow charts that have only limited empirical support. One approach to traffic safety that is rapidly gaining attention is the attempt to use computational technology. This is based on the assumption that because computers are faster and more reliable than humans, they can eliminate or reduce costly human errors. Designing systems to assist or automate driving is a fascinating challenge and may ultimately justify the investment of massive resources (Green and Brand, 1992). However, the successful development of such systems requires a fuller understanding of the fundamental characteristics of driving (Owens et al., 1993).

Researchers seeking a useful understanding of proficient driving and traffic safety skills should pursue multiple levels of conceptualization. They must address the regulatory difficulties that have persisted in licensing, training, and evaluating drivers as well as the problems associated with new technological systems. Many of these issues are widely recognized and repeatedly investigated, but systematic analyses are lacking. It is doubtful that useful general theories will follow from a focus on specific regulatory questions or from basic research on isolated or narrowly defined aspects of human performance. Rather, progress in understanding driving behavior will require coordination and integration of work at multiple levels of analysis, ranging from assessment of individual performance through specification of broad behavioral and informational requirements of the driving task to social aspects of communication among motorists in a changing traffic environment.

The theoretical requirements are challenging, and they offer exciting opportunities for gaining basic insights into human behavior in complex systems. The resulting agenda, however, should not be purely theoretical. Indeed, theory can receive an invaluable impetus from systematic attention to immediate practical problems. If investigators of current problems aim from the start to integrate their findings into a larger account of proficient driving performance in a complex traffic environment, their research will be doubly fruitful.

Screening and Licensing

Licensing examinations are universally administered. In the United States, they are typically based on an in-vehicle road test, a visual examination, and a series of questions testing the applicant's knowledge of driving regulations. The central problem with driver licensing is the poor predictive power of current test procedures. This issue is gaining in importance as a consequence of the 1990 Americans With Disabilities Act, which requires that exclusionary regulations be justifiable. We need to base licensing decisions on valid predictors of safe driving. Several states have already begun to relax visual acuity requirements as part of graduated licensing programs. Human factors research will be critically important in paving the way to valid and fair licensing procedures. At the same time, development of more reliable and efficient test procedures can contribute to the development of a comprehensive theory of driving.

In-Vehicle Testing and the Possible Benefits of Simulation

The current driving test typically evaluates the ability to negotiate a vehicle at low speeds on a specially constructed test track involving numerous turns and traffic signals. Such tests rarely, if ever, involve conditions under which accidents are most likely to occur, for example, under long-term demands, high speeds, competition for attention, emergencies, nighttime, or adverse weather driving conditions. If candidates for a driving license are to be tested under more realistic conditions, the method of choice is a driving simulator (McKnight and Stewart, 1990). Simulators are still expensive so they will probably be used only in selected cases. This situation appears to be changing, however, as increasingly sophisticated video and computing systems emerge at dramatically reduced costs. Such simulators may offer the most practical method to evaluate drivers under conditions that correspond more closely to those encountered in the driving environment. Although driving simulators have been available for many years, few are interactive, that is, with the visual environment changing to conform with vehicle movements. Interactive displays are the rule in aviation simulators, but making them available for driving simulation will require the development of less expensive display equipment.

A major problem in all simulators is the coordination of visual and vestibular simulation. A moving vehicle stimulates the vestibular system. In a fixed-base simulator, however, only the visual component of motion is available; the resulting mismatch between the visual and vestibular (and to a lesser extent proprioceptive) systems can induce nausea. This phenomenon, simulator sickness, has been of interest in aviation, but primarily for simulations of high-performance aircraft. Simulator sickness sometimes causes complex symptoms similar to motion sickness and has been reported to produce disorientation inside and outside the simulator. It is not yet clear how great a problem this will be for driving simulators; this must be evaluated as test simulators are developed. One possible solution is to impart motion to the operator (moving base simulator), but this would entail considerable additional expense.

Current computational technology is proceeding at a dazzling pace and it is not unreasonable to entertain the possibility of having driving simulators with realistic interactive displays and at a reasonable cost in the not-distant future.

Vision Testing

It is often said that the driving task is 90 percent visual. While the original source of this proposition is obscure, most would agree that good vision is necessary. The problem is that we do not yet know what good vision means in the context of driving. The conventional standard, which calls for a corrected visual acuity of 20/40, is not empirically justified and may be virtually irrelevant. The ineffectiveness of present visual testing procedures was dramatically demonstrated by Burg (1967, 1968, 1971), who correlated the results of both standard and nonstandard visual tests with the frequency of accidents for 17,000 drivers. Of all the tests employed, only dynamic visual acuity, a test not included (to our knowledge) in any licensing procedure currently in use in the United States, correlated with accidents, and even there the correlation was weak.

The basis for this low predictive power is probably related to the discrepancy between visual demands encountered during testing and those encountered while driving. Visual acuity tests evaluate only the threshold of resolution for high-contrast optotypes, that is, the ability to read fine print in bright illumination. Such tests have proven to be valuable for prescribing reading glasses. However, driving usually depends on large fields of relatively coarse visual structure, and it frequently requires recognizing low-contrast objects under low (mesopic) illumination. Therefore, it should not be surprising that research shows little or no correlation between standard acuity tests and accidents.

There is, however, a growing body of literature, both empirical and theoretical, suggesting that tests of peripheral vision, contrast sensitivity, and motion perception may be more useful. Johnson and Keltner (1983) obtained visual fields on 10,000 eyes and determined that individuals with binocular scotomas were more likely to be involved in accidents. Testing of visual fields on a large scale will require the development of automated perimetric devices. This area is being actively investigated, not because visual fields are critical for driving but because visual field examinations can detect early visual pathology.

Another promising approach is to assess vision for moving targets. Burg (1971) found that dynamic acuity, which requires the observer to resolve a moving optotype, was the only subtest related to accidents. Because motion is pervasive outside the laboratory, both the driver and the stimuli of interest are moving. The introduction of a gaze stability requirement for licensure is a logical step toward greater ecological validity.

One should also note that most critical visual stimuli encountered during driving have low luminance contrast. During the past several decades, the vision community has accumulated impressive evidence that contrast sensitivity is more informative than visual acuity for high-contrast optotypes, especially in predicting performance for visual tasks other than reading. For example, an observer with incipient cataracts can demonstrate normal acuity even though his or her ability to recognize low-contrast objects has been severely impaired. Several inexpensive contrast-sensitivity tests are now available.

In view of the poor predictive ability of the static optotypes now being used in visual testing, and the empirical and theoretical evidence for the importance of both a contrast-sensitivity criterion and motion, the development of a contrast-sensitive test of dynamic spatial vision is both wanted and wanting.

Perhaps the most significant concomitant of normal aging is the loss of ability to receive light. With age the pupil of the eye becomes progressively smaller. Typically, the maximum pupil diameter changes from 8 mm in youth to 4 mm in old age, a fourfold reduction in pupil area and, therefore, in light transmission. In addition, with increasing age, the optical media gradually become less transparent. On the average, the effective light transmitted to the retina is reduced by a factor of 2 every 15 years. Since the accident rate per mile driven is approximately 3.5 times higher at night than during the day, some test of nighttime vision is clearly to be recommended.

Despite these facts, and despite the extensive literature and instrumentation on night vision (developed primarily during World War II), to our knowledge no tests of night vision are currently in use, and, except for student drivers, there is no proscription for driving at night. What is needed is a test of nighttime vision that can be shown to be related to driving performance under low illumination levels. Whether this should evaluate recognition vision, visual guidance, or both is an empirical question. On the basis of information currently available, it would appear reasonable to restrict night driving for some individuals. Such graduated licensing regulations will require that reliable and valid testing procedures be developed.

The possibility of innovative tests of driving proficiency should also be considered. Based on the pioneering work of Gibson (1950, 1966, 1968), it has recently been suggested that the ability to guide a vehicle based on optical flow is critical in driving (e.g., Warren et al., 1991; Royden et al., 1992; Crowell and Banks, 1993). With the low-cost computers and displays available, such tests, if they demonstrate predictive power, are now feasible. Other visual abilities that might be considered include distance perception, ability to judge closing rates, glare sensitivity, velocity perception, and the ability to handle divided attention. The low predictive power of present procedures is a cogent argument for utilizing the wealth of information already available about vision and vision tests to develop new and better procedures.

Alcohol

Alcohol is universally recognized as a major contributor to traffic accidents (Shinar, 1978). It has been estimated that approximately 50 percent of automobile accidents involve alcohol (Evans, 1991). Depending on state driving laws, the critical level of blood alcohol, above which the driver is assumed to be legally intoxicated, is on the order of 0.08 or 0.10 mg per 100 mg. However, the behavioral data do not support a fixed blood-alcohol criterion as a threshold of impairment. There is extensive evidence that many individuals are impaired at blood-alcohol levels that are half the values being used (Moskowitz, 1985). In addition, the limited performance data available demonstrate strikingly high intersubject variability (Wilson and Plomin, 1985). Some people with blood-alcohol levels below the current legal limits frequently perform less well on dynamic contrast sensitivity tests than others whose levels would classify them as legally intoxicated (Andre et al., 1992). Data are urgently needed to resolve this dilemma. If there is a blood-alcohol level that does not impair performance, it must be supported by performance data. Alternatively, the combination of drinking and driving should be prohibited and the legal limit set at the measurement error of the evaluation procedure.

Attention and Automaticity

A major problem in human performance is vigilance, that is, the need to remain alert during a repetitive task that induces drowsiness and lack of attention. The deficit may be general or may involve a narrowing of the effective field of attention. There are marked individual differences in the "useful field of view" (UFOV); these are correlated with accident records and, therefore, have significant implications for driving proficiency (Ball and Owsley, 1992). Recent studies indicate that restriction of the field of attention or UFOV poses a serious problem for older drivers. This important finding raises several research questions about how drivers allocate attention. What are the effects of fatigue, expectation, traffic demands, experience, and age? To what extent or in what mode is the UFOV test related to specific components of the driving task (Ball et al., 1993)? The relationship of the UFOV test to driving performance provides yet another example of the need for theoretical development. Much needs to be learned about the relevant attentional and automatic processes in driving.

Lack of alertness seems to be an obvious factor in transportation safety. Depending upon the state of the driver, reaction times can vary from less than 1 second to 2.5 seconds or more (e.g., Summala, 1981; Triggs and Harris, 1982). Unfortunately, it is difficult for drivers to monitor their own state of alertness, particularly during prolonged driving. A system that would inform drivers of waning alertness would be extremely valuable.

Training New Drivers

Understanding the skills involved in driving will provide insights about how these skills are acquired. The effectiveness of current driver training is problematic, and it is important to learn better ways of teaching the skills involved in driving.

We suspect that the basic problem stems from the hierarchical nature of the driving task. The operational component—starting, steering, and stopping a vehicle—is relatively easy to learn and, once it has been learned, gives the novice driver the feeling of being in control. New drivers can also learn much of the "knowledge" (i.e., rules of the road) in a short time. Indeed, such rule-based knowledge is strongly emphasized in most driver tests. However, the driving task involves much more than vehicle control and familiarity with rules: for example, perceiving and anticipating the actions of other motorists and the potential of other roadway hazards. The identification and teaching of such abilities should be the focus of research.

Misestimated Risk

We suggested above that the fundamental skills involved in driving are easy to acquire and that accidents may arise from failure to perceive and compensate for risk. In effect, the driver is aware of the danger and has the ability to take the necessary precautions but does not always anticipate or recognize the relevant hazards (Leibowitz and Owens, 1986). The higher accident rate among young drivers and the effects of alcohol, which typically increase risk-taking behavior, are consistent with this observation. Wilde (1988) has proposed that drivers adjust their risk level so as to compensate for engineering innovations in driver safety such as seat belts (risk homeostasis). In recent years, there has been an increase in the research literature devoted to risk-taking behavior, and its application to driving is clearly indicated (Wagenaar, 1992).

Displays

In contrast to the extensive research in aviation, relatively little attention has been given to displays for vehicles. It has been suggested that displays that convey information to the driver—such as stopping distance, momentum, and distance to road obstructions—would be useful. It has also been proposed that display information be optically superimposed on the road ahead so as to reduce the need for scanning the instrument panel (Weihrauch et al., 1989; Okabayashi et al., 1989). ("Heads-up" displays are now common in military aircraft.) Such proposals should be empirically tested. An adequate theory of driving will help specify the types of information that are most useful for safe driving. It will also help determine when the information is useful and what format of presentation (display design) is most effective (Weintraub and Ensing, 1992).

Aging

A common thread relevant to all of these problems is the effect of age on driving skills (Barr and Eberhard, 1991; Eberhard and Barr, 1992). It is well known that some major abilities relevant to driving—such as night vision, smooth pursuit eye movements, and reaction time—all decline systematically with age and that the elderly frequently experience difficulty with cluttered environments. It is also well established that individual differences in aging are pronounced and that age per se is not an unequivocal predictor of performance. Furthermore, to a great extent, the aging driver probably compensates behaviorally for perceptual and cognitive deficiencies, thus minimizing or delaying costs to safety.

Given the increasing longevity of our population and the central role of mobility as a factor in the quality of life, research in this area is particularly critical (Waller, 1991). Again, we see the need for a comprehensive theory of driving behavior that will enable clearer delineation of the specific skills and situations in which older drivers are most likely to face difficulties. As we noted earlier, we already have valuable information about some age-related changes (e.g., night vision and the UFOV), but we do not clearly understand the impact of such changes on safe driving (e.g., to what extent the older driver can or does compensate). In order to evaluate the capabilities of elderly drivers and provide appropriate advice and restrictions, we need a scientific foundation for graduated licensing criteria. In effect, for all of the categories mentioned above, special attention should be given to the role of aging.

Intelligent Vehicle Highway

Efforts are currently being directed toward the application of sophisticated control systems that would take over some of the driver's responsibilities. For example, the difficulty of estimating distance and velocity is assumed to account for the frequency of accidents involving a vehicle turning left in the face of oncoming traffic. To prevent such errors, it has been proposed that each vehicle be equipped with a system that senses how far the vehicle is from other vehicles and that informs the driver when it is unsafe to attempt a left turn. Assuming such systems are technically feasible, the critical question is whether drivers accustomed to observation outside the vehicle would divert their eyes from the road in order to obtain information from a display. The application of sophisticated information systems assumes that the driver will be willing to trust and act on such information. Solving these problems will require extensive human engineering testing to determine the optimum method of presenting the information, for example, heads-up displays and auditory warnings. The human factors community has had extensive experience with similar problems in aviation, but they are by no means resolved. For example, ground proximity warning systems provide urgent auditory warning of impending flight into terrain, but in many instances, crews have disregarded the warning and have flown into the ground. It is critical that human factors analysis be incorporated in these programs (Weintraub and Ensing, 1992; Ervin, 1993; Owens et al., 1993).

Legal Applications

A potentially fruitful area for research is the interaction between the law and transportation. Specifically, many laws require operators of vehicles to accomplish tasks that are not within their capabilities. This leads to unnecessary litigation and appellate reviews and creates a disrespect for laws. If statutes such as the Assured Clear Distance Ahead rule and regulations governing the use of alcohol were examined in relation to the behavioral sciences literature on human capabilities and limitations while operating a vehicle, the findings could lead to more rational laws and codes (Leibowitz et al., submitted for publication). Efforts directed at this problem represent a potentially valuable contribution.

General Comments

The automobile plays a critical role in the quality of life in the United States as well as in other technologically advanced countries. Without the mobility it provides, the lives of many millions would be severely degraded. For this reason, legislatures have been reluctant to impose restrictions on driving licenses. From the data presently available, it is not readily apparent whether the unprecedented costs in human suffering and economic loss that our society tolerates as the price of mobility are justifiable. However, there is no doubt that measures to reduce this burden—whether they involve training, engineering innovations, alternative sources of transportation, or licensing restrictions—must be based on behavioral data.

TEAM PERFORMANCE AND AVIATION SAFETY

In stark contrast with automobile travel, commercial aviation is the safest form of mass transportation. Traditional ergonomics has played a substantial role in the development of all modern transport aircraft and remains an integral part of the design process (Wiener and Nagel, 1988). Commercial aviation remains one of the few vocations in which retirement at a fixed age is still required. Thus, age-related accidents are not a current problem, and, similarly, little evidence has been found to suggest that alcohol and drugs pose major safety problems. (These problems do exist, however, in general aviation.)

Despite the focus on human factors in aviation, more than two-thirds of those accidents and incidents that do occur in the system include human error/human factors as causal elements (Lautman and Gallimore, 1987; Cooper et al., 1980). Although the total number of casualties is small relative to those produced by the automobile, aviation accidents are highly visible and elicit strong demands for action. The aviation community has responded to these data with both research and training addressing the human factors identified in accidents and incidents. These activities reflect a broader conception of human factors as it intersects social, personality, and organizational psychology. The results of these interventions, occurring worldwide and known generically as cockpit or crew resource management (CRM) training, represent one of the success stories of applied psychology and human factors. The programs have been shown to produce significant changes in the behavior and attitudes of flight crews and have been identified as preventing or mitigating serious accidents in both civil and military aviation (Diehl, 1991; Helmreich and Foushee, 1993; Helmreich and Wilhelm, 1991; Wiener et al., 1993). Despite the progress that has been made in CRM, a number of major research questions remain. In the following sections, we describe the CRM approach to aviation human factors and the open issues requiring research and committed action.

Human Factors Training Approaches

Flight crews operate within a system in which the individual functions as part of a team that functions within an organization that, in turn, is embedded in a regulatory and ambient environment. All components of the system influence group dynamics and the outcomes of a particular flight. As with many aspects of driver training, it is not clear that all relevant and needed training is provided or that the most appropriate aspects of performance are being evaluated. The Federal Aviation Administration defines pilot training specifications in the United States; these include extensive training in emergency procedures and technical maneuvers, many of which are seldom encountered in operational flight. Historically a single-pilot mentality has prevailed in regulation, which has ignored the need to perform effectively the team tasks that are required to pilot a modern transport in complex air space. For example, to maintain licensure, air transport pilots must currently undergo a proficiency check that focuses on individual technical expertise as demonstrated by execution of fixed maneuvers.

CRM training has evolved into a systematic approach to group communications, team coordination, leadership, decision making, and conflict resolution.

Effective training is based on diagnosis of organizational norms and problems, organizational restructuring (where needed), and experiential training that focuses on the development of specific behavioral skills (Federal Aviation Administration, 1993; Byrnes and Black, 1993; Taggart, 1994). Also integral to this approach is continuing (recurrent) training and reinforcement of effective behaviors, usually through the use of structured simulation (Butler, 1993). The critical difference between CRM training in aviation is the focus on the group rather than the individual as the unit of behavior and evaluation.

Research Needs in Aviation and Team Performance

Integrating Technical and Psychological Training

Although CRM training for flight crews has shown demonstrable, positive effects, it has not been successfully integrated with traditional technical instruction. The value of integrated instruction for complex technical operations—so that, for example, training in maintaining control after losing an engine on takeoff is integrated with training in the coordination, communication, and decision making required to deal with such a contingency—is widely recognized, and recently implemented Federal Aviation regulations will require such integration (Federal Aviation Administration, 1990; Helmreich and Foushee, 1993). The accomplishment of this goal, however, will provide a continuing challenge for human factors specialists.

A larger challenge will be found in the integration of human factors concepts with initial pilot training. To date, the emphasis has been on adding human relations skills to the repertoires of already qualified pilots, a process that often requires unlearning old behaviors and changing traditional practices. While the International Civil Aviation Organization, the component of the United Nations that regulates worldwide civil aviation, has introduced a requirement for human factors training as a condition of licensure, this does not equate to demonstrating and evaluating the human factors aspects of primary flight activities. A fully integrated curriculum that stresses interpersonal, as well as individual, skills from the outset should result in better-prepared pilots.

Understanding the Impact of Culture on Team Performance

One finding that emerged from research into the effectiveness of CRM training was the existence of differential reactions to training in different organizational and national settings (Helmreich and Foushee, 1993; Helmreich and Wilhelm, 1991). Many of the differences could be traced to the particular views that organizational cultures and subcultures held regarding appropriate roles and behaviors. In addition, comparative data on the impact of CRM training from various national cultures are in accord with Hofstede's (1980, 1991) conceptualization of the dimensions of national culture, particularly differences in individualism/collectivism and in attitudes toward authority defined as power distance (Merritt and Helmreich, in press). For example, a training emphasis on the importance of the *group* and its maintenance is highly congruent with values in collectivist cultures, such as many South American and Asian nations, but less readily acceptable in more individualistic societies, such as the United States and Ireland. Similarly, an approach that stresses reducing status differentials between leaders and team members and the need for juniors to question the actions and decisions of leaders when they threaten safety is compatible with U.S. and Australian notions of egalitarianism, but mystifying to those from cultures, such as China and Malaysia, that rank high on Hofstede's power distance dimension in the reluctance of subordinates to question the actions of leaders. CRM training has not been successfully exported from one culture to another without consideration of differing values; this shows the importance of cultural issues even for supposedly standardized tasks, such as managing the flight of a transport aircraft. One challenge for human factors is to understand how cultures, both national and organizational, influence attitudes and behaviors related to safety and efficiency and to adapt training to reflect cultural issues. It should be of great concern to the manufacturers of aircraft that are marketed throughout the world.

Optimizing Curricula and Instructional Methods

Related to the influence of culture on attitudes and receptivity to instruction is a need to more precisely define which issues need to be stressed in human factors training and how best to deliver instruction. In general, human factors specialists have concluded from experience that curricula need to have high levels of specificity, to avoid overreliance on psychological concepts and jargon, and to be experiential rather than didactic. Even in this area, however, cultural values may influence responsiveness. Cultures that place a high value on instruction by authority figures may be less receptive to group discussion and more favorably influenced by more authoritative presentations. Overall, the goal for human factors is to define curricula that are relevant, are understandable, and can be presented in a way that can be understood in the context of individual values.

Improving the Evaluation of Group and System Performance

It is axiomatic that behavior cannot be successfully taught unless instances of the desired behavior are reinforced and that reinforcement cannot occur unless the behavior can be precisely evaluated. Although the evaluation of individual technical performance in aviation has a long and successful history, this focus has stifled concern with developing more sophisticated approaches to the evaluation of group behavior and performance. Early attempts to improve group-level evaluation in aviation illustrated the difficulties of such assessments and the reluctance of operational personnel to expand the scope of evaluation (Gregorich and Wilhelm, 1993; Butler, 1993; Taggart, 1994). A more reliable and valid methodology for assessment of group-level phenomena should be widely beneficial, not only in aviation but also in other endeavors involving group-level tasks.

Also needed are improved methodologies for analyzing and understanding human error at the group and system levels. The study of complex determinants of error in technological environments has advanced significantly in recent years (Perrow, 1984; Rasmussen, 1993; Reason, 1990). Detailed analyses of aircraft accidents, in which pilot error is clearly the proximal cause, typically uncover an array of contributing factors that influenced the decision making and group dynamics on the flight deck. For example, the crash of a Canadian airliner on takeoff during a snowstorm resulted from the crew's flawed decision to take off with ice contaminating the wings. However, the findings of a commission of inquiry that investigated all aspects of the aviation system isolated a number of contributing factors at the regulatory, organizational, and group levels that created an environment with inadequate safeguards against a fatal decision (Helmreich, 1992; Moshansky, 1992). A taxonomy of human factors problems that can be applied to the analysis of accidents and incidents would be invaluable for researchers and for those charged with safety (Jones, 1993).

Automation and the "Electronic" Crew Member

As increasingly sophisticated computer systems characterize the flight decks of modern aircraft, crews face the new dilemma of how to integrate an "electronic" crew member into team operations. When many activities are shifted from human to computer control, issues of maintaining competencies, vigilance, and awareness arise, and shifts may occur in the dynamics of crew interaction. Defining a coherent, research-based philosophy of automation is a critical task that involves human factors experts participating in the design and manufacture of aircraft as well as in the organizations that operate them (Billings, 1989; Wiener, 1993). The philosophy of automation needs to address what should be automated as well as how automation should be accomplished. The principle that what can be automated should be automated has not demonstrated marked success in reducing human error or workload, but has resulted in enormous capital expenditures. An excellent discussion of issues of automation design is found in the report of an extensive investigation of the fatal flight into terrain of a highly automated aircraft in 1992 (Ministère de L'Équipement, des Transportes, et du Tourisme, 1993). A philosophy of automation also needs to address the use of automated systems: when they should be used and when control should revert to the human operator. At the user level, this needs to be incorporated into training and organizational norms.

Recent research suggests that there are large cultural differences in attitudes regarding the acceptance and use of automated systems, including, for example, willingness to disengage automated systems and revert to human control when conditions change (e.g., Sherman and Law, 1994). Human factors specialists and airframe manufacturers must recognize that human-computer interfaces are highly varied and culturally determined; they can then undertake research to define these differences and training strategies to deal with them.

Advanced technology aircraft are now operated with a crew of two pilots. However, crews are augmented with relief pilots for extremely long intercontinental flights. Extended routine cruising raises questions of maintaining proficiency, combatting fatigue and complacency, and transferring control among extended crews. Many of these issues have not been systematically addressed in research.

Dealing with Training Failures

The success of human factors training in aviation is diminished to some extent by the fact that the concepts and behaviors taught are not universally accepted, even within a particular organization or culture. In all programs evaluated, a small subset of individuals fails to respond positively to training efforts, and some may even become less accepting of and more resistant to team coordination (Chidester et al., 1990a, 1990b; Helmreich and Wilhelm, 1989, 1991). These failures in training are necessarily of concern because those who actively reject strategies to enhance performance are likely to pose the greatest threat to safety. One source of resistance may lie in personality characteristics of those who react negatively. For example, Chidester et al. (1990a, 1990b) found that those lacking in attributes associated with effective interpersonal behavior were more prone to reject training that stressed the importance of interpersonal communication. Given the unlikelihood of effecting basic changes in personality except by extended psychotherapy, improved selection in terms of performance-related personality traits may be the last line of defense against ineffective crew coordination and teamwork. However, the human factors community should exhaust all avenues of research to determine if there are means of remediation for those deficient in interpersonal communication skills.

Extending the Concepts Beyond the Cockpit Door

Human factors research in aviation has grown from its original focus on enhancing communication and coordination among the basic flight team into an awareness that many of the problems in the system reside in interfaces between teams and other components of the aviation system, including, for example, air traffic control and ground operations (Taggart, 1993, 1994). New strategies and associated research will be required to develop and validate training methodologies that effectively address intergroup as well as intragroup coordination and cooperation.

EXTENDING HUMAN FACTORS TRAINING INTO OTHER DOMAINS

It is becoming clear that the human factors approaches encompassed by CRM in aviation apply more generally to endeavors in which teams function with technology under demanding conditions (Helmreich et al., 1993). For some years, the National Transportation Safety Board (NTSB) has been advocating the extension of CRM from the cockpit to the bridge of maritime vessels (e.g., National Transportation Safety Board, 1993). More recently, the medical profession has noted similarities in human factors problems in the operating room and in the cockpit and has begun to adapt training and evaluation strategies from aviation (Ewell and Adams, 1993; Howard et al., 1992; Helmreich and Schaefer, 1994). Similar issues should exist in control rooms of nuclear power plants, petrochemical operations, and other manufacturing enterprises (Helmreich et al., 1993). It is essential that the human factors community avoid trying to export training from one domain to another in a simplistic manner; the pitfalls of such an approach were discovered in attempts to move training across national boundaries. The characteristics and cultures of each domain must be investigated and understood before common concepts can be applied effectively.

Maritime Operations

A few examples may illustrate extensions of human factors concerns from aviation to seafaring. Although events play out at a much slower pace than in an automobile or aircraft, ships are operated by teams interacting with technology in a regulated environment. In the majority of marine disasters, as in aviation, human error is also implicated. Automation is also becoming a common shipboard characteristic, associated with large decreases in personnel and again raising issues of vigilance and reliance on computer solutions. Issues of national culture are also of great importance in maritime operations, and many ships are operated by multinational crews for economic reasons. The traditional culture of the sea, in which the captain is the unchallenged (and frequently autocratic) master, can inhibit the effective communication and utilization of available resources, especially human ones. Team interface issues have also been implicated in a number of marine accidents, specifically flawed communications between pilots and masters while operating in restricted waters (e.g., National Transportation Safety Board, 1993).

As noted, the NTSB has concluded that many of these problems could be alleviated through effective human factors interventions and the adoption of approaches successfully employed in aviation. This will, however, require a substantial research endeavor to define common and divergent factors and to design human factors programs that will be accepted within the general maritime and specific organizational cultures involved.

CONCLUSIONS

The field of transportation provides rich opportunities to expand the scope of human factors in areas for which outcomes have major consequences. Optimizing the interface between individuals and groups with complex technology and systems requires a multidisciplinary approach that embraces the full range of concerns of human factors specialists. Concern with a particular problem, whether vehicular or aviation safety, should not blind researchers to concepts that transcend problem areas and that reflect more broadly on human capabilities and limitations. At the same time, research and solutions must be sensitive to the characteristics of the particular endeavor.

REFERENCES

Andre, J.T., R.A. Tyrrell, M.E. Nicholson, M. Wang, and H.W. Leibowitz
    1992   Measuring and predicting the effects of alcohol on contrast sensitivity for static and dynamic gratings. *Investigative Ophthalmology and Visual Science* (ARVO Abstract) 33(4):1416.

Ball, K., and C. Owsley
    1992   The useful field of view: a new technique for evaluating age-related declines in visual function. *Journal of the American Optometric Association* 63:71-79.

Ball, K., C. Owsley, M. Sloan, D.L. Roenker, and J.R. Bruni
    1993   Visual attention problems as a predictor of vehicle crashes among older drivers. *Investigative Ophthalmology and Visual Science* 34(11):3110-3123.

Barr, R.A., and J.W. Eberhard
    1991   Safety and mobility of elderly drivers, Part I. Special Issue. *Human Factors* 33(5):497-603.

Billings, C.E.
    1989   Toward a human centered aircraft automation philosophy. Pp. 1-8 in *Proceedings of the Fifth International Symposium on Aviation Psychology*. Columbus: Ohio State University.

Burg, A.
1967   *The Relationship Between Vision Test Scores and Driving Record: General Findings.* Department of Engineering Report No. 67-24. Los Angeles, Calif.: University of California.
1968   *The Relationship Between Vision Test Scores and Driving Record: Additional Findings.* Department of Engineering Report No. 68-27. Los Angeles, Calif.: University of California.
1971   Vision and driving: a report on research. *Human Factors* 13(1):79-87.

Butler, R.E.
1993   LOFT: full-mission simulation as crew resource management training. Pp. 231-259 in E. Wiener, B. Kanki, and R. Helmreich, eds., *Crew Resource Management.* San Diego, Calif.: Academic Press.

Byrnes, R.E., and R. Black
1993   Developing and implementing CRM programs: the delta experience. Pp. 421-443 in E. Wiener, B. Kanki, and R. Helmreich, eds., *Crew Resource Management.* San Diego, Calif.: Academic Press.

Chidester, T.R., R.L. Helmreich, S.E. Gregorich, and C.E. Geis
1990a   Pilot personality and crew coordination: implications for training and selection. *International Journal of Aviation Psychology* 1:23-42.

Chidester, T.R., B.G. Kanki, H.C. Foushee, C.L. Dickinson, and S.V. Bowles
1990b   *Personality Factors in Flight Operations,* Vol. 1. *Leader Characteristics and Crew Performance in Full-Mission Air Transport Simulation.* NASA Technical Memorandum No. 102259. Moffett-Field, Calif.: NASA-Ames Research Center.

Cooper, G.E., M.D. White, and J.K. Lauber, eds.
1980   *Resource Management on the Flightdeck: Proceedings of a NASA/Industry Workshop.* NASA CP-2120. Moffett Field, Calif.: NASA-Ames Research Center.

Crowell, J.A., and M.S. Banks
1993   Perceived heading with different retinal regions and types of optic flow. *Perception and Psychophysics* 53(3):325-337.

Diehl, A.E.
1991   The effectiveness of training programs for preventing aircrew "error." In *Proceedings of the Sixth International Symposium of Aviation Psychology.* Columbus: Ohio State University.

Eberhard, J.W., and R.A. Barr, eds.
1992   Safety and mobility of elderly drivers, Part II. Special Issue. *Human Factors* 34(1):1-65.

Ervin, R.
1993   Bringing Human Factors Expertise to the Conceptualization of Active-Safety Technology. Paper presented at the UMTRI Human Factors Festival. Ann Arbor: University of Michigan Transportation Research Institute.

Evans, L.
1991   *Traffic Safety and the Drive.* New York: Van Nostrand Reinhold.

Ewell, M.G., and R. Adams
1993   Aviation psychology, group dynamics and human performance issues in anesthesiology. Pp. 499-504 in *Proceedings of the Seventh International Symposium on Aviation Psychology.* Columbus: Ohio State University.

Federal Aviation Administration
1990   *Advanced Qualification Program.* Washington, D.C.: Federal Aviation Administration.
1993   *Crew Resource Management Training.* Advisory Circular AC 120-51A. Washington, D.C.: Federal Aviation Administration.

Gibson, J.J.
1950 *The Perception of the Visual World.* Boston, Mass.: Houghton Mifflin.
1966 *The Senses Considered as Perceptual Systems.* Prospect Heights, Ill.: Waveland Press.
1968 What gives rise to the perception of motion? *Psychological Review* 75:335-346.

Green, P., and J. Brand
1992 *Future In-Car Information Systems: Input from Focus Groups.* SAE Technical Paper No. 920614. Warrendale, Penn.: Society of Automotive Engineers.

Gregorich, S.E., and J.A. Wilhelm
1993 Crew resource management training assessment. Pp. 173-198 in E. Wiener, B. Kanki, and R. Helmreich, eds., *Crew Resource Management.* San Diego, Calif.: Academic Press.

Helmreich, R.L.
1992 Human factors aspects of the Air Ontario crash at Dryden, Ontario. In V.P. Moshansky, ed., *Commission of Inquiry Into the Air Ontario Accident at Dryden, Ontario: Final Report.* Ottawa, Ontario: Minister of Supply and Services, Canada.

Helmreich, R.L., and H.C. Foushee
1993 Why crew resource management? Empirical and theoretical bases of human factors training in aviation. In E.L. Wiener, B.G. Kanki, and R.L. Helmreich, eds., *Cockpit Resource Management.* San Diego, Calif.: Academic Press.

Helmreich, R.L., and H.-G. Schaefer
1994 Team performance in the operating room. Pp. 225-254 in M.S. Bogner, ed., *Human Error in Medicine.* Hillsdale, N.J.: Erlbaum.

Helmreich, R.L., and J.A. Wilhelm
1989 When training boomerangs: negative outcomes associated with cockpit resource management programs. In R.S. Jensen, ed., *Proceedings of the Fifth International Symposium on Aviation Psychology.* Columbus: Ohio State University.
1991 Outcomes of crew resource management training. *International Journal of Aviation Psychology* 1(4):287-300.

Helmreich, R.L., E. Wiener, and B.G. Kanki
1993 The future of crew resource management in the cockpit and elsewhere. Pp. 479-501 in E. Wiener, B. Kanki, and R. Helmreich, eds., *Crew Resource Management.* San Diego, Calif.: Academic Press.

Hofstede, G.
1980 *Culture's Consequences: International Differences in Work-Related Values.* Beverly Hills, Calif.: Sage.
1991 *Cultures and Organizations: Software of the Mind.* Maidenhead, England: McGraw-Hill.

Howard, S.K., D.M. Gaba, K.J. Fish, G. Yang, and F.H. Sarnquist
1992 Anesthesia crisis resource management: teaching anesthesiologists to handle critical incidents. *Aviation, Space, and Environmental Medicine* 63:763-770.

Johnson, C.A., and J.L. Keltner
1983 Incidence of visual field loss and its relation to driving performance. *Archives of Ophthalmology* 101:371-375.

Jones, S.G.
1993 Human factors in incident reporting. Pp. 567-572 in *Proceedings of the Seventh International Symposium on Aviation Psychology.* Columbus: Ohio State University.

Lautman, L.G., and P.L. Gallimore
1987 Control of the crew-caused accident. *Airliner Magazine*, April-June:1-7. Seattle, Wash.: Boeing Commercial Airplane.

Leibowitz, H.W., and D.A. Owens
  1986   We drive by night. *Psychology Today* 54-58.
Leibowitz, H.W., D.A. Owens, and R.A. Tyrrell
           The Assured Clear Distance Ahead Rule: Implications for Traffic Safety and the Law. (Submitted for Pulication)
McKnight, A.J., and M.A. Stewart
  1990   *Development of a Competency Based Driver License Testing System*. Final Report for the California Department of Transportation. Landover, Md.: National Public Services Research Institute.
Merritt, A.C., and R.L. Helmreich
  In Press   Human factors on the flightdeck: the influence of national culture. *Journal of Cross-Cultural Psychology*.
Ministère de L'Équipement, des Transportes, et du Tourisme
  1993   Rapport de la Commission d'Enquête sur l'Accident Survenu le 20 Janvier 1992 Près du Mont Sainte Odile (Bas Rhin) a l'Airbus A.320 Immatricule F-GGED Exploité par la Compagnie Air Inter. Paris, France: Ministère de L'Équipement, des Transportes, et du Tourisme.
Moshansky, V.P.
  1992   *Commission of Inquiry Into the Air Ontario Accident at Dryden, Ontario: Final Report*, Vols. 1-4. Ottawa, Ontario, Canada. Minister of Supply and Services.
Moskowitz, H., M.M. Burns, and A.F. Williams
  1985   Skills performance at low blood alcohol levels. *Journal of Studies on Alcohol* 46(6):482-485.
National Safety Council
  1993   *Accident Facts*, 1993 edition. Itasca, Ill.: National Safety Council.
National Transportation Safety Board
  1993   Grounding of the UK passenger vessel *RMS Queen Elizabeth 2* near Cuttyhunk Island, Vineyard Sound, Massachussetts. NTSB/MAR-93/01. Washington, D.C.: National Transportation Safety Board.
Okabayashi, S., M. Sakata, J. Fukano, S. Daidoji, C. Hashimoto, and T. Ishikawa
  1989   *Development of Practical Heads-Up Display for Production Vehicle Application*. SAE Technical Paper Series No. 890559. Warrendale, Penn.: Society of Automotive Engineers.
Owens, D.A., G. Helmers, and M. Sivak
  1993   Intelligent vehicle highway systems: a call for user-centered design. *Ergonomics* 36(4):363-369.
Perrow, C.
  1984   *Normal Accidents: Living With High Risk Technologies*. New York: Basic Books.
Rasmussen, J.
  1993   Deciding and doing: decision making in natural context. In G. Klein, J. Orasanu, R. Calderwood, and C. Zsambok, eds., *Decision Making in Action: Models and Methods*. Norwood, N.J.: Ablex.
Reason, J.
  1990   *Human Error*. Cambridge, England: Cambridge University Press.
Royden, C.S., M.S. Banks, and J.A. Crowell
  1992   The perception of heading during eye movement. *Nature* 360:583-585.
Sherman, P.J., and J.R. Law
  1994   Are there differences for standard vs. automated aircraft? Within-airline comparisons of aircraft types. Pp. 110-115 in *Proceedings of the 14th Biennial Meeting of Applied Behavioral Sciences Symposium*. Colorado Springs, Colo.: United States Air Force Academy.

Shinar, D.
1978 *Psychology on the Road*. New York: Wiley.
Sivak, M.
1993 Some Speculations About How to Determine Traffic-Safety Priorities. Unpublished presentation at the UMTRI Human Factors Festival, Ann Arbor, Michigan. University of Michigan Transportation Research Institute, March.
Summala, H.
1981 Driver/vehicle steering response latencies. *Human Factors* 23:683-692.
Taggart, W.R.
1993 How to kill off a good CRM program. *The CRM Advocate* 93(1):11-12.
1994 Crew resource management: achieving enhanced flight operations. In N. Johnston, N. McDonald, and R. Fuller, eds., *Aviation Psychology in Practice*. Brookfield, Vt.: Ashgate.
Triggs, T.J., and W.G. Harris
1982 Reaction time of drivers to road stimuli. Report No. 0-86746-1470 for Commonwealth Department of Transportation, Australian Office of Road Safety.
Wagenaar, A.C.
1992 Risk taking and accident causation. Pp. 257-281 in J.F. Yates, ed., *Risk-Taking Behavior*. New York: John Wiley & Sons.
Waller, P.F.
1991 The older driver. *Human Factors* 33(5):499-505.
Warren, W.H., Jr., D.R. Mestre, A.W. Blackwell, and M.W. Morris
1991 Perception of circular heading from optical flow. *Journal of Experimental Psychology: Human Perception and Performance* 17(1):28-43.
Weihrauch, M., T.C. Goesch, and G.G. Meloeny
1989 *The First Head Up Display Introduced by General Motors*. SAE Technical Paper No. 890288. Warrendale, Penn.: Society of Automotive Engineers.
Weintraub, D.J., and M. Ensing
1992 *Human Factors Issues in Head-Up Displays: The Book of HUD*. Wright-Patterson Air Force Base, Ohio: Crew Systems Ergonomics Information Center.
Wiener, E.
1993 Crew coordination and training in the advanced-technology cockpit. Pp. 199-230 in E. Wiener, B. Kanki, and R. Helmreich, eds., *Crew Resource Management*. San Diego, Calif.: Academic Press.
Wiener, E., B. Kanki, and R. Helmreich, eds.
1993 *Crew Resource Management*. San Diego, Calif.: Academic Press.
Wiener, E.L., and D.C. Nagel
1988 *Human Factors in Aviation*. San Diego, Calif.: Academic Press.
Wilde, G.J.S.
1988 Risk homeostasis theory and traffic accidents: propositions, deductions and discussion of dissension in recent reactions. *Ergonomics* 31(4):441-468.
Wilson, J.R., and R. Plomin
1985 Individual differences in sensitivity and tolerance to alcohol. *Social Biology* 32(3-4):162-184.

10

Cognitive Performance Under Stress

J. Frank Yates, Roberta L. Klatzky, and Carolynn A. Young

The media have given considerable attention to the stress in contemporary society. For instance, the topic was the cover theme of a *Time* magazine issue in 1983 (June 6). Within the past few years, there have been numerous highly publicized incidents in which U.S. Postal Service employees have attacked and killed their supervisors and coworkers. The "headline" segment on a recent ABC News *20/20* program was devoted to these events. The program speculated that job stress, including the stress associated with the Postal Service's efforts to improve its highly automated sorting systems, was a significant contributor to the violence.

The current focus on stress extends beyond the popular press. Scholarly analyses of several prominent tragedies have cited stress as a factor:

• Wickens (1992) conjectures that stress may have impaired the performance of a control room operator on duty during the nuclear reactor accident at Three Mile Island, making the situation worse than it might have been otherwise. Wickens suggests that stress may have played a similar role in the downing of an Iranian airliner by the crew of the U.S. missile frigate *Vincennes*, who thought they were under attack by a military aircraft.

• In March 1989, Air Ontario Flight 363 crashed and burned on takeoff from Dryden, Ontario. The pilot, copilot, a flight attendant, and 21 passengers were killed. The immediate cause of the accident was snow and ice on the wings. Helmreich (1990) performed a human factors study of the incident. He concluded that time pressure and other stressors quite likely induced the pilots to make faulty—and ultimately fatal—decisions about deicing and departure.

• In every armed conflict, military personnel sometimes mistakenly fire on their own confederates. There were numerous instances of such "friendly fire" among American troops in the Persian Gulf War of 1991. Military analysts acknowledge a host of causes of these incidents, including cognitive ones like poor situational awareness (U.S. Army, 1992a, 1992b); these causes are also thought to include factors commonly acknowledged as forms or concomitants of stress, such as anxiety.

The stress literature is not restricted to high-profile occurrences like those just described. There is considerable published work on the stresses of everyday life, including—or perhaps especially—in the ordinary workplace (Manuso, 1983). In 1988, the National Research Council published a highly influential report by its Committee on Techniques for the Enhancement of Human Performance (Druckman and Swets, 1988). The committee did a critical review of research on a host of performance-related topics, ranging from motor skills to paranormal phenomena. Subsequent work by the committee (Druckman and Bjork, 1991) extended the range of topics even further. The 1988 report (p. 115) acknowledged that "none of the topics . . . has received more attention than the management of stress." This impression seems accurate; the scholarly literature on stress is enormous. The American Psychological Association's PsycINFO database lists over 35,000 entries under the keyword *stress*. Using the joint keyword query *stress and performance*, a recent search directed toward topics more closely associated with human factors yielded 2,565 entries.

Everyone seems to acknowledge that stress is a serious problem, and the research community has appeared to respond by devoting a great deal of energy to solving that problem. Why, then, is the topic being proposed as an area of additional emphasis in human factors research? There are several responses to this question, which we will develop more fully below. Briefly, however, the thrust of our argument is not that we need to increase the sheer volume of stress research (although that may well be true), but that future research should direct more attention to particular cognitive aspects of stress-performance interactions, aspects that have managed to escape close scrutiny. (By convention, the term *cognition* refers to acts of perception and knowing.) A hint of this oversight is suggested by a search of the PsycINFO database using the query *stress and cognition*. This search produced only 223 items, a yield of fewer than 9 items per year over the 1967-94 coverage of the database. In their comprehensive survey, Mross and Hammond (1989) have also noted the sparseness of the stress-cognition literature.

A good case also can be made that stress and the significance of its consequences will increase over the coming years:

- Global economic competition has increased dramatically over the past couple of decades, and it shows little sign of slackening in the near future. Thus, both public organizations, such as the U.S. Postal Service, and private organizations will undoubtedly find themselves severely challenged simply to survive. These challenges are virtually guaranteed to cause every person in those organizations to experience greater stress. For instance, polls suggest that there is an emerging sense among the public that job security is rarer now than in the past and that this condition will last a long time (Church, 1993).
- Economic and technological developments have led to greater interdependence among individuals and organizations, often accompanied by centralization. This implies that an action by any element in the contemporary workplace has wider—and hence more serious—consequences than in the past (see also Driskell and Salas, 1991b). Take the case of financial markets. On a dollar volume basis, the bulk of the activity in these markets is undertaken by the managers of large institutions, such as pension funds, not by individual investors (Siconols, 1992). Hence, as the market crash of 1987 demonstrated, buying and selling by a remarkably small number of people can have dramatic and far-reaching effects. So, to the extent that stress influences those individuals' choices, its ramifications extend far beyond their personal welfare and affect most people.
- The potential effects of stress on cognition are further amplified by the changing nature of work itself. In almost every arena, today's jobs place greater emphasis on cognitive than on motor performance. For example, as noted by Hartzell (1992), at one time a good helicopter pilot had to exercise extraordinary manual control. These days, however, a good pilot depends more on the ability to select appropriate procedural sequences, which in turn are executed by computer-controlled devices.
- There is also the role of new technology itself. For instance, it is now possible for employers to continuously monitor the performance of workers such as telephone operators and telemarketers. There is some evidence that this practice itself exacerbates the stress experienced by workers (e.g., Rogers et al., 1990).

A PERSPECTIVE

Hans Selye (1956) is the scientist who first called special attention to the concept of stress in humans. Selye (1983) noted that the body's physiological reactions to an injury or intense emotion can be divided into those that are peculiar to the particular stimulus in question and those that appear in response to any noxious stimulus. He called the latter constellation of responses the general adaptation syndrome (GAS). It includes three stages (Selye, 1983:4-5): (a) an "alarm reaction," representing a "general call to arms of the body's defensive forces," including changes in body temperature, blood pressure, and hormonal secretions; (b) a "stage of resistance," in which there is adaptation to the stressor and the reduction or even disappearance of the elements of the alarm reaction; and (c) a "stage of exhaustion," when the compensatory measures in the resistance stage deplete the available resources. Selye essentially equated stress with this nonspecific physiological response pattern. For some time, most work on stress and psychological functioning continued to emphasize this biological, response-oriented characterization of stress. Thus, for instance, Broadbent (1963) and others sought to account for stress effects in terms of fluctuations in arousal.

The concept of stress has evolved over the years. Stress researchers are far from unanimous in their use of the term (see, e.g., Everly and Sobelman, 1987:Chapter 1). And this situation has prompted at least some authors to suggest that the very concept of stress has outlived its usefulness (e.g., Hammond and Doyle, 1991). Nevertheless, there is reasonable consensus about the meaning of the expression in work on cognition, and we will interpret it here according to that consensus (see Hancock and Warm, 1989). Therefore, by *stress* we mean an individual's reactions to apparent significant threats to his or her welfare, reactions that often entail heightened emotion (see Keinan et al., 1987; Yates, 1990). As articulated by Novaco (1988:4), a *threat* is a self-perceived imbalance between the demands made on a person and the resources he or she can apply to satisfying those demands. That is, the person doubts his or her ability to meet the challenge imposed by the circumstances. Several implications of this stress conception can be highlighted with the aid of Figure 10.1.

To start with, stress is a response to situational conditions. If these conditions induce the perception of threat, they achieve the status of *stressors*—i.e., "that which causes stress" (Selye, 1983:9).

The *personal circumstances* element in Figure 10.1 calls attention to the fact that identical conditions can have different threat implications at different times. That is, a given individual's varying resources, physical and otherwise, alter that person's vulnerability. A problem-solving task that is daunting when one is exhausted at day's end is "no sweat" after a good night's sleep.

The *personal characteristics* feature highlights two kinds of stable individual differences. First are the differences in people's abilities to respond to challenges. Second are differences in the intensity with which people experience (or anticipate) the losses that would ensue should they fail to

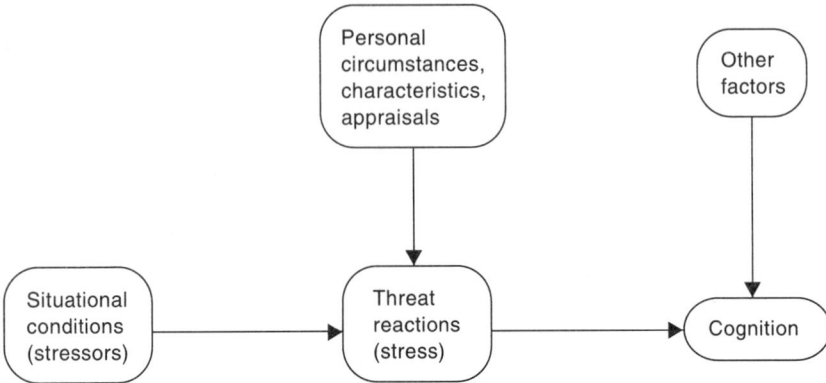

FIGURE 10.1 A schematic representation of stress and its connections with other entities.

meet a challenge. These differences are well illustrated in the discussion of medical devices used in the home in Chapter 4. Even a trained outpatient is likely to see himself or herself as less competent at using the equipment than a professional and is thus more likely to experience stress in an emergency. And since the patient's own health or life—not someone else's—is at stake, stress should be even more intense.

The *personal appraisals* element in Figure 10.1 is an acknowledgment that ignorance or special sensitivity to situational conditions can preclude or exacerbate the experience of threat (Lazarus and Folkman, 1984; Paterson and Neufeld, 1987). As an extreme case, if a surgical attendant is completely oblivious of a cardiac monitor's malfunctioning, he or she will perceive no threat to the patient's survival and hence experience no stress.

Finally, the *other factors* part of Figure 10.1 is a recognition that, in virtually any situation, the cognitive activities in question will be affected by other things besides stress. And the effects of these factors may or may not be independent of stress effects. Lighting quality has an obvious influence on machine assembly performance. But within bounds, that influence may or may not differ for people who are subjected to large rather than small amounts of stress.

The conceptual framework depicted in Figure 10.1 is minimal. Nevertheless, it does provide a useful perspective for the ensuing discussion, and it will be elaborated in the context of that discussion.

MANAGING STRESS-COGNITION EFFECTS

Human factors as a discipline focuses on the interactions between the people and the artifacts that together compose human-technology systems for accomplishing various tasks. Thus, a nuclear power plant has a human factors problem if displays of reactor status induce high probabilities of operator error, and an anesthesiology protocol has poor human factors features if it encourages physicians and technicians to administer inappropriate dosages of gas. Typical activities that people undertake in human-technology systems have both cognitive and noncognitive aspects. Implicit in the earlier discussion is the assumption that stress often has adverse effects on the cognitive elements of those activities. This is a common and plausible assumption, but there may be conditions under which stress has beneficial influences on cognition. Indeed, several theories imply what some of those conditions might be (e.g., Easterbrook, 1959; Yates, 1990:Chapter 13). We thus submit that a major theme in human factors research should be deepening our fundamental understanding of stress-cognition relationships. A complementary aim should be the development of practical techniques for counteracting negative effects and exploiting positive ones.

A PREREQUISITE: METHODOLOGICAL ADVANCES

Consider an experiment by Rothstein (1986). On any given trial in that experiment, the subject's task was to predict a criterion variable, C. The subject was shown vertical bars labeled A and B. The heights of these bars served as numerical cues for C, which was statistically related to A and B in a particular way. For instance, in one condition, the optimal rule for predicting C might have the form

$$C' = f(A) + g(B),$$

where f and g are specific linear or inverted-U functions. Via a block of subject-controlled trials with feedback, each subject first learned to predict the criterion to a satisfactory level of accuracy. Half of the subjects then performed test trials under time pressure. On each trial, the subject had to predict C from A and B within a deadline of six seconds. One of Rothstein's conclusions was that time pressure induced the subjects to apply more widely differentiated weights to the cues. So, if a subject initially tended to predict C according to a rule similar to

$$C^* = f^*(A) + g^*(B),$$

then time pressure altered that rule to one more like

$$C^* = 2f^*(A) + g^*(B)$$

or

$$C^* = f^*(A) + 2g^*(B).$$

The tentative generalization was that time pressure tends to make judgment policies less evenhanded.

Rothstein's (1986) results are interesting in themselves. But we highlight his experiment here because in one important respect it is typical of laboratory studies of stress and cognition. Recall that the consensus definition of stress requires a person to doubt his or her ability to meet a task demand. Missing-response data suggest that Rothstein's six-second time limit did indeed engender such doubt. However, another stress requirement is that the affected person should recognize task failure as a significant threat to his or her welfare. For average college students (as were Rothstein's subjects), failing to complete any assigned intellectual task is probably an ego threat. But in Rothstein's study, that seemed to be the *only* plausible threat; subjects were simply "required" (p. 85) to respond within the time limit. Therefore, in terms of threat, the stressfulness of Rothstein's time pressure was probably minimal. Rothstein did not explicitly apply the term *stress* to his time pressure conditions. Nevertheless, his manipulations were actually quite representative of how researchers have often sought to induce stress in controlled settings.

The problem illustrated by this example is that the stressors used in laboratory studies are essentially benign (almost a contradiction in terms); they constitute almost no threat to the subject. From an ethical, as well as a strictly legal, perspective (in the United States, at least), this is how it should and must be. After all, no responsible researcher would want to risk causing the serious harm implicit in the "significant threat" element of the stress construct.

Standard informed consent protocols have another consequence, too. Not only must the potential consequences of a subject's actions be mild; the conditions confronting the subject cannot be surprising. As numerous stress specialists (e.g., Levine, 1988) have noted, uncertainty, including complete unexpectedness, is a critical element in many truly stressful situations. As a fanciful example, consider the popular science fiction film *Alien*. Perhaps the film's most terrifying moment occurs shortly after the protagonists conclude that their embryonic extraterrestrial tormentor had inexplicably but surely disappeared, leaving them safe and secure. Then, with shocking suddenness, they learn that the monster has been with them all along. It had been maturing parasitically within a crew member's body. Everyone would agree that the crew's stress level—to say nothing of the audience's—is then much higher than had the alien been the usual, garden-variety, Godzilla-type monster.

Most experimentalists are surely aware that the stress levels they induce in the laboratory are far milder than those in the real-world situations to which they wish to generalize. So, implicit in their work is the hopeful assumption that, although the effects of laboratory stressors are markedly weaker than those of severe natural stressors, they are qualitatively the same. But not all stress investigators are so optimistic. Take the situation in which the stakes in a real-world decision problem can be a significant stressor in themselves (Janis and Mann, 1977). The prospect of losing a 50-cent bonus because of a bad decision in a laboratory experiment might seem so inconsequential that the subject can treat the decision task as simply an exercise in pure reasoning, on a par with solving a math problem. But suppose the situation entails the possibility of losing a daughter's life because of choosing the wrong medical treatment (see Ritov and Baron, 1990). Mann (1992) speculates that the very nature of the decision process is fundamentally different in this kind of distinctly stressful circumstance.

Right now it is impossible to say whether what we have learned from laboratory studies of stress and cognition does, in fact, generalize to more extreme levels of the stressors involved. But, as was implied above, we do know that those experiments can tell us virtually nothing about the role of uncertainty in stress effects. There is yet another limitation: the fact that experiments are short-term affairs. *Acute stressors* are present only briefly, whereas *chronic stressors* are sustained for extended periods (e.g., on long space flights or submarine tours of duty). For the most part (there are exceptions, of course), human factors and basic cognition researchers have experimentally examined the contemporaneous effects of acute stressors on, for instance, task performance in the presence of noise or extreme temperatures (see Hancock, 1986b). In contrast, clinical psychologists and other health care providers (e.g., Newberry et al., 1987) have more often addressed the long-term consequences of extended stress exposure, consequences that might occur in contexts completely removed from the locus of the stressor. Post-traumatic stress syndrome, such as combat fatigue, is perhaps the best-known example (Levine, 1988). As suggested by the work of Cohen (1980; Cohen and Spacapan, 1978), the effect of chronic stress can present surprises. These surprises seem inaccessible via standard laboratory techniques.

Given the above observations, it is essential that human factors investigators of stress and cognition seek to create ethical yet effective means of studying the influences of stressors that more closely approximate those in real-world human-technology systems. Indeed, it is plausible that a major reason for the very sparseness of the stress and cognition literature is that researchers often find it prohibitively difficult to create appropriate conditions.

How might this methodological research challenge be approached? Numerous avenues could be explored. One point of departure is to note that the issues are similar to those that confront investigators in education and medicine. In these fields, the stakes can be the difference between a child's academic success and failure or even between life and death. (Should this child participate in a new, experimental learning method? Should this patient receive a new drug or a placebo?) Methodologists and ethicists in education and medicine have responded to the challenge with numerous innovations (e.g., Cook and Campbell, 1979). Human factors researchers would do well to see whether some of those innovations can be adapted to the demands of stress research. At a minimum, human factors methodologists might find it instructive to examine the process by which their counterparts in education and medicine have addressed analogous dilemmas.

It is impossible to anticipate which, if any, research strategies developed in education and medicine warrant close examination for emulation in stress and cognition research. However, a good case can be made for investigating the feasibility of methods that entail central roles for two specific techniques: (a) the analysis of naturally occurring incidents and (b) the study of simulations, including competitive games.

Incident Analyses

Helmreich's (1990) inquiry into the Air Ontario crash at Dryden is a good illustration of incident analysis. The investigator sifted through the available records of a significant, catastrophic event, trying to reconstruct conditions that may have led or contributed to it. Among those conditions were ones that fit prevailing stress theories. It so happened that, because of his previous work, Helmreich was aware of stress effects and hence would have been on the lookout for their possible involvement in the Dryden incident. This might not always be the case whenever incident analyses are routinely performed (e.g., in high-risk domains such as aviation, nuclear power, and surgery). We therefore recommend that protocols for regularly commissioned incident analyses (e.g., by the Nuclear Regulatory Commission, the Federal Aviation Administration, and various surgery review boards) be designed so that the possible role of stress can be evaluated critically.

Among the inherent weaknesses of full-blown mishap analyses are their expense and the (fortunate) rareness of the incidents. One variation on the incident analysis approach relies on statistical explorations of richer (but necessarily less detailed) routine archival data and could thus circumvent these drawbacks. An example of the requisite kind of information source is the database maintained by the National Aeronautics and Space Administration's Aviation Safety Reporting System (ASRS). This is an archive of reports submitted by individuals throughout the U.S. civilian aviation system. Such reports, which are recorded anonymously, describe any conditions or activities the reporter thinks might be hazardous. Williams et al. (1992) successfully analyzed ASRS reports about incidents involving resource management and geographic disorientation. There is nothing to preclude similar analyses of stressed-related incidents.

Another variation on incident analysis entails embedding experiments or quasi-experiments (Cook and Campbell, 1979) within naturally occurring activities that are likely to be stressful, but to which certain individuals have already voluntarily committed themselves. One example (which was not considered particularly successful by the investigators themselves) was a study by Idzikowski and Baddeley (1983) of the reactions of inexperienced public speakers prior to delivering colloquium papers. Another widely cited pair of studies were those in which Fenz (1974; Fenz and Jones, 1972) traced arousal patterns in parachutists preparing to make jumps. The embedding studies that have perhaps been least susceptible to subject self-selection biases are those that tested ways of teaching surgery patients to manage the stress associated with their impending operations (e.g., Langer et al., 1975).

Incident analyses have been performed for years (see, e.g., Woods, 1993). The reports resulting from such analyses, as well as the manuals that sometimes guide them, suggest a remarkable degree of methodological sophistication and thoroughness. For example, the National Transportation Safety Board's *Investigator's Manual* prescribes in great detail who is to participate in the investigation of airplane crashes and what their roles should be. An especially attractive feature of the manual is a checklist of human performance factors that should be examined; for example, life habit patterns, training, and control design. There has been some discussion of critical issues underlying incident analyses (e.g., Reason, 1990). Nevertheless, it appears that the scholarly literature on such issues is surprisingly scant. The rarity of relevant reports highlights both an opportunity and a need for methodological developments to which human factors stress researchers could be major contributors.

In collaboration with practicing incident analysts from diverse backgrounds, human factors specialists should aggressively seek to extend the public literature on incident analysis techniques. There are numerous specific issues such a literature should address, as our illustrations have highlighted. Take the case of Helmreich's (1990) examination of the Air Ontario crash at Dryden. We conjectured that Helmreich's awareness of stress theories probably sensitized him to the possible role of stress in the crash. But critics might say that such awareness actually threatens the integrity of incident analyses. For example, research has shown (e.g., Chapman and Chapman, 1967) that clinical psychologists who hold theories that certain signs ought to be indicative of patients' true conditions tend to see "illusory correlations" between those signs and conditions even when in reality no connections exist. A strong literature on incident analysis methods would address techniques for precluding such validity threats.

Simulations

In any simulation, the aim is to duplicate some, but not all, of the essential features of a particular experience (see Jones et al., 1985). For instance, flight simulators are used in pilot training mainly because they are much cheaper to build and operate than real aircraft; they closely approximate the behavior of an aircraft but not its costs. The present research recommendation is that the same approach be attempted in the domain of stress and cognition. That is, an effort should be made to develop stress simulators as research tools that would induce realistic stress reactions but would not actually expose subjects to the threat of serious harm that is an essential element of stress.

Passable stress simulators should in principle be more difficult to construct than physical simulators. History has shown that a functional flight simulator needs to emulate only the physical response to an operator's manipulation of a control device (including what is seen through the cockpit windshield). The details of actually building such a simulator might be quite involved, but the underlying physical (and perceptual) principles are well known. The builder of a stress simulator, however, must bring about psychological illusions; the subject should feel stress without actually being subject to the hazards that are normally a prerequisite for it. Unfortunately, what we know about stress principles is not enough to guide us in creating those illusions in a systematic way. Thus, to the extent that stress simulators are feasible, they must emanate from researchers' intuitions and everyday experiences.

One starting point for developing stress simulators might be commonplace activities that appear to already approximate what we would like the simulators to do. Several examples come to mind: (a) films, (b) amusement park rides, (c) role-playing exercises, and (d) competitive games. We can all recall encounters with one or more of these activities in which we experienced what felt like serious stress. (Think of your most memorable horror movie, fun house, or roller coaster ride.) At one level of consciousness, we were fully cognizant that we were in no real danger; otherwise, we would never have agreed to participate. However, that assessment was temporarily suspended and put completely out of mind. Effectively, we allowed ourselves to be encapsulated within a small, insular world where gut-wrenching threats abounded.

It is unclear exactly what makes an effective horror movie, for instance. And perhaps that is the genius of artistry. But it should be possible to systematically review good and poor films, rides, role-playing exercises, and the like in order to glean hints of techniques that could be mimicked in building useful stress simulators. For example, one plausible working hypothesis is that a good simulator must insulate the subject from external stimulation that would encourage thoughts like, "This is only a game," "I realize that nothing bad can actually happen to me," or "In the larger scheme of things, this just doesn't matter."

In the abstract, the kinds of stress simulators suggested here seem almost far-fetched or frivolous. They are not, however. Instantiations of similar ideas already exist and have proven their value. Take the case of role-playing. Armstrong (1987) has reviewed research (including some of his own) about forecasting the outcomes of conflict situations, such as union negotiations. He specifically compared the relative accuracy of forecasts generated by expert opinion and those that relied on role plays of the conflict situations. Forecasts based on role-playing tend to be far better.

Or take competitive sports. There is an active subarea of sports psychology that focuses on stress and performance among athletes (see Jones and Hardy, 1990). Work in this area should be examined not only for methodological ideas but also for possible generalizations of its substantive findings beyond the domain of sports.

There is currently considerable activity in the development of virtual reality technology. Much of that work is proceeding at a breakneck pace in the private entertainment industry, to the point that several virtual reality games and "rides" are currently available to consumers in special arcades and amusement parks (Corliss, 1993). One perspective on these developments is that they are expected to be so lucrative precisely because they promise to be so much better than competing technologies (e.g., conventional video games and films) at simulating extreme stressors. We suggest that their potential as stress and cognition research platforms be thoroughly explored. Should the anticipated commercial success of virtual reality entertainment devices be realized, researchers might also find themselves with a ready-made volunteer subject pool analogous to the skydivers studied by Fenz (1974) and his colleagues. In fact, researchers might even want to explore developing research simulators that are in fact "games" included among other games in commercial arcades and parks. Decision researchers have sometimes tested the generalizability of behavioral decision-making principles in the high-stakes, real-world environment of Las Vegas casinos (e.g., Goodman et al., 1979; Lichtenstein and Slovic, 1973).

One paradoxical complication of stress simulation should be acknowledged and addressed by stress research methodologists and ethicists. Our discussion has emphasized the simulation of stress without the real danger of direct physical harm. For instance, in a simulator, there is no chance of a subject's being killed in a high-speed crash. But to the extent that real stress is induced, there is still the potential for psychological and indirect physical harm from the stress itself (see, e.g., Cox, 1988). In the vernacular, the simulation might "scare the subject out of his wits." Screening subjects on appropriate risk factors is one seemingly reasonable approach to such possibilities (e.g., Shanteau and Dino, 1993).

STRESS-HANDLING APPROACHES

We have noted that existing stress and cognition theories suggest that there are conditions in which stress can be expected to enhance performance. However, the overarching theme of our recommendations for substantive research is that human factors investigators should develop practical means for counteracting or handling adverse stress-cognition interactions. Five approaches to solving human factors stress problems can be distinguished:

Approach 1: Eliminate or weaken the stressor.
Approach 2: Reduce stress reactions by the current system participants.
Approach 3: Select system participants who are stress resistant if not stress immune.
Approach 4: Train system participants to function effectively even when stressed.
Approach 5: Design system features and procedures so that system goals are achieved despite the presence of high stress levels among system participants-—stress-proofing, as it were.

To varying degrees, all five of these approaches are discernible in the stress-related literature (most of which does not explicitly address human factors). Nevertheless, some of them—approaches 1 and 2, in particular—are emphasized more than others. In fact, approaches 1 and 2 can be seen as the standard strategies for dealing with stress. The underlying idea is that if stress is a bad thing, get rid of it. In approach 1, the work situation is designed to minimize the presence and intensity of stressors, for instance, by eliminating noise or altering workload schedules (see Chiles, 1982). In approach 2, an effort is made to change how system personnel react to stressors. A classic example of this approach is "stress inoculation training (SIT)" (Meichenbaum, 1985). When a person is inoculated against an infectious disease (e.g., chicken pox), he or she is exposed to mild forms of the disease with the expectation that the body will develop the ability to resist the more severe natural forms of that disease. Similarly, a key element of SIT is that an individual is exposed to progressively more intense forms of a stressor and is assisted in developing tools for managing that stressor, to the point that stress reactions no longer occur.

The logic of stress-elimination approaches is not unreasonable. However, as noted by Druckman and Swets (1988), it is impossible to eliminate or even reduce stress all the time. No matter what we do, some crises will inevitably occur. We should not simply concede defeat and allow nature (in this instance, dysfunctional stress-cognition interactions) to take its course. Approaches 3 to 5 are specific tacks for engineering effective system performance no matter what circumstances might arise. We will now make research recommendations for each of these strategies.

Selecting Stress-Resistant Personnel

Implicit in the conceptual scheme of Figure 10.1—and consistent with common experience—is the assumption that individuals tend to differ in how stressful they find particular circumstances and how well they perform under those conditions. Also as suggested by the conceptual scheme, part of the reason for this variation is that people have different competencies. Thus, Operator A is less stressed by task demands than Operator B because he or she knows the job better; it constitutes no threat. The more interesting situation is that in which equally competent individuals differ in their stress reactions. Our conceptual scheme implies that this could occur for either of two reasons. First, these individuals might appraise situations and their competencies differently; the stressed person could see tasks as more demanding and his or her skills as weaker, thereby anticipating less adequate performance. Alternatively, individuals might vary in the intensity with which they feel the pain of task failure. Such appraisal and reactivity differences are a major focus of work on decision and motivation processes (see Arkes, 1982; Yates, 1990). Consider, for example, why not all consumers choose the same car or why not all students pursue their studies vigorously.

Fortunately, stress and cognition researchers have begun seeking to understand the nature of individual differences and the bases for them (see Hockey et al., 1986, especially Section IV: "Individual Differences, Adaptation, and Coping"). However, for our purposes, the essential issues are somewhat different. In particular, the primary question is whether and how we can accurately and economically predict which individuals are likely to be unaffected by severe stressors in given situations.

Consider air traffic control. To the lay observer, the responsibilities and the pace of air traffic control seem crushingly stressful. Airplanes appear to converge on airports unceasingly, and a single mistake could be fatal for hundreds of people. There is some evidence that air traffic control specialists (ATCSs) do exhibit unusually high stress symptoms (e.g., Cobb and Rose, 1973), but the literature is by no means unanimous on this conclusion (Crump, 1979; Smith, 1985). Numerous studies indicate that ATCSs do experience stress levels comparable to those of other workers but they regard stress as among the least of their concerns. It is conceivable—in fact, plausible—that the stress resistance of the individuals who compose the corps of working ATCSs is higher than that of the initial pool of trainees. Somehow, because of self-selection and dismissals for correlated factors, stress-prone trainees leave the program. This natural filtering process might be an effective way to choose stress-resistant participants for a system in which stressors are inevitable. However, that process is undoubtedly expensive and perhaps risky and inefficient. It thus seems appropriate that a priority for stress and cognition researchers is to explore other ways to identify people who are stress resistant.

Do people, in fact, exhibit behaviors that could be used to predict stress resistance? Perhaps so. For instance, there are indications that ATCSs are reliably distinct from the general population with respect to several easily assessed personal characteristics, such as low trait anxiety and tendencies toward nonconformity (Smith, 1985). Research in the more general personality literature reveals other encouraging signs (but see Driskell and Salas, 1991b). For example, Allred and Smith (1989) have shown that individuals who score high on so-called hardiness scales tend to find pleasant and inviting challenges in what others regard as unpleasant stressors (see also Kobasa, 1979). Parkes (1986) similarly found that extraversion scores among nurses were correlated with their reactions to work stressors. Finkelman and Kirschner (1980) have reported data consistent with their proposal that laboratory measures of information processing, such as delayed digit recall, could be used as specific predictors of stress resistance in human-technology contexts such as air traffic control. And work reviewed by Reason (1988) indicates that stress vulnerability is correlated with simple questionnaire reports of everyday "cognitive failures." We previously recommended the development of various forms of stress simulators, such as competitive games and virtual reality experiences. If that effort succeeds, then it seems only natural that performance in the simulations be examined for their ability to predict individuals' performance in the work situations of ultimate interest: if trainee Smith thinks clearly in tight game situations, is it reasonable to expect him or her to do the same when a catastrophe strikes the workroom?

Training for Stressful Conditions

Keinan and Friedland (1984) performed an interesting experiment in which Israeli military personnel served as subjects. The focal task required the subject to search a page for randomly dispersed instances of a designated digit, say, "3." At the end of the session, the subject's skill at the task was evaluated by performance on three criterion trials. (Surprisingly, there are actually better and worse ways of performing such tasks.) After each of the first two test trials, the subject received a 1.5 mA shock to his fingers. The five conditions in the experiment were distinguished according to the context in which the subject was trained to perform the task. During high-fidelity training, the subject received a 1.5 mA shock after 4 of the 10 training trials. In low-fidelity training, these shocks were of 0.2 mA intensity. Shocks of 0.2, 0.75, 1.0, and 1.5 mA intensity were administered in succession during graduated-intensity training. Shocks at these same levels of intensity were administered in random sequence during random-intensity training. Finally, subjects in a no-stress condition received no shocks during training. Figure 10.2 shows the mean numbers of correct target identifications, by condition, during the three criterion trials and the last three training trials. We recommend that a priority for future research be the development of procedures for training personnel to perform well even under stressful conditions (see also Means et al., 1993). Keinan and Friedland's results speak directly to how such development might proceed.

First, note that the performance of the group trained under no stress was superior to that of any other group. Moreover, the performance of the graduated-fidelity and random-intensity subjects was especially poor. This suggests that stress can be expected to interfere significantly with skill acquisition (but see Christianson, 1992, for a contrasting view). Moreover,

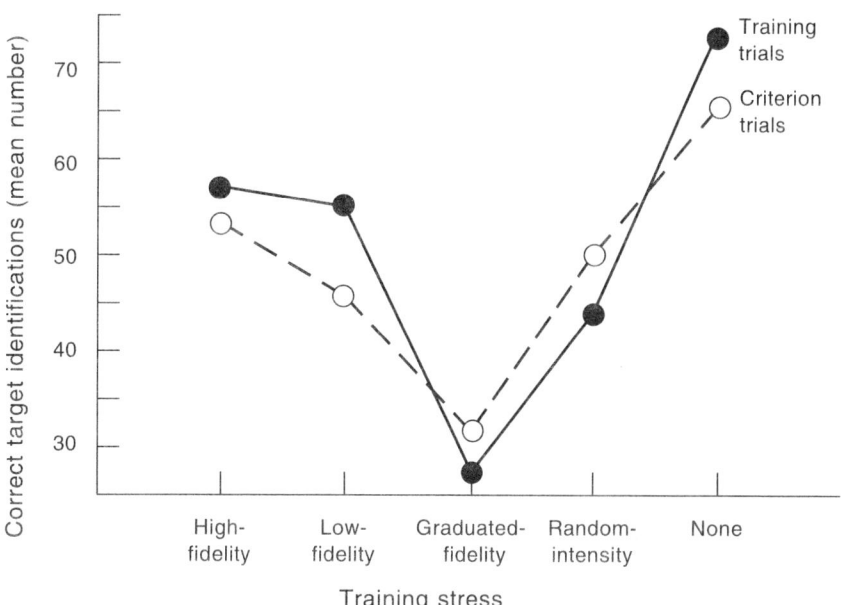

FIGURE 10.2 Mean number of correct target identifications as a function of training session stress. SOURCE: Keinan and Friedland (1984:Table 1).

it seems that the worse the stress, the worse the interference. (Consistent with previous discussions about stress effects, the apprehension and uncertainty in the graduated-fidelity and random-intensity conditions should amplify the associated stress.) The implied conclusion is that personnel should be trained to perform required tasks under relaxed conditions. This contrasts with the strategy behind the various stress conditions in the Keinan and Friedland experiment. In that view, training should be most effective when it occurs under conditions similar to those in which trainees must eventually perform.

Keinan and Friedland's results indicate that we should expect a relaxed-training approach to yield some performance decrement when a stressor is eventually encountered. (Observe that criterion performance was worse than terminal training performance for the no-stress group.) But there might be a way to preclude this decrement. A plausibly effective hybrid training strategy would train in basic skills under relaxed conditions and then allow trainees to "overlearn" those skills under stressful conditions. Yet another potentially effective strategy would be to concede that skill acquisition is slower under stress, but insist that it simply needs to be extended for a longer period of time. Neither of these strategies was tried directly by Keinan and Friedland, but these and others should be high on the research agenda.

Leaving aside the use of electrical shock, research like that of Keinan and Friedland is clearly the logical starting point for the kind of work that is required to inform the development of techniques for training people to perform well under stress. But as we have suggested, this work needs to be extended in various ways. There are also numerous matters of detail. The Keinan and Friedland experiment addressed only one training issue: When and in what form should stress be introduced into a training regimen? Consider some of the other issues, which are closely tied to fundamental questions about stress and cognition.

Task Type: Well Defined Versus Free Form

Visual search similar to that required of Keinan and Friedland's subjects is clearly an important part of many duties in human-technology systems. However, there are many other activities in which such well-defined tasks play a minimal role. In fact, we can expect that in crisis situations (e.g., during catastrophic weather conditions or equipment breakdowns) the required tasks will be anything but well defined; instead, they will be free form in the extreme. As was argued above, the inherent uncertainty in such unique situations can be expected to intensify the stress that is experienced (Harris, 1987). A close examination of the stress literature suggests that almost all experimental stress research has studied performance on well-defined tasks. Therefore, our beliefs about stress influences, including those indicated by Keinan and Friedland's results, might not be as generalizable as we would like. Future studies should determine whether or not this is the case.

Strategic Versus Reactive Stress Effects

The notion of controlled versus automatic information processing is widely recognized (Schneider and Shiffrin, 1977; Shiffrin and Schneider, 1977). This notion is different from the distinction between strategic and reactive stress responses but is related to it. In a strategic response, the individual executes a deliberate plan for coping with stressful conditions. For instance, faced with a stringent time limit for making a choice, a decision maker might rationally elect to alter the decision process, giving temporal priority to those aspects of the options that are deemed most important (e.g., Ben Zur and Breznitz, 1981). The decision maker realizes that there might be insufficient time to deliberate all the factors he or she would ideally weigh and thus concludes that he or she must ignore some of them (see also Payne et al., 1988). In contrast, a reactive response is automatically evoked by a stressor; the individual cannot help himself or herself. An example is suggested by Janis and Mann's (1977) theory of decision behavior. This theory proposes that the nature of the decision process is conditioned by the intensity of the stress involved. When stress is at its zenith, the theory suggests that the decision maker enters a stage of hypervigilance characterized in part by the kind of loss of control associated with pure reactivity (see also Janis et al., 1983).

It seems that the nature of stress performance training should be dictated by whether a stressful situation tends to evoke strategic or reactive responses. If responses tend to be strategic, then we might want to focus on things like incentives for selecting one type of action rather than others. Such tactics would be ineffective when responses are normally reactive. In the latter case, we would have to "reprogram" the subject to spontaneously perform different actions given appropriate conditions, a process that must rely on seemingly countless repetitions. A good analogy is what occurs in sports training. Tennis coaches sometimes have their students repeat a maneuver over and over again. That way, when the corresponding conditions occur during a tense match, the player does the right thing without having to think; it just happens. Thus, part of the stress training research agenda should be directed toward distinguishing which stress responses are strategic and which are reactive. A related issue that must be addressed is deciding which responses an organization might *want* to be strategic rather than reactive, or vice versa.

Specific Countermeasures

Exactly what should personnel be trained to do when performing their duties under stress? Suppose that an operator is instructed to do A, then B, and then C under normal conditions. On one hand, we might expect stress to induce the operator to deviate from the A-B-C routine when he or she ought to adhere to it. If so, training must be directed to maintaining that adherence. But how? On the other hand, an emergency might require that ideally the operator act differently from what would be appropriate under normal conditions. How should the operator be taught to respond most adequately to the altered circumstances? To the best of our knowledge, these kinds of essential issues have not been directly addressed, and clearly they must be. One way researchers might proceed is to begin with what previous research has taught us about adverse stress-cognition interactions. Stress-performance training protocols can then be pointedly designed to counteract those interactions, as we will illustrate below.

We noted previously—and hence this call for more research—that the stress and cognition literature has always been sparse. Nevertheless, that literature indicates consensus about several fundamental principles that we hope will be verified by future work (see, e.g., Hamilton and Warburton, 1979; Hockey, 1983, 1986; Hockey and Hamilton, 1983). For instance, there seems to be general agreement that stress tends to reduce short-term memory (STM) performance. It is not clear how and why this occurs (see also Christianson, 1992). There might be a reduction in sheer capacity. Alternatively, capacity might remain constant, but must now be divided between a focal task and other tasks, such as monitoring the stressor (Cohen, 1978; Reason 1988). It is important for stress researchers to determine the basis for this decrement in STM performance. Suppose that worsened STM performance is due to attention being shared with monitoring and managing the stressor. It should be easier to recover the attention if that stressor is an ambient stressor than if it is a task stressor. An ambient stressor (e.g., heat, noise, close confinement) is one whose presence is independent of the person's task performance, whereas the opposite is true of a task stressor (e.g., time pressure or high performance stakes) (Yates, 1990).

Regardless of the basis for the reduction of STM performance with stress, there are direct training implications. For example, suppose a task analysis indicates that an operator's duties require him or her to perform mental operations that rely heavily on STM (e.g., operations like those required in the decision algorithms Zakay and Wooler, 1984, trained their subjects to perform). Then the operator should be taught to distrust his or her ability to carry out those operations unaided when under duress; instead, the operator might use paper and pencil and double-check.

Another consensus conclusion is that stress reduces the scope of perceptual attention (Baddeley, 1972), a phenomenon sometimes called tunneling. Thus, stress reduces the range of elements in the environment a person sees or hears. An adjustable-beam spotlight provides a useful metaphor for perceptual narrowing. The claim is that stress reduces the width of the beam. A plausible alternative hypothesis is that the width of the beam remains constant but the beam simply develops a more erratic sweep (see Keinan, 1987; Keinan et al., 1987).

As in the case of STM, there are several fundamental issues that need to be resolved. Assuming that perceptual narrowing does indeed occur, along what gradient does it proceed? As noted by Yates (1990), in the context of decision making, at least three different gradients have been proposed in the literature. In the Easterbrook (1959) hypothesis, attention is restricted according to the objective, functional significance of various items of information for decision quality. The thesis implicit in the work of Ben Zur and Breznitz (1981), however, says that attention is reduced according to the personal evaluative significance of various features of the choice alternatives. And then there is the idea that (in one form, originated in drive theory) stress reduces attention according to the inverse of the probabilities of information being taken into account under nonstressful conditions (see Zajonc, 1968). Another attention-narrowing question concerns memory. Much of the information used in cognitive tasks is retrieved from memory. Does stress induce retrieval-narrowing phenomena analogous to perceptual restrictions?

The ultimate resolution of fine-grained theoretical issues like perceptual-narrowing gradients will indeed have practical implications. Nevertheless, stress training designers need not wait for a complete understanding of the theoretical issues. As suggested in the case of STM decrements, trainees might be warned that perceptual narrowing is likely to occur. But if the experience of debiasing in judgment and decision making is a guide (see Fischhoff, 1982), this is unlikely to decrease inappropriate narrowing to any great extent. Instead, the trainee must be given specific prescriptions and must be taught when and how to apply them, or perhaps even "conditioned" to implement them essentially automatically. Exactly what those prescriptions might be will depend on the designers' ingenuity.

There have been some indications that stress alters people's reliance on information from different sources. For instance, Wickens et al. (1989) cite evidence that stress encourages people to depend more heavily on information retrieved directly from long-term memory (LTM), presumably at the expense of information that is perceived directly from the environment or synthesized anew in STM (see also Klein, 1989). It is as if stress increases the significance of a person's prior experiences. Perhaps this is why, independently of competency differences, stress is sometimes found to have greater effects on performance for novices than for experts (Baddeley, 1972;

Hancock, 1986a). Greater dependence on LTM might be advantageous in some stressful situations. However, the kind of rigidity it implies might well be dysfunctional in crisis situations involving novel circumstances (e.g., a breakdown in a mechanism that has never broken down). There is at least some evidence that stress does in fact reduce certain forms of creativity that are required in such circumstances (Shanteau and Dino, 1993; Voss, 1977). Once again, training designers must be inventive in devising ways to counteract these effects.

Stress-Proofing Human-Technology Systems

Our final recommendation is similar to the previous one, but it applies to human-technology systems rather than to individual participants. Specifically, we suggest that researchers seek to develop system features and routines that allow the systems to achieve their goals even when participants are stressed. This should begin with an effort to understand how stressors affect both individual and group behavior. We have already discussed some of what is known about stress and cognition interactions at the individual level, including the fact that the literature is so sparse. For various reasons the literature on stress at the group level is even more limited (see Davis et al., 1992; Park, 1990). Thus, researchers who take up the challenge posed by this final proposal will need to make contributions to fundamental understanding of the problem as well as to practical innovations.

The reactions of the individual are key to the design of systems that function well in stressful circumstances. In our previous discussion, we suggested that training efforts be directed toward equipping system participants with personal skills to counteract the adverse effects of stress on cognition. The system could also be redesigned in a way that gives participants tools and procedures to counteract those effects. For example, recall that one effect of stress is to restrict the range of attention and consider crisis situations, such as the medical emergencies involving outpatients' operating sophisticated equipment discussed in Chapter 4. A specially constructed expert system might force operators to explicitly acknowledge that they have reviewed checklists presented to them by the system.

Or consider the creativity limitations we should expect in crises, when creativity is at a premium. Although the kinds of options that would occur to any single individual during stress are likely to be restricted, different individuals should be expected to generate somewhat different ideas. Thus, systems might be designed so that under stressful conditions, procedures require that inputs from multiple participants be elicited and synthesized. ("Two heads are better than one," and so on.) Not surprisingly, such activities can be done in ways that are more of a hindrance than a help (see Hill,

1982). Fortunately, however, there is a literature that discusses how such group "process losses" can be avoided (e.g., Dennis and Valacich, 1993).

As we have indicated, stress influences on group processes have been studied less extensively than we might like. Nevertheless, the literature that does exist suggests that stress is likely to alter communication and perhaps authority patterns among human-technology system participants and that not all of those changes are for the better (e.g., Driskell and Salas, 1991a; Gladstein and Reilly, 1985). For instance, Janis (1972) has proposed that stress is a major antecedent of the dysfunctional form of organizational decision making called "groupthink." In groupthink, a policy-making group becomes so preoccupied with achieving concurrence that other aspects of the decision task suffer from sometimes-fatal neglect. As researchers attempt to develop the kinds of stress-resistant systems envisioned, they should include features that anticipate and circumvent such dysfunctional social patterns.

SUMMARY

As we noted at the outset, there are numerous indications that the stresses experienced by the people involved in human-technology systems have intensified in recent years and are likely to continue doing so for some time in the future. Human roles in these systems increasingly place greater demands on cognitive than on, say, motor activities. This implies that good human factors design will increasingly rely on our understanding of the connections between stress and cognition. An examination of the literature reveals that the requisite knowledge base is surprisingly skimpy, given broad-based interest in stress. We concluded that a major contributor to this state of affairs is probably the inherent difficulty of inducing significant stress in subjects while protecting those subjects from the potential harm inherent in the stress construct.

Our analysis implied that a prerequisite for meeting the challenge is the development of better methods of doing stress and cognition research. Hence, our first class of research recommendations is methodological. Perhaps modeling their efforts on those of methodologists and ethicists in education and medicine, human factors investigators should seek to develop techniques that are practically and ethically feasible yet capable of revealing how significant stress does indeed affect cognition. Any and all promising approaches should be explored. But we argued that two especially promising methodological avenues should have priority. One of them relies on analyses of naturally occurring incidents involving stressful conditions, such as accidents. The other uses various forms of simulation, including competitive games and virtual reality technologies.

We concluded that the pressing needs for substantive research on stress and cognition are most productively conceptualized in terms of alternative approaches for handling adverse effects of stress on cognitive performance. Three particular approaches seem to demand special attention:

- Selecting stress-resistant personnel. The critical assumption underlying this approach is that people differ reliably in their tendencies to perform cognitive tasks well or poorly under stressful conditions. The aim of the research recommended in this area is identifying easy-to-assess predictors of such individual differences.
- Training people for stressful conditions. A fundamental yet unresolved training issue that needs to be addressed is when and in what form stress should be introduced into training programs. Other recommendations address questions that need to be resolved in order to provide specific guidance to stress trainers: How, if at all, do stress effects on cognition depend on whether the cognitive task is well defined as opposed to free form? Which cognitive responses to stress are strategic and which are reactive, and what are the implications of the distinction? What specific countermeasures for such adverse stress responses as diminished short-term memory performance are effective—and why?
- Stress-proofing human-technology systems. Most discussions of stress have emphasized the behavior of individuals. However, a good case can be made that much is to be gained by designing human-technology systems, including protocols for social interactions, so that the systems function well even during times of heightened stress. Our broad recommendation is that human factors researchers seek out principles that would inform such design objectives. Specific recommendations call for the development of expert systems and organization schemes that compensate for anticipated human difficulties, such as restricted attention and reduced creativity.

The practical challenges posed by research on stress and cognition are formidable and have undoubtedly contributed to the paucity of such work in the past. However, there is reason to be optimistic that developments along the lines suggested here will meet with greater success than in the past. That work promises more than immediate practical benefits. It should also lead to significant advances in our understanding of fundamental principles of cognition and stress.

REFERENCES

Allred, K.D., and T.W. Smith
    1989   The hardy personality: cognitive and physiological responses to evaluative threat. *Journal of Personality and Social Psychology* 56:257-266.

Arkes, H.R.
1982   *Psychological Theories of Motivation*, 2nd ed. Monterey, Calif.: Brooks/Cole.
Armstrong, J.S.
1987   Forecasting methods for conflict situations. Pp. 157-176 in G. Wright and P. Ayton, eds., *Judgmental Forecasting*. Chichester, England: Wiley.
Baddeley, A.D.
1972   Selective attention and performance in dangerous environments. *British Journal of Psychology* 63:537-546.
Ben Zur, H., and S.J. Breznitz
1981   The effect of time pressure on risky choice behavior. *Acta Psychologica* 47:89-104.
Broadbent, D.E.
1963   Differences and interactions between stresses. *Quarterly Journal of Experimental Psychology* 15:205-211.
Chapman, L.J., and J.P. Chapman
1967   Genesis of popular but erroneous psychodiagnostic observations. *Journal of Abnormal Psychology* 72:193-204.
Chiles, W.D.
1982   Workload, task, and situational factors as modifiers of complex human performance. Pp. 11-56 in E.A. Alluisi and E.A. Fleishman, eds., *Human Performance and Productivity*. Hillsdale, N.J.: Erlbaum.
Christianson, S.-A.
1992   Emotional stress and eyewitness memory: a critical review. *Psychological Bulletin* 112:284-309.
Church, G.J.
1993   Jobs in an age of insecurity. *Time* November 22:32-39.
Cobb, S., and R.M. Rose
1973   Hypertension, peptic ulcer, and diabetes in air traffic controllers. *Journal of the American Medical Association* 224:489-492.
Cohen, S.
1978   Environmental load and the allocation of attention. Pp. 1-29 in A. Baum, J.E. Singer, and S. Vallins, eds., *Advances in Environmental Psychology*. Hillsdale, N.J.: Erlbaum.
1980   Aftereffects of stress on human performance and social behavior: a review of research and theory. *Psychological Bulletin* 88:82-108.
Cohen, S., and S. Spacapan
1978   The aftereffects of stress: an attentional interpretation. *Environmental Psychology and Nonverbal Behavior* 3:43-57.
Cook, T.D., and D.T. Campbell
1979   *Quasi-Experimentation: Design and Analysis Issues for Field Settings*. Chicago, Ill.: Rand McNally.
Corliss, R.
1993   Virtual, man! *Time* 1(November):80-83.
Cox, T.
1988   Psychobiological factors in stress and health. Pp. 603-628 in S. Fisher and J. Reason, eds., *Handbook of Life Stress, Cognition and Health*. New York: Wiley.
Crump, J.H.
1979   Review of stress in air traffic control: its measurement and effects. *Aviation, Space, and Environmental Medicine* 50:243-248.

Davis, J.H., T. Kameda, and M.F. Stasson
1992 Group risk taking: selected topics. Pp. 163-199 in J.F. Yates, ed., *Risk-Taking Behavior*. Chichester, England: Wiley.

Dennis, A.R., and J.S. Valacich
1993 Computer brainstorms: more heads are better than one. *Journal of Applied Psychology* 78:531-537.

Driskell, J.E., and E. Salas
1991a Group decision making under stress. *Journal of Applied Psychology* 76:473-478.
1991b Overcoming the effects of stress on military performance: human factors, training, and selection strategies. Pp. 183-193 in R. Gal and A.D. Mangelsdorff, eds., *Handbook of Military Psychology*. Chichester, England: Wiley.

Druckman, D., and R.A. Bjork, eds.
1991 *In the Mind's Eye: Enhancing Human Performance*. Committee on Techniques for the Enhancement of Human Performance, National Research Council. Washington, D.C.: National Academy Press.

Druckman, D., and J.A. Swets, eds.
1988 *Enhancing Human Performance: Issues, Theories, and Techniques*. Committee on Techniques for the Enhancement of Human Performance, National Research Council. Washington, D.C.: National Academy Press.

Easterbrook, J.A.
1959 The effect of emotion on cue utilization and the organization of behavior. *Psychological Review* 66:183-201.

Everly, G.S., Jr., and S.A. Sobelman
1987 *Assessment of the Human Stress Response*. New York: AMS Press.

Fenz, W.D.
1974 Arousal and performance of novice parachutists to multiple sources of conflict and stress. *Studia Psychologica* 16:133-144.

Fenz, W.D., and G.B. Jones
1972 The effect of uncertainty on mastery of stress: a case study. *Psychophysiology* 9:615-619.

Finkelman, J.M., and C. Kirschner
1980 An information-processing interpretation of air traffic control stress. *Human Factors* 22:561-567.

Fischhoff, B.
1982 Debiasing. Pp. 422-444 in D. Kahneman, P. Slovic, and A. Tversky, eds., *Judgment Under Uncertainty: Heuristics and Biases*. New York: Cambridge University Press.

Gladstein, D.L., and N.P. Reilly
1985 Group decision making under threat: the tycoon game. *Academy of Management Journal* 28:613-627.

Goodman, B., M. Saltzman, W. Edwards, and D.H. Krantz
1979 Prediction of bids for two-outcome gambles in a casino setting. *Organizational Behavior and Human Performance* 24:382-399.

Hamilton, P., and D. Warburton, eds.
1979 *Human Stress and Cognition*. Chichester, England: Wiley.

Hammond, K.R., and J.K. Doyle
1991 *Effects of Stress on Judgment and Decision Making, Part II*. Final Report for Contract No. DAAL03-86-D-001. Research Triangle Park, N.C.: U.S. Army Research Office.

Hancock, P.A.
1986a   The effect of skill on performance under an environmental stressor. *Aviation, Space, and Environmental Medicine* 57:59-64.
1986b   Sustained attention under thermal stress. *Psychological Bulletin* 99:263-281.
Hancock, P.A., and J.A. Warm
1989   A dynamic model of stress and sustained attention. *Human Factors* 31:519-537.
Harris, J.H.
1987   Prior task experience and psychological stress: an investigation. Pp. 45-53 in J.H. Humphrey, ed., *Human Stress: Current Selected Research*, Vol. 2. New York: Ames Press.
Hartzell, J.
1992   Army-NASA Aircrew-Aircraft Integration (A^3I) Program. Presentation (in October) to the Committee on Human Factors, National Research Council, Atlanta, Ga.
Helmreich, R.L.
1990   Human factors aspects of the Air Ontario crash at Dryden, Ontario: analysis and recommendations to the Commission of Inquiry. Pp. 319-348 in V.P. Moshansky, ed., *Technical Appendices to the Final Report of the Commission of Inquiry Into the Air Ontario Crash at Dryden, Ontario.*
Hill, G.W.
1982   Group versus individual performance: are N + 1 heads better than one? *Psychological Bulletin* 91:517-539.
Hockey, G.R.J., ed.
1983   *Stress and Fatigue in Human Performance.* Chichester, England: Wiley.
Hockey, G.R.J.
1986   Changes in operator efficiency. In K. Boff, L. Kaufman, and J. Thomas, eds., *Handbook of Perception and Performance*, Vol. II. New York: Wiley.
Hockey, R., and P. Hamilton
1983   The cognitive patterning of stress states. Pp. 331-362 in G.R.J. Hockey, ed., *Stress and Fatigue in Human Performance.* Chichester, England: Wiley.
Hockey, G.R.J., A.W.K. Gaillard, and M.F.H. Coles, eds.
1986   *Energetics and Human Information Processing.* Dordrecht, Netherlands: Martinus Nijhoff.
Idzikowski, C., and A. Baddeley
1983   Waiting in the wings: apprehension, public speaking and performance. *Ergonomics* 26:575-583.
Janis, I.L.
1972   *Victims of Groupthink.* Boston, Mass.: Houghton Mifflin.
Janis, I.L., and L. Mann
1977   *Decision Making.* New York: Free Press.
Janis, I.L., P. Defares, and P. Grossman
1983   Hypervigilant reactions to threat. Pp. 1-42 in H. Selye, ed., *Selye's Guide to Stress Research*, Vol. 3. New York: Scientific and Academic Editions.
Jones, J.G., and L. Hardy, eds.
1990   *Stress and Performance in Sport.* Chichester, England: Wiley.
Jones, E.R., R.T. Hennessy, and S. Deutsch, eds.
1985   *Human Factors Aspects of Simulation.* Working Group on Simulation, Committee on Human Factors, National Research Council. Washington, D.C.: National Academy Press.

Keinan, G.
1987 Decision making under stress: scanning of alternatives under controllable and uncontrollable threats. *Journal of Personality and Social Psychology* 52:639-644.

Keinan, G., and N. Friedland
1984 Dilemmas concerning the training of individuals for task performance under stress. *Journal of Human Stress* 10:185-190.

Keinan, G., N. Friedland, and Y. Ben-Porath
1987 Decision making under stress: scanning of alternatives under physical threat. *Acta Psychologica* 64:219-228.

Klein, G.A.
1989 Recognition-primed decisions. *Advances in Machine Systems Research* 5:47-92.

Kobasa, S.C.
1979 Stressful life events, personality, and health: an inquiry into hardiness. *Journal of Personality and Social Psychology* 37:1-11.

Langer, E.J., I.L. Janis, and J.A. Wolfer
1975 Reduction of psychological stress in surgical patients. *Journal of Experimental Social Psychology* 11:155-165.

Lazarus, R.S., and S. Folkman
1984 *Stress, Appraisal, and Coping.* New York: Springer.

Levine, S.
1988 Stress and Performance. Background paper prepared for the Committee on Techniques for the Enhancement of Human Performance, National Research Council, Washington, D.C.: National Academy Press. Available through the National Academy Press's Publication-on-Demand Program.

Lichtenstein, S., and P. Slovic
1973 Response-induced reversals of preference in gambling: an extended replication in Las Vegas. *Journal of Experimental Psychology* 101:16-20.

Mann, L.
1992 Stress, affect, and risk taking. Pp. 201-230 in J.F. Yates, ed., *Risk-Taking Behavior.* Chichester, England: Wiley.

Manuso, J.
1983 The Equitable Life Assurance Society program. *Preventive Medicine* 12:658-662.

Means, B., E. Salas, B. Crandall, and T.O. Jacobs
1993 Training decision makers for the real world. Pp. 306-326 in G.A. Klein, J. Orasunu, R. Calderwood, and C.E. Zsambok, eds., *Decision Making in Action: Models and Methods.* Norwood, N.J.: Ablex.

Meichenbaum, D.
1985 *Stress Inoculation Training.* New York: Pergamon.

Mross, E.F., and K.R. Hammond
1989 *Annotated Bibliography for Cognition and Stress.* Report No. 295. Center for Research on Judgment and Policy. Boulder: University of Colorado.

Newberry, B.H., D.R. Baldwin, J.E. Madden, and T.J. Gerstenberger
1987 Stress and disease: an assessment. Pp. 123-151 in J.H. Humphrey, ed., *Human Stress: Current Selected Research*, Vol. 2. New York: Ames Press.

Novaco, R.W.
1988 Stress Reduction Programs. Background paper prepared for the Committee on Techniques for the Enhancement of Human Performance, National Research Council. Washington, D.C.: National Academy Press. Available through the National Academy Press's Publication-on-Demand Program.

Park, W.-W.
 1990 A review of research on groupthink. *Journal of Behavioral Decision Making* 3:229-245.
Parkes, K.R.
 1986 Coping in stressful episodes: the role of individual differences, environmental factors, and situational characteristics. *Journal of Personality and Social Psychology* 51:1277-1292.
Paterson, R.J., and R. Neufeld
 1987 Clear danger: situational determinants of the appraisal of threat. *Psychological Bulletin* 101:404-416.
Payne, J.W., J.R. Bettman, and E.J. Johnson
 1988 Adaptive strategy selection in decision making. *Journal of Experimental Psychology: Learning, Memory, and Cognition* 14:534-552.
Reason, J.
 1988 Stress and cognitive failures. Pp. 405-421 in S. Fisher and J. Reason, eds., *Handbook of Life Stress, Cognition and Health*. New York: Wiley.
 1990 *Human Error*. New York: Cambridge University Press.
Ritov, I., and J. Baron
 1990 Reluctance to vaccinate: omission bias and ambiguity. *Journal of Behavioral Decision Making* 3:263-277.
Rogers, K.J.S., M.J. Smith, and P.C. Sainfort
 1990 Electronic performance monitoring, job design and psychological stress. Pp. 854-858 in *Proceedings of the Human Factors Society 34th Annual Meeting*. Santa Monica, Calif.: Human Factors Society.
Rothstein, H.G.
 1986 The effects of time pressure on judgment in multiple cue probability learning. *Organizational Behavior and Human Decision Processes* 37:83-92.
Schneider, W., and R.M. Shiffrin
 1977 Controlled and automatic human information processing: I. detection, search, and attention. *Psychological Review* 84:1-66.
Selye, H.
 1956 *The Stress of Life*. New York: McGraw-Hill.
 1983 The stress concept: past, present, and future. Pp. 1-20 in C.L. Cooper, ed., *Stress Research*. New York: Wiley.
Shanteau, J., and G.A. Dino
 1993 Environmental stressor effects on creativity and decision making. Pp. 293-308 in O. Svenson and A.J. Maule, eds., *Time Pressure and Stress in Human Judgment and Decision Making*. New York: Plenum.
Shiffrin, R.M., and W. Schneider
 1977 Controlled and automatic human information processing: II. perceptual learning, automatic attending, and a general theory. *Psychological Review* 84:127-190.
Siconols, M.
 1992 Individual investors' holdings of U.S. stocks fall below 50% of total market for the first time. *The Wall Street Journal* 13(November):C1.
Smith, R.C.
 1985 Stress, anxiety, and the air traffic control specialist. Pp. 337-358 in I.G. Sarason and C.D. Spielberger, eds., *Stress and Anxiety*, Vol. 9. New York: Hemisphere Publishing.
Time
 1983 June 6 issue.

U.S. Army
  1992a   Fratricide: reducing self-inflicted losses. *CALL Newsletter* 92(April):4. Fort Leavenworth, Kans.: U.S. Army Combined Arms Command.
  1992b   Fratricide risk assessment for company leadership. *CALL Handbook* 92(March):3. Fort Leavenworth, Kans.: U.S. Army Combined Arms Command.

Voss, H.-G.
  1977   The effect of experimentally induced activation on creativity. *Journal of Psychology* 96:3-9.

Wickens, C.D.
  1992   *Engineering Psychology and Human Performance*, 2nd ed. New York: Harper Collins.

Wickens, C.D., A. Stokes, B. Barnett, and F. Hyman
  1989   *The Effects of Stress on Pilot Judgment in a MIDIS Simulator (U)*. Armstrong Aerospace Medical Research Laboratory Report No. AAMRL-TR-88-057. Wright-Patterson Air Force Base, Ohio: Army Aerospace Medical Research Laboratory.

Williams, H.P., M. Tham, and C.D. Wickens
  1992   *Resource Management and Geographic Disorientation in Aviation Incidents: A Review of the ASRS Data Base*. Aviation Research Laboratory Report No. ARL-92-3/NASA-92-2. Savoy: University of Illinois.

Woods, D.D.
  1993   Process-tracing methods for the study of cognition outside of the experimental psychology laboratory. Pp. 228-251 in G.A. Klein, J. Orasunu, R. Calderwood, and C.E. Zsambok, eds., *Decision Making in Action: Models and Methods*. Norwood, N.J.: Ablex.

Yates, J.F.
  1990   *Judgment and Decision Making*. Englewood Cliffs, N.J.: Prentice Hall.

Zajonc, R.B.
  1968   Social facilitation. *Science* 149:269-274.

Zakay, D., and S. Wooler
  1984   Time pressure, training and decision effectiveness. *Ergonomics* 27:273-284.

11

Aiding Intellectual Work

John D. Gould

INTRODUCTION

Nearly all human intellectual work is aided by technology. In the next decade, new technology will make possible many new potential aids. The main thrust of this chapter is to structure the problem of aiding intellectual work in such a way that human factors researchers can contribute toward the development and evaluation of electronic systems that will indeed aid their users.

The Problem

The problem focused on in this chapter is how to carry out human factors research aimed at augmenting human intellectual work with electronic aids. Aids include artifacts or tools, as well as procedures, techniques, organizational structures, and facilitated, encouraged communication patterns. Most human intellectual work tends to be mentally demanding, with a priority on timeliness and with a high premium on getting the "right information" (i.e., not just the right answer to a question asked, but information that is relevant to questions that should have been asked but were not).

Today, all human intellectual work is augmented to some degree. Electronic aids for intellectual work include ubiquitous, simple "single-function" tools, some of which are largely electronic versions of established books (e.g., dictionaries, thesauruses, foreign language translators) and some of which are more (e.g., calculators, spelling verifiers). There are spreadsheets for businesspeople, high-level programming languages and tools for computer application developers, Cadam and very large-scale integrated (VLSI) tools for hardware designers, and conferencing systems for group work. To help users transfer learning from one computer application to another, software developers have attempted to standardize user interface styles (e.g., Apple Corporation's Macintosh style and IBM's CUA style). Many jobs simply could not be performed without electronic aids (e.g., work related to forecasting, reservations, large databases, simulations, and much medical practice).

Aiding intellectual work performance is a broad topic. Several domains have been addressed by recent National Research Council committees and are not explicitly covered here. One report summarized evidence on whether human performance is enhanced through techniques that have their roots in academic research, as well as techniques that have their roots in commercial enterprise. The various techniques used to enhance human performance include training, pain management, stress management, expert guidance, meditation, self-help subliminal audiotapes, self-assessment techniques, mind-altering drugs, and sports-psychology techniques to aid performance under pressure (see Druckman and Bjork, 1991; Druckman and Swets, 1988). Another National Research Council report (Ferber et al., 1991) discusses how human intellectual work performance can be affected by the linkages between work and such family and personal areas as finances, marital status, child care, the need to support elderly relatives, marital stress, illness, and working at home. Other studies have reviewed intellectual work aids for (a) people with special needs (e.g., sensory and motor impairments, communication difficulties) (see Chapter 3) and (b) students in primary, secondary, and collegiate schools.

Why Is Aiding Intellectual Work Important?

There are three main reasons for understanding how to better aid human intellectual work.

It Increases Productivity and Creativity

Productivity is a national problem: many studies report that the productivity of the U.S. labor force is relatively stagnant. Some widely used electronic aids have not helped much. Nevertheless, there is the belief, supported by informal evidence, that electronic aids, if developed with users and their work organizations in mind, do enhance productivity and creativity and can allow people to work in new ways. For example, using electronic spreadsheets, business professionals routinely run "what if" analyses in seconds, a task that previously took hours or days or that would simply not have been done a few years ago. A variety of intellectual workers use simulations to provide quick studies of situations that previously could be studied only by actually building an object itself (if indeed it could be built). Scientists move single atoms, study their effects, and produce new combinations with awesome possibilities. With new visualization systems, engineers and scientists can see how wings are affected by wind or how a thunderstorm forms and thus develop an understanding that cannot be gained from studying the thermodynamics and the underlying isolated equations themselves. Powerful visualization machines have led to a change in the design of drugs: "scientists use graphics workstations to visualize a protein in three dimensions and determine which drugs will fit best into its active sites. Such visualization is so powerful that no pharmaceutical researcher would try to do the job without it" (Pool, 1992). An entire new field (experimental mathematics) and a way of thinking for mathematicians has developed based upon these new visualization tools (Pool, 1992).

It Is Socially Responsible

The abilities of people to cope with increasing intellectual work demands and their resulting self-images are affected by tools they are given (e.g., see the section below on "Organizational Impact of a New Technology"; see also Kraut et al., 1989). Some workers are deficient in problem-solving skills, spelling, composing, calculating, language skills (including second language problems), giving instructions, and so forth. Intellectual aids have the potential to reduce these deficiencies. Intellectual aids also have the potential to help people deal with problems outside work (e.g., financial planning, finding support groups, interacting with their children), and these, in turn, impact on their work lives (Ferber et al., 1991). Human factors researchers can help identify which intellectual tasks, if aided, could lead to the most benefits (for individuals and society), and they can help identify aids that can effectively help people with specific intellectual deficiencies compensate for what is latent or missing.

Our Expanding Technology Makes It Easier to Do

Rapid advances in electronic technology provide opportunities to improve existing intellectual aids, create effective and unforeseen new ones, and reduce costs, thereby bringing heretofore expensive aids to larger audiences. The productivity and creativity of individuals and work groups can be dramatically affected, in unforeseen ways, by the faster, cheaper processors with larger storage that are becoming available. In the next decade, the costs of electronic storage and of computer processing are expected to continue decreasing at about the same rate as they have been (about 10 to 20 times per decade). The speed of computer processing is expected to continue increasing at about the same rate as it has been (about 10 to 20 times per decade). Communications bandwidth is expected to grow even more dramatically—about 100 times during the next decade.

Human Factors Special Interest

The evaluation and creation of intellectual aids draw on a variety of disciplines involved in human factors work, including assessment, cognitive science, applied and experimental psychology, work science, social science, organizational theory, and systems theory. It is the human-centered focus and the assessment and system development methodologies pioneered by these disciplines that provide the potential contributions.

THREE RECOMMENDED HUMAN FACTORS RESEARCH NEEDS AND STRATEGIES

Three human factors research efforts are recommended to improve human intellectual aids. Although they differ in emphasis, the three research needs overlap, as do their respective research strategies and methodologies. Several studies used as examples to illustrate one human factors research strategy can, from a different angle, be used to illustrate one of the other research strategies.

Assessing Existing Intellectual Aids

Today, little experimental and formal empirical evidence exists as to the impact of various electronic aids on individual worker and organizational productivity. Human factors research should be directed at empirically and experimentally assessing the value of existing intellectual aids, particularly as related to individual and organizational productivity.

Exploring the Nature of the Tasks to be Aided

The fundamental nature of the intellectual tasks to be aided must be understood. Stated differently, the characteristics of aids that will improve worker and organizational productivity and workers' quality of life must be discovered. Human factors research should empirically study ongoing intellectual work to identify which work tasks are "aidable," the cognitive characteristics of these tasks, and the characteristics that the aids should have. There exist a variety of human factors methodologies that can be used to conduct empirical investigations of intellectual work in creative jobs and work organizations in order to understand what types of intellectual aids would be valuable. These methods include serious observational visits to work locations, task analyses, interviews, and thinking-aloud protocols.

Participating in Interdisciplinary Efforts

The disciplines that underlie human factors research are necessary to improve existing intellectual aids and to create new useful ones. But they are not sufficient. Human factors researchers should participate in serious interdisciplinary efforts with other scientists and engineers. This is necessary both because human factors researchers have the potential to make a contribution and to help ensure that human factors research has a significant impact. This chapter provides frameworks for thinking about how to address these three research needs, suggestions for following the recommended research strategies, and examples of existing research.

RESEARCH NEED AND STRATEGY 1: ASSESS EXISTING AIDS

It is important to know the effectiveness of the various intellectual work aids. It would seem reasonable to assume from observation that, to augment their intellectual work, people are successful in using some aids (e.g., actuarial tables) but not others (e.g., whatever aids many mutual fund managers used in the last decade). Although free enterprise market forces and differential commercial success of various intellectual work aids would seem to be a good indicator of their effectiveness, this may not always be the case. The experimental results that do exist sometimes question the value of the generally accepted intellectual aid studied (see the examples of studies below; see also Attewell, 1994).

Future human factors research should assess existing, ubiquitous intellectual aids to learn how well, if at all, they aid human intellectual performance and productivity; to understand more deeply how they affect people's minds and the productivity of their work organizations; and to use the results as insights in defining new aids that would be valuable. In addition to increasing productivity, knowledge about the effectiveness of intellectual work aids might help identify their common characteristics that have long-term positive effects (e.g., making workers more productive, improving the quality of work life). These discoveries might center on characteristics of the aids themselves or on characteristics of the processes used to develop them. For example, a number of questions might be asked about aid development processes. What design processes lead to the best aids? Do the system design processes recommended by human factors people lead to good aids (e.g., iterative design, user-centered design; see Helander, 1988)? If so, then are there more case study data in support of these design and development processes? If not, then these recommended processes can be modified.

Experimental assessment of intellectual aids could help identify the cognitive implications of widely used aids. For example, does a calculator lead students to become poorer at basic math skills, as is often suspected, or better at math skills, which one could argue on the basis of accurate feedback? Intellectual aids can have surprising, unpredictable effects and intellectual implications. A personal example is presented here: Prior to the availability of copiers, I used to make notes and know a relatively small number of journal articles in detail; now I have a more peripheral knowledge of a larger number of journal articles and, when I need to know any details, I rely on remembering the existence and location of the copies that I have stacked in my office. Experimental studies of existing aids would focus energy on tasks for which the aids are designed to help, rather than on convenient lab or toy tasks that are often used in psychological investigations. This may lead to developing new methodology and a worthwhile theory about the work under study. It also can lead either to the identification of ways to improve an aid or to ideas about the design of new aids.

Figure 11.1 summarizes key points to be kept in mind when designing experiments or empirical field studies to assess an existing intellectual aid. Although a few readers may think that some of these points are obvious, many of us are familiar with studies that fail to follow "obvious" suggestions.

Examples of Studies

A few studies are described here to illustrate how to successfully assess a potential aid.

Composing Aids

Composing by Dictation In the 1970s dictation was often used to produce office correspondence. Vendors and employers encouraged their professional employees to compose letters and memos this way; office dictation systems, including hardware, software, and required procedures, were being designed, purchased, and used. There were several generally accepted notions about dictation at that time, as summarized in Figure 11.2. To test these commonly held views, in a series of about 20 laboratory experiments, participants were given a specific topic on which to compose a letter and a specific composition method with which to do it. They composed similar

> **Key Methodological Points in Assessing Existing Aids**
>
> - Try to pick a ubiquitous, important task and a ubiquitous, important aid.
> - Continually ask the question: does this aid do what it claims to do? Does it do it in the SHORT run and in the LONG run?
> - Figure out how the task can be studied experimentally in the laboratory or empirically in the field. Often short-term laboratory experiments are not as informative as longer-term empirical studies conducted in context.
> - In laboratory experiments, include appropriate experimental controls, e.g., have participants do the same work with and without the aid (if this is possible). In empirical field studies, strive to do comparative studies of multiple groups or longitudinal studies of the "aided" group, especially if newly "aided."
> - Use tasks, equipment, material, procedures which relevant users and vendors would agree, a priori, provide a fair evaluation and could lead to their altering their behavior depending upon the outcome.
> - Design experiments so that results generalize to intended users (e.g., use appropriate people as experimental participants) and to a useful range of variation in the intellectual task under study (e.g., if it is querying a database make sure enough different types of queries are studied).
> - Choose meaningful dependent variables, including performance ones and attitudinal ones.
> - Try to identify and understand theoretically the underlying mental processes involved in the tasks and use of the aid.
> - Provide limits of generalization of the results in the study report.
> - Suggest how to improve the aid studied, or the possible use of alternative ones.
> - Prior to starting the experimental work, write conclusions sections that vary depending upon what is found. Show them to a few friends and get their reaction. This will help you determine whether it will be possible to draw any conclusions from the study design under consideration.

FIGURE 11.1 Key points in designing experiments to assess existing intellectual aids.

letters with each method studied. Their performance and behavior were measured while they composed, and the quality of their work was measured in various ways.

The results showed that, contrary to what was generally supposed, it does not take a long time to learn to dictate with dictating equipment (Gould, 1978a). Eight college graduates who had never dictated spent part of one day learning to dictate to a machine. The next day, they dictated four business letters and handwrote four similar business letters. The experimental results showed that they dictated and wrote letters in about the same time and with about the same resulting quality. Quality was rated by judges who viewed typed copies of the letters and who did not know which method was used to compose them.

> **Common Notions about Dictation in the 1970s**
>
> 1. "Dictating requires a long time to learn."
> 2. "Eventually, dictating is much faster than writing. That is, potential maximum output rates, as measured informally, are 200 words per minute (wpm) and 40 wpm, respectively."
> 3. "Dictating may be qualitatively superior to writing because the higher potential output rate permits faster transfer of ideas from limited capacity working memory, thus reducing forgetting through interference or decay."
> 4. On the other hand, handwriting was thought to have "an advantage over dictating because it is easier to review and modify."

FIGURE 11.2 Hypotheses about different composition methods in the 1970s. SOURCE: Gould (1980:101).

A second key result was that, contrary to what was generally supposed, (Figure 11.2), people experienced at using dictation equipment did not dictate several times faster than they wrote. Indeed, eight business executives who dictated regularly for years dictated routine business letters about 60 percent faster than they wrote them ($p < .001$), and they dictated more complex letters (e.g., essays on the Bicentennial celebration, on the issue of capital punishment, on their favorite teacher) about 25 percent faster than they wrote them ($p < .01$) (Gould, 1978b). These experienced dictators dictated routine business letters about 20 percent faster than did novice dictators ($p < .05$), and they dictated more complex letters in about the same time as did the novice dictators. Dictated and handwritten letters, when typed, were rated as similar by judges blind as to the method of composition and the experience of the authors. Part of the aid assessment strategy recommended in this chapter is to develop a theoretical understanding of users' tasks and to relate the results of experiments to that understanding. These dictation studies found that the reason people do not dictate five times faster than they write, even though they can say words five times faster than they can write them (40 wpm vs. 200 wpm), is that the generation component of composition (i.e., the actual moving of one's hand or mouth) takes only a small fraction (25 percent or less) of the total time. Planning (i.e., pausing and thinking) takes two-thirds of the total time.

A third key result was the discovery that inexperienced dictators *feel* it takes them longer to learn to dictate than it actually does because they *think* the quality of their dictation is poorer than it actually is. Novice dictators, just after composing a letter, rated their dictated letters as significantly poorer than their written letters (Gould and Boies, 1978). Upon receiving a typed copy of their dictated and written letters and incorporating proofreading and editing changes, they elevated the ratings of their dictated letters to those of their handwritten letters. Experienced dictators, on the other hand, rated their dictated and written letters equivalently at both stages. Judges later rated the quality of the final versions of the dictated and written letters as equivalent.

A fourth key result had to do with the just-emerging method of composing by speaking "voice documents," which recipients listen to (rather than read). Experimental results showed speaking to be a much faster method of composing than dictating or writing. Authors could use syntax, phraseology, and organization that they had learned over the years to be appropriate for listening but not necessarily for reading. In addition, in contrast to dictating, authors did not have to give typing instructions, which is potentially a cognitively disrupting "secondary" task (Gould, 1980). This information on voice documents had an impact on technology. In the middle 1970s, researchers at IBM had begun work on a "super dictating system," one that allowed users to apply great editing power to their aural documents. The results of the research project described here were a key factor in IBM's decision to shift the emphasis away from a super dictation system and toward a voice messaging system (Gould and Boies, 1983). Since then, voice messaging has grown dramatically as an industry (e.g., voice mail, telephone answering machines, etc.), whereas dictation has declined.

Composing With Text Editors In the 1970s, office professionals began using text editors to compose their own memos and documents—and to create the final text versions without the aid of secretaries. This trend led to research in the late 1970s to experimentally assess the productivity of professionals using text editors to compose documents (Gould, 1982), the method of composition that has since become dominant in much American industry. In the first study, 10 professionals, who regularly used a computer-based, mainframe, line-oriented text editor to compose their own correspondence and longer documents, composed four letters with a text-editor (T letters) and four letters with handwriting (W letters). A secretary typed each W letter, and then the participants revised and proofread the typed copy.

The results questioned the productivity advantage of using the line-oriented text editors of the 1970s. The basic result was that participants spent significantly more time composing T letters (29.5 minutes) than they did composing W letters (19.2 minutes) ($p < .001$). There was no difference in the judged quality and effectiveness of the T and W letters (based upon ratings by judges afterwards). Participants made many more modifications to T letters (41.3) than to W letters (8.5). There was no difference in the length of T and W letters, and there were no observable differences in the style of T and W letters. The report concluded that no productivity advantages were found for professionals who composed their own letters with their own text editor.

These results seemed questionable to some, particularly because of the lack of differences in quality between the T letters, with all their additional modifications, and the W letters. A few years later, Card et al. (1984) repeated this study, using the same tasks, materials, and procedure, but using a "display-oriented" editor (called Bravo) rather than a line-oriented editor. Again, participants' T and W letters were of the same length and were judged to be of the same quality and style. Again, participants made several times more modifications to T letters than to W letters, and again these did not lead to higher quality in the T letters. However, in this experiment the composition times for the two methods were about equivalent (mean times for T and W, respectively, were 22.8 and 21.7 minutes versus 29.5 and 19.2 minutes in the Gould study). That is, the mean time for T letters was reduced to about that of W letters. Besides extending the findings to other editors, an important point for this chapter is that results showed that experiments aimed at assessing aids can lead to cumulative knowledge.

Comment Note how these experiments reflect the recommended methodological points of Figure 11.1. Important aids were studied. Participants were experienced with *all* methods of composition. This helps generalization to the intended audience. Appropriate tasks, that is, letters rather than, for example, sentences or minor utterances, were used. A variety of letter types were studied, which leads to greater generalization of the results. Meaningful dependent variables—time and effectiveness—were measured in a variety of ways. In real work, these variables reflect productivity. Since the goal of communication is often to persuade, having judges select the most persuasive letters in some of the studies was better than simply judging spelling, syntax, and stylistic considerations. The experimental results were repeatable, extendable, and cumulative.

Some of the work had an impact on composing technology. Human factors research should not just question the value of existing aids but, whenever possible and perhaps aided by theory, find something that is a good replacement. Speaking documents as a method of composition was so identified.

Each report contained a theoretical explanation of the results, a serious discussion of limitations, and a pointed conclusion paragraph. For example, the Gould (1982) report on text-editing systems pointed out that only one text editor was studied; that even though there were no other experimental data on the use of text editors in composing tasks, it was known that the text editor used can affect the speed of routine secretarial editing tasks (Roberts, 1979); that line-oriented editors can lead to longer editing times for very simple editing tasks than display-oriented editors (Card et al., 1979); that the results are limited to one-to-two-page letters, typical of those composed in industry; that these were not letters in which participants were deeply ego-involved or did an extensive number of drafts.

Missing from these studies, however, was a serious discussion, based upon interviews with participants afterwards, about why they were not going to modify their behavior after learning about the results. In other words, a powerful dependent variable in assessment studies is to try to take some "aid" away from people and see how much they resist this or fight to get it back. Related to this is the observation that many developers do not, even when appropriate, themselves use the aid that they are making for others to use.

Everyday Aids

Norman (1988) analyzed the way many "everyday" aids (e.g., doorknobs, VCRs, stoves, faucets, automobile controls) work in his popular book on the subject. The book is valuable for its motivational, theoretical, and methodological content. It is largely anecdotal, empirical, and analytical (rather than experimental) in its approach. It shows an excitement to get started, a motivation to focus on existing aids. It articulates theories of mental mechanisms involved in using common aids. It correctly describes how successful artifact design proceeds. The book finds fault with many existing implementations of intended aids, but sometimes points to successful ones as well. Norman's methods can be extrapolated to aiding intellectual work.

Behavioral and Motivational Effects of Automation

One work setting that has become increasingly automated is the flight deck of modern jet transports. Many intellectual tasks, such as flight planning and navigation, have been delegated to flight management computers, with the tasks of flying becoming increasingly centered on entering data and monitoring electronic displays. This, of course, has been done to "aid" the crew and improve airline safety. But some researchers have raised questions about unintended consequences of this automation, such as complacency, overreliance on malfunctioning automated systems, reduced job satisfaction, changes in authority relationships, and loss of skills. One recent study found no differences in technical performance (errors) between crews flying an automated version and those flying a standard version of the same aircraft on a simulated flight involving mechanical malfunctions and diversions from planned routing (Wiener et al., 1991). There are probably many broad, longer-term issues surrounding the impact of using "electronic colleagues" that go beyond performance results from laboratory studies.

Organizational Impact of a New Technology

The studies described above have focused more on individuals than on groups and more on outcomes than on processes. The next study shows how to focus on all four. Methodology derived from experimental psychology is not the only way to assess the effect aids have on productivity. Empirical field studies, especially if they are aimed at organizations of people, rather than just individuals, can provide another powerful approach. While lacking some of the control of laboratory experiments, empirical field studies can assess whether individual productivity, measured in the laboratory, translates into organizational productivity and can also take into account contextual variables.

Kraut et al. (1989) studied what happened to productivity and quality of working life when a large company introduced a computerized record system to replace a microfiche record system in their customer service department. Management's goal was to increase efficiency and cut labor costs. Customer service representatives, who were the highest-paid nonmanagement employees in the company, were the primary contact between the company and customers. They attempted to solve customers' problems, collect overdue bills, and sell new company services.

Kraut et al. studied 10 customer service offices, collecting survey data from each (a) prior to the introduction of the new technology, (b) one month after its introduction, and (c) three months after its introduction. This approach allowed pre/post comparisons within an office. Each month the new system was introduced into another office. Thus, Kraut et al. started studying each office at successive monthly intervals (e.g., office 1 in November, office 2 in December, etc.). This "lagged, time-series" design also allowed comparison among offices in various stages of automation.

Management's main goal was realized. Productivity of service representatives, as estimated by the service representatives themselves (more objective data were not available), increased by 50 percent. They reported that their jobs got easier. However, they also reported (Kraut et al., 1989:230) that their jobs:

> became less satisfying, less interesting, and generally of poorer quality following the introduction of the computerized record system. Furthermore, contact with work colleagues became a less frequent and less satisfying component of service representatives' work life. . . . Interview data suggest[ed] that these changes resulted partly from changes in social interaction, based upon new seating arrangements, new privacy panels, and service representatives' limited physical movement that came from coordinating information through a database rather than through word of mouth and transfer of documents.

The supervisors of these service representatives, on the other hand, reported their jobs became much more difficult after the introduction of the computerized system. They reported that they now worked harder. Evaluation became more time-consuming, largely because obtaining the relevant data from the new system was harder than before. Overflow work could no longer be done at home; it now had to be squeezed into business hours. Supervisors no longer felt expert in the details of the work that service representatives did, although service representatives reported that their supervisors were good sources of information and responded well to pressure and uncertainty during the transition.

Comment This study by Kraut et al. (1989) is a valuable example of how to conceptualize the issues, design an empirical study of the organizational impact of a new technology, and summarize the experience. It demonstrates the value of being guided by a general theoretical view of human work that takes into account a multiplicity of factors. The results showed that the new system's effects depended on workers' jobs, the tasks they performed, and the types of offices in which they worked. The study illustrates that impact on productivity is not the only key measurement of an aid. "Both economic and social theory as well as a rich case study literature suggest that while information technology may increase productivity it can degrade the work lives of those who use it" (Kraut et al., 1989:220). Quality of work life can sometimes be neglected in formulating a study, because nearly all human factors people work for management (not labor); this has powerful consequences, often subtle, on the goals of the work.

In developing and refining group aids, it is important to consider not only how they affect group processes and team efforts, but also how these affect individual motivation, skills, and job satisfaction in the long term and how they shape organizational cultures and norms. These goals would probably require that even longer studies be carried out.

Group Work Aids

Attempts to aid collaborative group work have taken many forms, including new systems, technologies, facilities, and procedures (and the tools and infrastructure to create them) and have been aimed at many tasks, including collaborative writing, design, problem solving, consulting, and programming. The emphasis has been upon tools for interactive, real-time work, although there have been some attempts to help people who are separated in time as well as in space.

A combination of experimental laboratory work and empirical field studies to assess various aids for group work is a valuable human factors approach. One such approach was carried out by Judy and Gary Olson (Olson et al., 1992:91) at the University of Michigan:

Our research strategy began by understanding group work in the field where we noted important baseline features of group work and gathered ideas about how various kinds of technology can help. We have now moved to the laboratory to study specific phenomena such as how technologies change the flow of activities in meetings or affect the quality of the work. From these laboratory studies, we will move back to the field, assessing the use of these new technologies in the full organizational context. This strategy has the benefit that we are informed by actual practice and can eventually test ideas in the field, but also puts the burden on us to construct laboratory situations that mirror the field in important ways.

In one experiment, Olson et al. (1992) compared the work processes and performance of small groups whose members were mainly business degree holders with business experience. Some groups used a group editor that allowed all participants to see and edit the same document at the same time, and other groups used paper and pencil, white-boards, and similar traditional aids. Each group worked face to face and collaborated for 90 minutes to draft the initial requirements statement for an automated post office. The designs produced by the "aided" groups were of higher quality than those produced by the groups working with traditional methods. The "aided" groups focused more on the core issues of the problem (in other words, they did less extensive exploration of the design space). However, the "aided" groups did not like their experience in the experiment as well as the control groups.

Research interest has been dramatic. In the most recent Association for Computing Machinery conference on computer-supported cooperative work, 191 research papers were submitted (48 were selected for inclusion) (Turner and Kraut, 1992:1). The conference papers appear to reinforce the research strategies recommended above. In the area of assessment, there was a feeling that empirical studies of the impact of aids on *group* work need to be longer-term than studies of individual work. According to Cool et al. (1992:31), "while the basic cycle of build-study-redesign is still valid, the implementation should be more complete and robust, and the study should extend over a longer time period before one trusts conclusions about ultimate value of a new communications system." Regarding the second recommendation—study existing work situations in order to determine how best to aid them—Berlin and Jeffries (1992) did precisely that for consultants helping apprentices. Bentley et al. (1992) carried out an ethnographic study of air traffic controllers in the United Kingdom. This methodology, relatively new to systems design, uses observational methods and puts emphasis upon identifying the social and cooperative processes involved in group work.

RESEARCH NEED AND STRATEGY 2: EMPIRICALLY STUDY ONGOING INTELLECTUAL WORK

In contrast to the first strategy (assessment), in which the emphasis is upon studying both an aid and a task, here the emphasis is upon studying an intellectual work domain or task itself (which, of course, is already aided in some way, since all tasks are). This means identifying the social, organizational, physiological, and cognitive mechanisms involved in the task. The goal is to use these observations to gain insights into what would make good new aids. There is no cookbook for this transition; one hopes the studies cited below will help generate ideas. Particularly in the case of group work, deciding which task to study can be a major challenge. Issues include whether to study long-term tasks (such as effectiveness in increasing sales, morale, safety, and student learning) or short-term tasks (such as teacher-student contact hours), entire work assignments (such as changes in productivity of a group) or subtasks (such as effectiveness in communicating). The underlying belief is that mental analysis and task analysis can help in aiding intellectual work more effectively.

To this end, it is probably more important to study the work of groups and organizations than the work of individuals because (1) it is harder to do, (2) most people work in organizations, and (3) there is uncertainty about how increases in individual performance affect organizational performance (see Chapter 1).

Many formal and informal human factors methods are available for carrying out empirical studies of ongoing intellectual work. Figure 11.3

Methods for Observing On-Going Intellectual Work

- Visit worker locations.
- Study the work organization—understand the target group's role.
- Observe workers working—with and without existing aids they may use or wish they had.
- Talk with workers—about their work and any existing aids.
- Have workers think aloud while they work.
- Videotape workers doing their jobs with and without intellectual aids.
- Try the job yourself—with and without intellectual aids.
- Collect opinions from associated workers outside the target group.
- Study worker-made notes attached to equipment, documentation, bulletin boards for insights into what would be a better aid.
- Surveys and questionnaires.
- Use computer forums to learn about other possible aids.

FIGURE 11.3 Methods for collecting data on ongoing intellectual work. SOURCE: Gould (1988).

> **Key Methodological Points in Identifying the Characteristics of Useful Intellectual Aids**
>
> - Understand the work that you want to aid, so that the proposed aid will likely have the right characteristics.
> - Focus on the work itself and the workers themselves: this keeps you close to the tasks you are trying to aid.
> - Pay attention to the social and organizational interactions and needs.
> - Analyze at a sufficiently abstract level for the purpose of identifying new ways for people to work.
> - Understand the *context of the work* (e.g., what policemen actually do) so that a proposed aid (e.g., a geographic database system) will be appropriate.
> - Go beyond activity analyses; involve workers' personal and organizational intentions, motivations, socialization.
> - Will the aid reduce manufacturing and development cycle times ("design for manufacturing")?
> - Consider the trade-offs between integrating several intellectual aids, rather than using intellectual separate aids.

FIGURE 11.4 Key points in studying ongoing intellectual work for the purposes of identifying the characteristics of useful intellectual aids.

lists some informal ones; these are described more fully in Gould (1988), with the emphasis upon designing computer applications. In addition, there is a substantial sociological and anthropological literature on participant observation and ethnography. All of these methods emphasize coming into direct contact with workers, their organizations, and the intellectual tasks they do or could perform. Failure to do this will lead to misunderstanding the characteristics of potentially valuable aids as surely as if one attempted to define an important aid for intellectual workers who live on other planets. Figure 11.4 provides further methodological guidelines.

The studies cited below are more valuable even than the following methods and results might indicate. Nearly every one provides a thoughtful discussion of the motivation, methods, results, and their implications.

Example of Studies

Work Activity Analyses

Observational studies, using a variety of methodologies, have summarized how office workers spend their time. For example, Klemmer and Snyder (1972) had observers note specific types of work behavior exhibited in a large research and development laboratory in offices, hallways, and other areas. In addition, employees answered questionnaires on how they spent their time. They found that office professionals in the late 1960s spent most of their work time (42 percent) communicating interactively (talking face to face or on the telephone) and less time writing (14 percent) and reading (12 percent). They spent only 3 percent of their time interacting with office machines—which is certainly less than the time office workers spend today interacting with computer terminals. Mintzberg (1973) followed five very-high-level executives for two weeks each, observing and categorizing their activities, and wrote a widely read book summarizing the findings. Panko (1992) recently reviewed over 50 of these use-of-time studies and made general suggestions for office automation technology consistent with the main results.

While these studies, and many others, have illuminated how white-collar workers spend their work time, few intellectual work aids have so far emerged based upon insights from them, especially as related to the work of groups and organizations. What has been needed for some years are studies that relate these observed activities to personal and organizational motivations and goals, trigger events, and interruptions. New methodology is needed to go beyond "activity counts" and gain insights into work processes that can be aided. Although such studies are very hard to carry out correctly, they could provide insights into potentially valuable aids.

A related approach to these activity analyses has involved detailed coding and classification of verbal communications at the level of individual utterances. This approach has been used to analyze group interactions in flight simulations and in the investigation of aircraft accidents (e.g., Kanki and Palmer, 1993; Predmore, 1991). A more holistic approach has used experienced raters to evaluate interactions in real time using specific behavioral markers to exemplify effective and ineffective processes (Helmreich et al., 1990). Both methodologies can be useful in evaluating aids since they appear to be transferable to a variety of work situations of which the researchers have a detailed knowledge.

Desk Organization

To learn more about office work, Malone (1983) interviewed 10 professional and clerical workers about the way they organized their desks and offices. Participants first gave a verbal tour of their offices, explaining what information was where and why they put it there. As an indicator of their information-retrieval ability, the participants were then asked to find certain documents (chosen by a knowledgeable co-worker as being either easy or hard to find). Malone's motivation was to obtain insights into designing natural and convenient computer-based information systems to aid office workers' productivity. His two main conclusions (1983:99), which have implications for the design of intellectual aids, were the following:

(1) A very important function of desk organization is to *remind* the user of things to do, not just to help the user *find* desired information. Failing to support this function may seriously impair the usefulness of electronic office systems, and explicitly facilitating it may provide an important advantage for automated office systems over their non-automated predecessors. (2) The cognitive difficulty of categorizing information is an important factor in explaining how people organize their desks. Computer-based systems may help with this difficulty by (a) doing as much *automatic classification* as possible (e.g., based on access dates), and (b) including untitled "piles" of information arranged by physical location as well as explicitly titled and logically arranged "files."

The point is not so much to focus on these particular implications of Malone's study, but rather to appreciate how he drew implications for system design from his observational methodology. A second, more general, point has to do with the distinction between the need to design electronic aids (e.g., to replace paper) that allow people to work in about the same way they now work (an approach Malone's conclusions seem to favor), versus the need to design electronic aids that allow people to work in relatively new ways.

Communication in Software Development Organizations

Probably millions of people are employed in system development work. To understand better how to aid the coordination of work activities in system development, Kraut and Streeter (1995) studied the communication practices of people in such organizations. Hundreds of software professionals in the software development divisions of one company—including managers, analysts, software engineers, programmers, testers, and documentation specialists—completed a written survey. They reported that they got much of their important information directly from other people. When they had a large network of personal contacts outside the project, information flow improved, especially when the project was uncertain. They said that interpersonal techniques for getting information from beyond their immediate work group were underused, while more formal procedures for tracking routine information were overused in terms of their value. The design implication seems clear: aids to facilitate informal communication are needed. Kraut and Streeter note that planners are especially ill-served by current coordination techniques. Importantly, Kraut and Streeter conclude with valuable organizational and technological suggestions for increasing communication across project boundaries. This type of well-designed study should be imitated by future researchers.

Software Development

Many studies have contributed to a shared understanding of the social, cognitive, and behavioral characteristics of the software development process (e.g., Boehm, 1988; Jones, 1986). The study by Kraut and Streeter (1995), just described, and one by Curtis et al. (1988) show how human factors researchers can investigate very large task domains through surveys, visits, observations, and interviews. Using their findings, they can then develop theoretical understandings of work organizations, which in turn can lead to ideas about the characteristics of helpful electronic aids. Curtis et al. (1988) carried out extensive visits to workplaces and interviewed many individuals involved in 17 large software development projects. They found three general classes of behavioral problems that limited development productivity: thin spread of application domain knowledge, communication and coordination breakdowns, and conflicting requirements. It is possible to design new tools for software developers that address these problems.

Diagnostic Judgments

Diagnostic judgments and decision making are intellectual activities that cut across most work domains. People frequently make diagnostic judgments to answer important questions in their work lives (e.g., which person to hire? which applicants to admit to law school? what is the likely response from this drug treatment?), as well in their personal lives, which in turn affect their work lives (e.g., should I marry this person? should we have children? what is the best form of discipline to use with my children? at which supermarket should I shop?).

Consider a clinician examining a patient who is feeling ill, an employer selecting salespeople, a teachers' committee deciding whether to refer a youngster to special classes for learning disabilities, and a graduate school admissions committee selecting students for scholarships. If these "judges" base such difficult decisions upon years of personal experience as successful diagnosticians, intuition about particular cases, and knowledge of the individuals involved, then they are using the clinical method of decision making. This is said even if they base their judgment upon one or more quantitative predictors as well.

On the other hand, sometimes people rely only on statistical predictors to make judgments (the actuarial method). "To be truly actuarial, interpretations must be automatic (that is, prespecified or routinized) and based on empirically established relations" (Dawes et al., 1989:1668). The insurance industry, for example, predicts people's life expectancy (and therefore insurance rates) based only on a combination of statistical predictors. No intuition or human element is involved. "In the clinical method the decision-maker combines or processes information in his or her head. In the actuarial or statistical method the human judge is eliminated and conclusions rest solely on empirically established relations between data and the condition or event of interest" (Dawes et al., 1989:1668).

Nearly all of over 100 studies comparing these two general approaches have shown that actuarial methods lead to more accurate diagnostic judgments than do clinical methods (Dawes et al., 1989). The topics studied included prediction of progressive brain damage, survival time following Hodgkin's disease, differential diagnosis of psychiatric disorders, length of psychiatric hospitalization, violent behavior, parolee's behavior after release, and graduate student performance. Even when clinicians are able to combine actuarial test results with their own interview data, the predictions based only upon the test results proved superior (Dawes et al., 1989). In other words, in this case, a third approach consisting of combining the two methods does not seem to help. This conclusion is not limited to clinical and social science situations. Murphy and Brown (1984) reported that "objective" (actuarial) weather forecasting is equal to or sometimes superior to "subjective" weather forecasting. The important point is that in straightforward ways, actuarial methods have the potential to be useful electronic aids, once the variables that correlate with successful predictions about the topics being judged are established.

There are elements of all three recommended research strategies in the aggregate of these studies. One problem for human factors research is how to put this conclusion into practice—where it presumably would have real value for society in many domains. To date, the results have had little impact. Dawes et al. (1989) cite several reasons, including lack of familiarity with the evidence, the continuation of previous practices in spite of the evidence, the feeling that an actuarial approach dehumanizes people, misconception, and inflated confidence about one's own ability to make accurate predictions (including situations in which the actuarial method, although superior to clinical judgment, achieves only modest results). Actuarial methods, when applied to new domains, can seem mechanistic, antidemocratic, inhuman. One task for human factors researchers is to figure out how to get people to internalize the research conclusion in this area so that it affects their behavior. McCauley (1991) notes that, when psychologists have to select National Science Fellows from a large list of applicants, they do not make use of available actuarial predictors.

A second problem for human factors research is how to extend these results to new domains (e.g., labor disputes, employee conflicts, system development planning). Here, one needs to identify the crucial predictors (which may already be known), carry out the empirical work when necessary, figure out ways to present the data effectively to potential users, assist them in using the data, and design and carry out the necessary follow-up studies to verify whether the actuarial method helps in the new domain of concern.

Additional Approaches

There are several other approaches to studying ongoing intellectual work. There are many variations of task analysis methodology (see Drury et al., 1987). One can apply task analysis to the work of an isolated individual or to that of individuals working within an organization. Task analysis can be limited to "logical analysis" of all the possible steps in a work process (as with an entirely new, nonexisting work system), or it can involve behavioral observations. Task analysis has been carried out mainly in military organizations, but there are many examples in industrial settings. The same handbook containing the Drury et al. paper contains chapters describing task analyses applied to process control systems (Woods et al., 1987) and to air traffic control systems (Lenorovitz and Phillips, 1987).

Another possible approach is to carefully study, through interviews, observations, task analyses, and other methods, experienced, successful workers performing crafts of high intellectual content (e.g., a physician diagnosing, a nurse treating, a musician composing, an architect designing, a teacher teaching). The goals would be to identify the intellectual skills involved, develop a theory about the task, and identify the characteristics of an aid that would presumably be helpful. Some designers find that, in the early stages of designing and developing a complex system, they want a list of classes of subtasks that most need augmenting or even automating. The challenge is to identify in general terms, if possible, these classes (e.g., spatial orientation, data interpretation, handling radioactive materials).

Another possible approach is to identify intellectual skills that cut across a large number of human intellectual tasks and then identify the characteristics of useful aids for that work. Existing examples of such aids are electronic verifiers for spelling and even grammar and electronic calculators for arithmetic. The idea is to determine not what the tasks have in common, but what common mental operations are carried out while working on a variety of tasks.

RESEARCH NEED AND STRATEGY 3: INTERDISCIPLINARY DEVELOPMENT WORK

Human factors researchers, to date, have been more likely to assess the value of an aid developed by others (e.g., the studies under strategy 1) than to join with others to develop an aid and then carefully assess its value and impact. There is a need for human factors researchers to become involved in interdisciplinary efforts to create superior intellectual aids. This work is needed for several reasons.

First, working alone is generally not enough. It is hard to find examples of human factors researchers who, *by themselves*, develop effective electronic aids. Making good electronic aids requires many disciplines. Second, many engineers and scientists have come to realize that "human factors" determine the difference between success and failure of most aids. Third, there is evidence that human factors involvement in development projects is cost-effective (Karat, 1992). Fourth, iterative testing and improvement of proposed aids are vital if the aids are to increase productivity and enhance the quality of work life.

If human factors researchers are involved, such testing and improvement are likely to be carried out much more thoroughly. Thus, the potential aid under development has a greater probability of improving organizational or individual productivity or other aspects of work lives. Chapters 4 and 5 have made similar recommendations for interdisciplinary work in, respectively, emerging medical and environmental domains, areas that offer great rewards for successful collaboration. Fifth, successfully instantiating ideas into technology is a much bigger problem than most human factors researchers uninitiated in this process realize, and it is something they often cannot carry out on their own. This is true no matter how well the ideas seem to have worked in the experimental psychology laboratory or with prototypes. The studies referred to above on decision making and those described below by Landauer et al. (1993) illustrate this. Finally, many of the disciplines underlying human factors, such as computer science and experimental psychology, have a much larger supply of techniques, procedures, and results than there is a demand for by others. This proposed advanced development strategy can identify, select, and shape the techniques that others want and the context in which they must fit.

Human factors researchers should become partners with others in designing, developing, and iteratively improving good aids. This recommended strategy is not just a plug for employment opportunities for human factors researchers or a profession-oriented expression of the importance of human factors. The idea is not just to work in support roles, as most industrial and government human factors workers now do, but to work as equal, long-term participating partners, with all the commitment, risk, and pressure that leadership demands. The idea is that as co-partners, human factors researchers should develop a sense of responsibility for the initiation and successful completion of an interdisciplinary project to produce a significant aid, with all the burdens this entails. For many human factors researchers, this requires new learning, motivations, work organizations, and professional payoff (particularly as related to the university promotion system). It requires a new conceptualization of what constitutes legitimate research. As outlined below, there are already some situations that can serve as organizational models of how to do the work and write it up for professional credit.

This recommended interdisciplinary work—wherein end users actually use, on a regular basis, the aid that human factors researchers jointly made with other necessary disciplines—may be hard to carry out because of personal and institutional barriers. It is longer-term work than is typical of many graduate school projects and may require longer-term commitments of funding than traditional research. To be successful, it will also require professional recognition and rewards, including ammunition for gaining tenure. There has been a start toward giving this work professional recognition. Case studies describing innovative research and development efforts involving human factors researchers in advanced development work exist, including Xerox's Star system (Smith et al., 1982), IBM's Audio Distribution System (Gould and Boies, 1984), and IBM's multimedia 1984 Olympic Message System (Gould et al., 1987). It is generally accepted that these systems were innovative and have influenced others who have made similar ones. Figure 11.5 summarizes key points to keep in mind carrying out interdisciplinary development work. These points are based upon personal experience and published case studies.

Key Characteristics about Effective Multidisciplinary Development Work

- Strong technical leader and mentor, not lacking greatly in people skills—a technical champion.
- Small group of talented people.
- Individual members do what they are good at.
- Commitment to colleagues.
- Drive to become famous as a group for making an outstanding aid.
- Willing to do what is necessary to be successful—no 80/20 or 90/10 rule, where the remainder is left to others.
- Use iterative engineering, including good development tools.
- Make early prototypes.
- Involve intended users from the outset.
- Use the aid that you are making yourselves whenever possible (if applicable).
- Follow up, and measure the value of the aid created.

FIGURE 11.5 Key points in interdisciplinary work aimed at making effective aids for intellectual work.

Examples of Studies

SuperBook

Many developers believe that electronic computers hold greater potential than printed books for helping people find desired information. Yet Landauer et al. (1993), working at Bellcore, noted that nearly every experimental study failed to find an advantage for various electronically presented text systems over comparable printed versions. They concluded that a significant human factors problem is to develop an electronic text system that people will use to better advantage than a standard printed book.

Because of experimental results on the use of command languages and navigational tools, as well as results related to other relevant cognitive issues they had been studying (summarized in Landauer et al., 1993), Landauer et al. felt that they had a chance to change this history. They observed that much information already exists in published books and that this is the favorite medium of most authors. So, in contrast to most development efforts, which focus on electronic books only, they focused on how to help people find information in already published books that are also stored in computers. They called this approach SuperBook. They combined "full text indexing" with a "dynamic, hierarchical table of contents" of the stored book. A review of some of their previous research is in order.

Two fundamental problems that users have in trying to find helpful information in electronic databases have to do with the command language they use and how they navigate through databases.

With respect to command languages, almost all electronic information retrieval systems—be they library subject catalogs, dial-up query systems, relational databases, or computer file systems—require users to select or enter some words that must match words used as identifiers by the authors or indexers. After many observations, Landauer et al. became convinced that most occasional users are usually unsuccessful in using standard query languages or in producing Boolean expressions from *ands* and *ors* to get desired information from databases. Thus, they rejected this approach as a way of helping casual users of electronic data bases.

Why not discover what words people naturally prefer to use? When trying to devise new command names that would help novices learn word processors, the Bellcore group (Landauer et al., 1993:76) decided:

> that any two people were unlikely to agree on a best term, so the common expedient of a one word command, file name or table label will usually fail to put a user in touch with data stored by someone else. . . . The chances that any two people would choose the same name for a command were less than twenty percent, and the most popular term was chosen by only thirty percent of potential users. This observation was extended to the use of

terms for queries in data bases of several kinds, from classified ad listings to recipe files and program names.

The Bellcore researchers built upon these observations by experimentally evaluating inventions of their own derived from their results. For example, unlimited aliasing (i.e., using every term that a sample of users apply to an item as an index to that item) led to a hit rate of over 70 percent in an interactive retrieval task on a small database of 188 recipes, with little increase in false positives. The same level of success was achieved with full text indexing (i.e., every word contained in each item in the database is part of the index for that item). Unfortunately, these results do not scale up to much larger databases because too many irrelevant items (i.e., false positives) are returned. Note how these researchers were not misled by optimistic lab results based upon limited tasks.

With respect to navigation, the Bellcore experimental studies (Landauer et al., 1993:80) found that people have fundamental problems when using menus to navigate through electronic databases:

> Typically menu selection schemes have a branching factor of four to ten options; more often than not, users stray from the path to their target by the third hierarchical level. Put differently, if there are more than about thirty items in the data base, the chances of finding the right one in a single error-free traversal drop below half.

They concluded provisionally that (Landauer et al., 1993:82):

> menu-traversal methods are fundamentally, and probably irremediably flawed as a primary method for information retrieval. As conceived by humans, almost all categories are essentially fuzzy; most items belong only partially or with only moderate probability to any one category, and can often fit reasonably well into several.

The Bellcore group also retained a "rather cautious use of links in the SuperBook system" (Landauer et al., 1993:129).

Thus, as the experiments showed, neither full text indexing nor a dynamic, hierarchical table of contents is sufficient by itself, but the Bellcore group thought a combination of the two might help. Full text indexing returns a superset of desired items, as they had already found, but the dynamic table of contents, they reasoned, should provide contexts to help users narrow this superset of items to those that are relevant to their intentions. So they built a prototype, and used it themselves (Landauer et al., 1993:98):

> We were . . . perfectly delighted with our system design. We were so overwhelmed with its obvious superiority to anything we had previously seen, and so pleased with our early in-house use that we dubbed it "SuperBook." We immediately began technology transfer efforts, efforts to make

this marvelous new tool available to other organizations at Bellcore. . . . Fortunately, [a new group member] reminded us of some of our public pronouncements about, and past experiences with the necessity of evaluation studies.

To judge how well the SuperBook actually aided people, the Bellcore group gave college students queries which they had to answer using either a published statistics book or its electronic clone.

Using the first version of the electronic aid (version 0), participants found slightly more correct answers with SuperBook than with the original printed book, but took significantly longer to do so. So the Bellcore group made many changes to its electronic aid, most based upon analyses of its experimental data. With the next version, participants answered their queries faster and more accurately with SuperBook than with the original printed book. Based upon what they learned in this experiment, the Bellcore researchers made more changes to their electronic aid and tested it again. This third evaluation study found that participants performed even better with SuperBook.

This interdisciplinary effort illustrates many important points. No one discipline could have done all the design, implementation, testing, and transfer efforts. Human factors researchers, like experts in other disciplines, can be overenthusiastic about their own inventions when empirical data are absent. Development of a successful aid usually takes a long time. The work shows the potential payoff of interdisciplinary effort directed at a very hard problem. The work reflects elements of all three recommended research strategies.

The experimental evaluations showed the status of the aid and suggested possible improvements. And the testing still goes on: Landauer et al. (1993) have carried out several other experimental evaluations with other materials and have made SuperBook available to over 40 research labs.

Aid for Detecting Drunk Drivers

Harris (1980) designed and carried out a study to develop an aid to help police officers spot drunk drivers. In the end, Harris provided officers in patrol cars with a low-tech, effective aid—a list of driver behaviors that best predict which motorists are driving while intoxicated (DWI). More important here is the process that led up to his designing, implementing, and evaluating this aid. His study is valuable for its methodology, completeness, impact, clear write-up, and (inexpensive) aid developed to help police select which drivers to stop.

In general, six percent of drivers at night have a blood-alcohol level of 0.10 or greater (most states define this as DWI), and another 9 percent have a blood-alcohol level between 0.05 and 0.10 (some states define this as "driving while impaired").

First, Harris studied 1,288 DWI arrests throughout the United States, analyzing the reported driver behaviors associated with each of these arrests. These driver behaviors could all be observed by police officers making the arrests (e.g., swerving, straddling) and could therefore serve as visual cues to aid in deciding whether or not to stop a potentially drunk driver. Harris also rode with officers in two states as they stopped drivers, and he noted the driving behaviors of those motorists and the conditions under which they were stopped. Nearly all of the stopped drivers (93 percent) were given a blood-alcohol test when they were stopped: 38 percent had a blood-alcohol level greater than 0.10 (DWI), and another 23 percent had a blood-alcohol level between 0.05 and 0.10. (This 61 percent is already 4 times higher than the 15 percent that could be expected by chance.) From these analyses he produced a drunk driver detection guide, a small card listing several visual cues (e.g., following too closely) and the percentage of drivers with this behavior who have a blood-alcohol level greater than .01 (e.g., 60%). The card was for police to carry with them or to attach wherever convenient. Harris then went on to study the value of this aid by studying 10 law enforcement agencies across the United States, before and after they used his guide. He found that DWI arrest rates increased significantly overall (from 66 to 74 percent; $p < .01$) for these 10 agencies. (No mention is made of false positives—arrests that did not lead to convictions.) Harris (1980:731) also reported:

> Experienced police officers . . . were generally skeptical that . . . the guide would enhance their own DWI detection ability. However, most officers considered the guide to be a valuable aid for increasing patrol awareness of useful detection cues, training inexperienced patrol officers, preparing DWI arrest reports, and supporting court testimony.

Other Approaches

The Landauer et al. (1993) studies provide an excellent example of extending the cognitive findings of one's own group to the design of a real electronic aid. Human factors researchers could take the initiative to extend other existing behavioral and cognitive knowledge to applied domains (a task much harder than many experimental and cognitive psychologists generally suppose). For example, for some years many secretaries and professionals have used commercially available computer-executable, rule-based styles to lay out and format their documents, for example, in prespecified fonts, margins, number of columns, line spacing, headings, and graphics. Now, if there exists behavioral knowledge of how to design textbooks and related reading material to enhance learning, human factors people, working with computer scientists, could develop a similar rule-based style to lay out documents according to this knowledge. Another example would be to extend the results on human decision making, described above, to real-world aids. The Landauer et al. (1993) studies provide an excellent example of extending one's own cognitive findings to the design of a real electronic aid.

Another approach might take existing technology or aids and combine them in behaviorally innovative ways. Perhaps the Xerox Star system (Smith et al., 1982), the highly influential graphic user interface developed at Xerox Parc in the 1970s, is an example of this. The developers took existing hardware (bit-mapped displays, mouse), existing software techniques, behavioral knowledge, and cognitive engineering models and integrated them into something much greater than the sum of the parts. Many observers agree that this work influenced subsequent graphic user interfaces made by other major corporations.

ACKNOWLEDGMENTS

I thank Liz Gould, Bob Helmreich, Bob Kraut, Tom Landauer, Ray Nickerson, Lynn Streeter, Jacob Ukelson, and Frank Yates, who provided helpful written reviews and suggestions to improve this chapter, and several other committee colleagues for additional suggestions.

REFERENCES

Attewell, P.A.
    1994   Information technology and the productivity paradox. Pp. 13-53 in D.H. Harris, ed., *Organizational Linkages: Understanding the Productivity Paradox*. Panel on Organizational Linkages, Committee on Human Factors, National Research Council. Washington, D.C.: National Academy Press.

Bentley, R., J.A. Hughes, D. Randall, T. Rodden, P. Sawyer, D. Shapiro, and I. Sommervile
    1992   Ethnographically-informed systems design for air traffic control. Pp. 123-129 in J. Turner and R. Kraut, eds., *Proceedings of the Conference on Computer-Supported Cooperative Work*. New York: ACM Press.

Berlin, L.M., and R. Jeffries
    1992   Consultants and apprentices: observations about learning and collaborative problem solving. Pp. 130-137 in J. Turner and R. Kraut, eds., *Proceedings of the Conference on Computer-Supported Cooperative Work*. New York: ACM Press.

Boehm, B.W.
    1988   Improving software productivity. *IEEE Computer* September:43-56.

Card, S.K., W.K. English, and B. Burr
    1979   *The Keystroke-Level Model for User Performance Time With Interactive Systems*. Report SSL-79-1. Palo Alto, Calif.: Xerox Palo Alto Research Center.

Card, S.K., J.M. Robert, and L.N. Keenan
    1984   On-line composition of text. Pp. 231-236 in *Proceedings of Interact'84, First IFIP Conference on Human-Computer Interaction*, Vol. 1. London, England: Elsevier North Holland.

Cool, C., R.S. Fish, R.E. Kraut, and C.M. Lowery
- 1992 Iterative design of video communication systems. Pp. 25-32 in J. Turner and R. Kraut, eds., *Proceedings of the Conference on Computer-Supported Cooperative Work.* New York: ACM Press.

Curtis, B., H. Krasner, and N. Iscoe
- 1988 A field study of the software design process for large systems. *Communications of the ACM* 31:1268-1287.

Dawes, R.M., D. Faust, and P.E. Meehl
- 1989 Clinical versus actuarial judgment. *Science* 243:1668-1673.

Druckman, D., and R.A. Bjork, eds.
- 1991 *In the Mind's Eye: Enhancing Human Performance.* Committee on Techniques for the Enhancement of Human Performance, National Research Council. Washington, D.C.: National Academy Press.

Druckman, D., and J.A. Swets, eds.
- 1988 *Enhancing Human Performance: Issues, Theories, and Techniques.* Committee on Techniques for the Enhancement of Human Performance, National Research Council. Washington, D.C.: National Academy Press.

Drury, C.G., B. Paramore, H.P. Van Cott, S.M. Grey, and E.P. Corlett
- 1987 Task analysis. Pp. 370-401 in G. Salvendy, ed., *Handbook of Human Factors.* New York: Wiley.

Ferber, M.S., B. O'Farrell, and L. Allen, eds.
- 1991 *Work and Family: Policies for a Changing Work Force.* Panel on Employer Policies and Working Families, Committee on Women's Employment and Related Social Issues, National Research Council. Washington, D.C.: National Academy Press.

Gould, J.D.
- 1978a An experimental study of writing, dictating, and speaking. Pp. 299-319 in J. Requin, ed., *Attention and Performance VII.* Hillsdale, N.J.: Lawrence Erlbaum Associates.
- 1978b How experts dictate. *Journal of Experimental Psychology: Human Perception and Performance* 4(4):648-661.
- 1980 Experiments on composing letters: some facts, some myths, and some observations. Pp. 97-118 in L. Gregg and E.R. Steinberg, eds., *Cognitive Processes in Writing.* Hillsdale, N.J.: Erlbaum and Associates.
- 1982 Writing and speaking letters and messages. *International Journal of Man-Machine Studies* 16:147-171.
- 1988 How to design usable systems. Pp. 757-789 in M. Helander, ed., *Handbook of Human-Computer Interaction.* Amsterdam, Netherlands: North-Holland, Elsevier Science Publishers.

Gould, J.D., and S.J. Boies
- 1978 How authors think about their writing, dictating, and speaking. *Human Factors* 20:495-505.
- 1983 Human factors challenges in creating a principal support office system: the speech filing system approach. *ACM Transactions of Office Information Systems* 4(1):273-298.
- 1984 Speech filing: an office system for principals. *IBM Systems Journal* 23:65-81.

Gould, J.D., S.J. Boies, S. Levy, J.T. Richards, and J.W. Schoonard
- 1987 The 1984 Olympic Message System: a test for behavioral principles of system design. *Communications of the ACM* 30(9):758-769.

Harris, D.H.
- 1980 Visual detection of driving while intoxicated. *Human Factors* 22(6):725-732.

Helander, M., ed.
1988   Handbook of Human-Computer Interaction. Amsterdam, Netherlands: North-Holland, Elsevier Science Publishers.

Helmreich, R.L., J.A. Wilhelm, S.E. Gregorich, and T.R. Chidester
1990   Preliminary results from the evaluation of Cockpit Resource Management training: performance ratings of flightcrews. *Aviation, Space and Environmental Medicine* 61(6):576-579.

Jones, T.C.
1986   *Programming Productivity.* New York: McGraw-Hill.

Kanki, B.G., and M.T. Palmer
1993   Communication and crew resource management. In E.L. Wiener, B.G. Kanki, and R.L. Helmreich, eds., *Cockpit Resource Management.* San Diego, Calif.: Academic Press.

Karat, C.
1992   Cost-justifying human factors support on software development projects. *Human Factors Society Bulletin* 35(11):1-4.

Klemmer, E.T., and F.W. Snyder
1972   Measurement of time spent communicating. *Journal of Communication* 22:148-158.

Kraut, R., S. Dumais, and S. Koch
1989   Computerization, productivity, and quality of worklife. *Communications of the ACM* 32(February):220-237.

Kraut, R.E., and L.A. Streeter
1995   Coordination in software development. *Communications of the ACM* 38(3):69-81.

Landauer, T., D. Egan, J. Remde, M. Lesk, C. Lochbaum, and D. Ketchum
1993   Enhancing the usability of text through computer delivery and formative evaluation: the SuperBook project. In C.M. McKnight, A. Dillon, and J. Richardson, eds., *Hypertext: A Psychological Perspective.* New York: Horwood.

Lenorovitz, D.R., and M.D. Phillips
1987   Human factors requirements engineering for air traffic control systems. Pp. 1724-1770 in G. Salvendy, ed., *Handbook of Human Factors.* New York: Wiley.

Malone, T.W.
1983   How do people organize their desks: implications for designing office automation systems. *ACM Transactions on Office Automation Systems* 1:99-112.

McCauley, C.
1991   Selection of National Science Foundation Graduate Fellows. A case study of psychologists failing to apply what they know about decision making. *American Psychologist* December:1287-1291.

Mintzberg, H.
1973   *The Nature of Managerial Work.* New York: Harper and Row.

Murphy, A.H., and B.G. Brown
1984   A comparative evaluation of objective and subjective weather forecasts in the United States. *Journal of Forecasting* 3:369-393.

Norman, D.
1988   *The Psychology of Everyday Things.* New York: Basic Books.

Olson, J.S., G.M. Olson, M. Storrosten, and M. Carter
1992   How a group-editor changes the character of a design meeting as well as its outcome. Pp. 91-98 in J. Turner and R. Kraut, eds., *Proceedings of the Conference on Computer-Supported Cooperative Work.* New York: ACM Press.

Panko, R.R.
  1992   Managerial communication patterns. *Journal of Organizational Computing* 2(1):95-122.
Pool, R.
  1992   Computing in science. *Science* 256(April 3):44-64.
Predmore, S.C.
  1991   Microcoding of communications in accident analyses: crew coordination in United 811 and United 232. Pp. 350-355 in *Proceedings of the Sixth International Symposium on Aviation Psychology*. Columbus: Ohio State University.
Roberts, T.L.
  1979   Evaluation of Computer Text Editors. Unpublished PhD dissertation. Computer Science Department, Stanford University.
Smith, D.C., C. Irby, R. Kimball, B. Verplank, and E. Harslem
  1982   Designing the star user interface. *Byte* 7(4):242-282.
Turner, J., and R. Kraut, eds.
  1992   *Proceedings of the Conference on Computer-Supported Cooperative Work*. New York: ACM Press.
Wiener, E.L., T.R. Chidester, B.G. Kanki, E.A. Palmer, R.E. Curry, and S.E. Gregorich
  1991   *The Impact of Cockpit Automation on Crew Coordination and Communication. I. Overview, LOFT Evaluations, Error Severity, and Questionnaire Data*. NASA Contractor Report No. 177587. Moffett Field, Calif.: NASA-Ames Research Center.
Woods, D., J.F. O'Brien, and L.F. Hanes
  1987   Human factors challenges in process control: the case of nuclear power plants. Pp. 1724-1770 in G. Salvendi, ed., *Handbook of Human Factors*. New York: Wiley.